Facets of Rationality

Facets of Rationality

Editors

DANIEL ANDLER
PARTHASARATHI BANERJEE
MAHASWETA CHAUDHURY
OLIVIER GUILLAUME

Sage Publications
New Delhi/Thousand Oaks/London

First published in 1995 by

Sage Publications India Pvt Ltd
M-32 Greater Kailash Market I
New Delhi 110 048

Sage Publications Inc
2455 Teller Road
Thousand Oaks, California 91320

Sage Publications Ltd
6 Bonhill Street
London EC2A 4PU

Published by Tejeshwar Singh for Sage Publications India Pvt Ltd, phototypeset by Pagewell Photosetters, Pondicherry, and printed at Chaman Enterprises, Delhi.

Library of Congress Cataloging-in-Publication Data

Facets of rationality / editors, Daniel Andler ... [et al.].
p. cm.
Proceedings of a workshop held Dec. 15–18, 1993, at the National Institute of Science, Technology, and Development Studies, New Delhi, India.
Includes bibliographical references.
1. Rationalism—Congresses. 2. Science—Methodology—Congresses. 3. Philosophy, Indic—20th century—Congresses. I. Andler, D. (Daniel) II. National Institute of Science, Technology, and Development Studies (India)
B833.F33 1995 149'.7—dc20 94–45238

ISBN: 0–8039–9225–4 (US-Hb)
81–7036–453–1 (India-Hb)

Sage Production Editors: *Sarita Vellani and Indiver Nagpal*

Contents

Introduction

Rationality is back. The reasons for its return are several.

First there is the joint assault of (scientific) relativism, pluralism, sociologism, anarchism etc. on the traditional claims of scientific rationalism; and the joint assault of (cultural and moral) relativism, historicism, hermeneuticism, post-modernism etc. on the very foundations of rationalism. These attacks were bound to trigger a call to arms among rationalists of all breeds.

Second, progress in the social sciences has uncovered the shortcomings of the received models of rational behaviour used in economics and social and political theory, as well as ethics: paradoxes abound, lack of realism, or downright implausibility is patent, and last but not least, many situations and behaviours simply don't fit the available models, no matter what degree of idealisation one is prepared to live with. The need was felt for vastly more sophisticated models, as well as systematic empirical inquiry.

Third, cognitive science has risen, at first quite slowly from its formal–logical roots, then at a brisker pace as neuroscience, mainstream psychology, linguistics and philosophy actually joined hands in an effort to draft and implement a thoroughly novel research programme covering all aspects and levels of the functioning mind/brain. This opened the possibility of connecting the abstract study of rationality to the scientific investigation of the mechanisms responsible for the production of belief and behaviour. It is also leading to a reappraisal of boundaries: in more than one way, the image of rationality as an island in the sea of irrationality seems no longer fitting.

Last, but not least, the Indian context provided both an additional source of insights—drawn from its philosophical tradition—and a sense of urgency caused by recent threats to the foundations of the Indian state guided by the lay ideal of progress and social justice (a threat which is equally felt, albeit with lesser strength, in France and other Western states).

Each of these developments has set new inquiries in motion. However, it is far from clear that these trends, old and new, converge. In fact, it often seems as if they were at best dimly aware of one another, and uncertain of the existence of some common core of interests. The time seems ripe, therefore, for exploring the possibility of bringing them together. Ours is an attempt in that direction. It is just one among many, shaped, to be sure, by somewhat unusual circumstances—or rather, it was shaped by circumstances in a way which is unusual for academic enterprises, and we wish to be quite candid about this, because it largely accounts for the wide scope of this volume and the variety of concerns its twenty chapters reflect.

Two of the organisers, one a philosopher from India visiting France (Mahasweta Chaudhury), the other a logician from France (Daniel Andler), first met in Paris and decided to try and promote both a defense of rationalism and Indo-French cooperation by setting up, under the umbrella of the Indo-French exchange programme in social sciences, a joint research programme on two of the above four major themes, *viz.* scientific method and cognitive science. An exploratory stage took the form of a seminar on 'Science, Rationality and Cognition' held at the 78th Indian Science Congress in Indore in January of 1991. Several among the contributors to the present volume gave papers and comments during that seminar, and an informal group was thus formed. Reinforcement was needed for furthering the project, and this came from two Delhi-based organisations, one Indian (NISTADS), the other French (Centre for Human Sciences). The former had provided support and institutional shelter to the Indore seminar, and now came forth with an offer for help and the contribution of a philosopher-scientist who was part of the Indore group (Parthasarathi Banerjee). The latter, under the guidance of its newly-appointed archaeologist director (Olivier Guillaume), was already involved in a project on rationality from the standpoint of social science and computer-assisted methodologies. With the two new members of the scientific committee, came the two other themes listed above: the issues of rationality in the social sciences and of rational action in moral and political philosophy, from an Indian perspective.

Thus was born the idea of holding in Delhi an international workshop on 'Rationality in Cognitive and Social Science', involving speakers of most of the disciplines concerned, coming mostly from India and France but from other countries as well.

The philosophy of our project was dictated as much by its genesis as by our convictions. Simply stated, as much as we recognise the necessity for intellectual progress, of focusing research on specific issues, it seemed arbitrary at the present stage and in the context of our meeting to gear the workshop in any one particular direction. Indeed, as we had recognised the variety of concerns leading to the present interest in rationality, and were aware of the need to explore their interconnections, it would have been downright irrational to proceed in any spirit other than openness.

It also seemed important to enhance the chances of a true meeting between two national academic communities whose previous contacts had been minimal. The workshop confirmed our feeling that they have much to share. In particular, they both harbour an active minority of thinkers working in the tradition of analytic philosophy and associated currents in the social sciences and cognitive science, yet in both countries the majority of philosophers pursue research conducted in a different framework. Ruling out simplistic views of (cultural) universalism and (scientific) revolutions, as well as their opposites, cultural particularism and scientific unanimism, the problem is posed of how to integrate and put to work, in each community, those ways, traditions and trends which can blend and bear fruit together—a challenge all the more exciting in the context of a general weakening, if not breakdown, of disciplinary barriers.

It must be left to the reader to decide the extent to which we have succeeded in our attempt to both prepare for the bridging of some gaps between intellectual traditions, disciplines, research programmes, countries and languages, and suggest directions for thought on our main topic, rationality. The present volume, however, is but a tiny part of an ongoing enterprise. It is also, perforce, incomplete: just as the workshop had to be finite in length and number of participants, the contributions gathered here cover but a fraction of the relevant research. By choosing the title *Facets of Rationality*, we wish to signal the bounds of our immediate ambitions. Our long-term outlook includes developing more specialised cooperative research programmes, setting up inter-university visits, as well as pursuing our quest for a more unified view of rationality.

It remains to acknowledge the financial, logistic and moral support we have received from many people and institutions. The workshop, which was held at NISTADS in New Delhi on 15–18

December 1993, was inaugurated by Shri P.R. Kumaramangalam, Minister for Science and Technology and H.E. Philippe Petit, Ambassador of France. About 40 scholars attended, and barring one from the United States, all of them were from India and France.

The administrative framework was provided by the Indo-French exchange programme in social sciences, run by the Maison des Sciences de l'Homme (MSH) in Paris and the Indian Council of Social Science Research (ICSSR) in New Delhi, with the assistance of the Office of the Counsellor for Cultural, Scientific and Technical Cooperation of the Embassy of France.

Funding was provided by the following agencies:

- Department of Science and Technology (DST), Government of India
- Embassy of France, New Delhi
- Extra-Mural Research, Council of Scientific & Industrial Research (CSIR), Government of India
- Indian Council of Philosophical Research (ICPR), New Delhi
- Indian Council of Social Science Research (ICSSR), New Delhi
- Université Charles de Gaulle, Lille.

Four research institutions were responsible for organising the workshop:

- Centre de recherche en épistémologie appliquée (CREA), Ecole polytechnique, Paris
- Centre de Sciences Humaines (CSH), Embassy of France, New Delhi
- Department of Philosophy, Calcutta University, Calcutta
- National Institute of Science, Technology and Development Studies (NISTADS), New Delhi.

The workshop was made possible by the wise and energetic action of a number of leaders in both countries: Dr Ashok Jain, Director, NISTADS; M. Pierre Barroux, Counsellor for Cultural, Scientific and Technical Cooperation of the Embassy of France; Professor D.N. Dhanagare, (then) Member Secretary, ICSSR; Dr Clemens Heller, Administrator (now retired), MSH; Professor

Maurice Aymard, Deputy Administrator (now Administrator), MSH; Paule Gentot, Deputy Administrator, MSH; Dr A.K. Chakrabarti, CSIR; and Professor (Mrs) Bhuban Chandel, Member Secretary, ICPR.

Finally, warm thanks are due to Sage Publications India for undertaking the publication of this work.

September, 1994

DANIEL ANDLER
PARTHASARATHI BANERJEE
MAHASWETA CHAUDHURY
OLIVIER GUILLAUME

I

Rationality, Social Choice and Moral Theory

Preface

Analysing a social situation, as contributors remind us here, is not only tackling a situation of discourse and injunction, imperatives and norms, liberty and preferences but also, at the same time, it is the speaking about the sentences being employed, the believing and intending, the causality and rationality. Rationality has been defined variously, in terms of sentences or consistency, as privately or socially. The research programme that draws most attention is the so called descriptive research programme, and all the contributors draw attention to causality.

Pierre Jacob brings in the issues of reference and belief ascription in the context of the ongoing debate between the two contending research programmes, viz. the descriptive research programme and the programme on direct reference (or causality). Jacob examines a view of belief-ascriptions due to Jerry Fodor, which is in close similarity with the theorists of direct reference. Direct reference demands that in all contexts, including contexts of propositional attitude attributions, the propositional content of a singular term is denotational. Citing Fodor's project on naturalisation of mental content, Jacob argues how causal asymmetric dependence condition and his views on the totality of epistemic liaisons push Fodor to reject content holism. As Fodor treats believing as a four-place relation, and as he argues that in the ascription of a belief, specification of the vehicle (mental symbol) is less constraining than the specification of its content, the present contributor shows how such a view on the content of an attribution has to concede to the non-semantic aspect of its constituents, or to go beyond the denotations. Jacob then argues that as the explanatory success of either a belief ascription or its consequent a desire-attribution may depend on the functional role of the attributed belief, besides being dependent on the syntactic properties of its vehicle and the reference to content, Fodor faces the dilemma that he has to consider the functional role either as a part of the causal role of the vehicle or as a part of the content. Therefore, according to Jacob, the causal role of a mental representation does not reduce to pure syntactic properties and a purely denotational theory of content must also be given up.

The next paper, by *Parthasarathi Banerjee*, brings in the notion of a complete sentence and a corresponding notion of rationality. This paper exclusively draws upon theories from Indian theoretical vantages, and apart from allusions does not directly refer to contemporary debates amongst the English speakers, though the problems addressed are close indeed notwithstanding their unique formulations. A sentence, as discussed here, is an action-sentence referring to a complete action, and even a descriptive sentence including propositions in science should be considered as part of an injunctional complete sentence. Such an injunction, it is argued, derives from a cognition where the cognition causes a desire causing an action for a desired fruit; and an action cannot be for attaining knowledge as such. Desire is not against the reason. The role of injunction and the sentence in the causal scheme of reason, desire and action are learnt from observing a discourse or through participating in it by the employment of cognitions of meanings syntactic over the property of being action.

In the following contribution *Allan Gibbard* explores the broad meanings that 'rational' might be given as being both normative and descriptive. Gibbard in fact rejects naturalism, non-naturalism and emotivism for normative languages in general, and says that this is a language that shares whatever special features it is that, in the bulk of their uses, makes and defines specially these terms. He defines, following Ewing, a primitive non-moral ought, which brings together the normative ought and the rational ought. Normative language expresses a state of mind, an acceptance of the norms, and which motivates one to certain kinds of actions. The term rational might then depend on the norms one accepts and its meaning might be distinguished from a purely causal/explanatory explanation. Meaning and judgement are part of the natural, causal order and therefore Gibbard maintains that while a judgement of rationality is in the causal order, its correctness or incorrectness is not. Gibbard argues that in fact, meaning may be given both normative and factual concepts. It is plausible that what a speaker means in the normative sense would supervene on what he means in a causal/explanatory sense. Similarly 'rational' can be given both non-factual, normative senses and causal/explanatory, factual senses.

In the next contribution by *Serge-Christophe Kolm*, questions of social good, social rationality as also several sets of individual preferences (with meaningful comparisons across) have been tackled primarily through an examination of Arrow's work on social choice. Kolm maintains that since the literature following Arrow is influenced by utilitarianism, it imposes *a priori* such questions as are not reflective of the actual possibilities, given the actual individual preferences. Social choice for a given possibility set is concerned with the world only through the individual's preferences and given the individual's poor moral judgement, social choice is reflective of the values of the very individuals it considers. In fact it does not explicitly consider the two important values of liberty and equality. Kolm cautions on many points. He states that the pluralities in this theory are not the successive social choices of the same individuals, neither does the theory refer to different populations, nor can it be a theory of constitutions. Kolm brings forward the necessity of a full-fledged theory of social ethical choice under uncertainty.

The last contribution to this section, by *Satish K. Jain*, distinguishes between two aspects of liberty—first, where it denotes freedom of choice for individuals in their protected spheres and the second, as a value judgement for determining the set of socially optimal alternatives. Jain maintains that though choice has no implication for the evaluation of social states, given the acceptance of liberty as a value judgement, the existence of rights coupled with individual preferences would have implication for social states. He argues that as the Gibbard paradox relates to inconsistency between liberty as a value judgement for the evaluation of social states and social rationality (meaning that social choice must not be non-optimal) rather than the existence of rights, it is possible to avoid the paradox by limiting the scope of the libertarian value judgements, without curtailing the rights. Again, through another theorem Jain shows that if there is at least one liberal individual, then the Sen paradox cannot arise. He shows further that both these paradoxes can be considered manifestations of divergence between individual rationality and social rationality, and these can be resolved if either one accepts a weaker notion of individual rationality that is closer to social rationality or accepts constraints on social rationality.

PARTHASARATHI BANERJEE

Belief-attribution and Rationality: A Dilemma for Jerry Fodor

PIERRE JACOB

Introduction: Beliefs and Belief-ascriptions

In this paper, I will assume that human beings do have beliefs and other propositional attitudes.[1] Along with token physicalists generally, I will assume that an individual's beliefs—particular mental-state tokens—are brain-state tokens of the individual. Furthermore, I will distinguish between beliefs, utterances whereby individuals express their beliefs, and belief-ascriptions or belief-reports whereby beliefs are attributed by one person to another.

Beliefs may represent states of affairs, and so may utterances whereby beliefs are expressed: Sarah's belief that London is pretty, as it might be, represents the fact that a particular city bears the property expressed by the English predicate 'pretty'. And so does the sentence she may utter to express her belief. The fact that Sarah has this belief—if she does—may explain why she does certain things—why, for example, she utters the French sentence 'Londres est jolie' which means that London is pretty.

A belief-report—such as 'Sarah believes that London is pretty'—is a (linguistic) representation of a belief (itself a mental representation of a state of affairs). Although Sarah may not speak a word of English, she may nevertheless hold a belief which an English speaker may report as the belief that London is pretty. The belief-ascriber would then use an English sentence to characterise the content of Sarah's belief. A belief-report is an utterance which expresses the belief-ascriber's belief about the ascribee's belief (about London, in this instance). Let us say that a person expresses her belief by making an assertion. So if Sarah, having learnt enough English and Roman history, were to utter the sentence, 'Cicero was a Roman orator', she would thereby express her belief that Cicero was a Roman orator.

Beliefs are paradigmatic psychological states with content ('propositional attitudes' as they have come to be called since Russell) and they play basically three roles. First, beliefs help explain what a person does intentionally. Second, a person *B* may indirectly derive information about aspects of the world not directly observable by him/her by forming a higher order belief about *A*'s belief about those aspects. Finally, a person's beliefs are subject to evaluation for consistency and/or rationality. Assessment of a person's rationality primarily involves an examination of the logical relationships between his or her beliefs.

Modes of Presentation and Frege's Constraint

On our unsophisticated conception of what it takes a person to be minimally rational, we would not consider Sarah to be irrational were she to express her belief that, unlike Tully, Cicero was a Roman orator, by uttering the sentence, 'Cicero was a Roman orator but Tully was not' or 'Cicero, not Tully, was a Roman orator', even though (we know that) Cicero was no other than Tully. What Sarah would thereby display is ignorance, not irrationality: she might well be rational and lack the knowledge that 'Cicero' and 'Tully' are two English names of one and the same Roman orator. Lacking a belief is not the same thing as holding two contradictory beliefs. Using the Fregean notion of a 'mode of presentation' (which Frege himself called a *Sinn*), we might say that Sarah believes of Cicero that he was a Roman orator under one mode of presentation and she disbelieves it under another mode of presentation. However, on the same unsophisticated conception of rationality, Sarah would display irrationality were she to hold the belief she would normally and sincerely express by means of her utterance, 'Cicero was and Cicero was not a Roman orator'. Her linguistic behaviour would then leave no room for a distinction between two modes of presentation of Cicero.

In the philosophical literature, there are two rival views of the proposition literally (or explicitly) expressed by the utterance of a sentence containing a singular term (such as a proper name, an indexical or a demonstrative pronoun): the Fregean view and the theory of direct reference (TDR). According to the former, the proposition explicitly expressed by the utterance of a sentence containing a singular term contains a mode of presentation of the referent of the singular term. According to the latter, the proposition contains the referent itself. On this latter view, such an utterance expresses what D. Kaplan (1979: 387) calls a Russellian 'singular proposition'.

The Fregean notion of a mode of presentation has been designed to account for the linguistic and psychological intuition according to which a person may believe of one and the same object both that it has and that it does not have a given property. Indeed, modes of presentation obey what Schiffer (1978: 180 and 1992: 502–3) calls 'Frege's Constraint':[2]

> First it says that a rational person *x* may both believe and disbelieve that a certain thing . . . *y* is such and such only if there are distinct modes *m* and *m'* such that *x* believes *y* to be such and such under *m* and disbelieves it to be such and such under *m'*. Then it says that there are distinct modes of presentation *m* and *m'* such that a rational person *x* believes *y* to be such and such under *m* and disbelieves *y* to be such and such under *m'* only if *x* fails to realise that *m* and *m'* are modes of presentation of one and the same thing. In other words, you cannot rationally believe and disbelieve something under one and the same mode of presentation, or under modes of presentation that you realise are modes of presentation of the same thing.

No philosopher I know of denies that something like Frege's notion of a mode of presentation of the referent of a singular term is involved in a person's thought about an object. However, what distinguishes the advocate of TDR from Frege and neo-Fregeans is that the former, unlike the latter, holds that the proposition explicitly expressed by the *utterance* of a sentence containing a singular term, unlike the thought expressed, contains no mode of presentation of the referent.

Fodor's Three Assumptions

My goal here is to examine a view of belief-ascriptions recently (1990b) defended by Jerry Fodor in a short paper, 'Substitution

Arguments and the Individuation of Beliefs' (SAIB). Fodor's view of belief-attribution bears a strong family resemblance to TDR according to which the propositional contribution of a singular term is *always* its referent, not some mode of presentation of the referent. *Always*, that is, in all contexts, including contexts of propositional attitude attributions such as 'Sarah believes that Cicero was a Roman orator'. I will argue that Fodor faces a dilemma: either he must revise his Formality Condition or he must weaken his rejection of Content Holism—or he must do both. In order to do so, I shall first lay out the three major assumptions which underlie Fodor's view of scientific psychology.

The Language of Thought Hypothesis

One of Fodor's great ambitions is to naturalise content (or intentionality), that is, to offer sufficient conditions for some symbol to be a representation or to have content. His strategy is twofold: first, he wishes to reduce all non-mental contents to mental contents. This first semantic reduction proceeds in two steps: first step, reduce the semantic properties of linguistic symbols to the contents of propositional attitudes; second step, reduce the contents of propositional attitudes to the semantic properties of mental symbols—which, of course, presupposes the existence of a language of thought.

Once non-mental content has been reduced to the semantic properties of mental symbols, then reduce the latter to the notion of information (or nomic dependencies between properties) via what he calls the Causal Asymmetric Dependence Condition. Suppose that the tokening of my mental symbol 'horse' (my concept of a horse) nomically depends on horses or on the instantiation of the property of being a horse. Then my 'horse' symbol would nomically depend on horses (or on the property of being a horse). Sometimes my 'horse' symbol is prompted by a donkey (something which instantiates the property of being a donkey). What the Causal Asymmetric Dependence Condition says is that all non-horse caused 'horse' symbols (such as my donkey–caused 'horse' symbol) asymmetrically depend on my horse-caused 'horse' symbols. If horses did not cause me to token my 'horse' symbol, donkeys (and other things) would not cause me to token my 'horse' symbol either. However, the converse is not true: my horse-caused 'horse' symbols do not so depend on my non-horse caused 'horse' symbols. The Causal Asymmetric Dependence Condition on content seems to state in non-semantic terms sufficient conditions for a symbol to possess some semantic property. It is, I think, clear how Fodor's approach to the naturalisation of mental content pushes him towards an atomistic theory of content and away from Content Holism.[3]

Intentional Realism

Common sense, I believe, assumes that propositional attitudes can be causes of intentional actions. Furthermore, my belief that the cup in front of me contains tea (together with my desire for a sip of tea) may cause me to pick up the cup by virtue of its content. I will call Intentional Realism the view that intentional behaviour falls under *ceteris paribus* psychological intentional laws: intentional behaviour can be nomologically explained by psychological intentional laws. The antecedent of such laws will typically refer to (or quantify over) the content of an agent's propositional attitude and its consequent will mention the agent's behaviour. So I do something intentionally because (among other things) of what I believe. For an advocate of Intentional Realism, the connection between what I believe and what I do (when I do what I do because I believe something) is covered by a *ceteris paribus* psychological intentional law.[4]

The Formality Condition

According to the second step of the semantic reduction, the content of a propositional attitude reduces to the semantic property of a symbol of the Language of Thought. Now, Fodor assumes that semantic properties of symbols are not causally efficacious *per se*. Only *syntactic* properties of symbols are. Mental processes are purely formal (or syntactic) operations on mental symbols. In other words, mental symbols undergo causal processes by virtue of their syntactic, not their semantic, properties. This assumption is known as the Formality Condition and is characteristic of the so-called Computational Theory of the Mind.[5]

You might, like Stich (1983: 188), wonder how the Formality Condition is consistent with Intentional Realism. The reason the two views are consistent is that psychological intentional laws and cognitive processes do not belong to the same level. The Formality Condition applies to mental processes, while Intentional Realism applies to psychological laws. The latter are supposed to be 'implemented' by the former in much the same way that biochemical processes implement Mendel's laws of inheritance of hereditary features.[6]

Now Fodor takes Intentional Realism to be inconsistent with what he calls (Fodor 1987: 56) Content Holism, 'the idea that the identity—specifically, the intentional content—of a propositional attitude is determined by the *totality* of its epistemic liaisons', where proposition *q* is an epistemic liaison of proposition *p* for an agent *A* if *A* takes the truth of *q* to be relevant to the semantic assessment of *p*. So he rejects Content Holism and espouses Content Atomism or what he calls the 'denotational' theory of content.

Frege's Puzzle and Belief-ascriptions

Consider utterances of sentences (1) and (2) which differ from each other in that in (2) 'Cicero' is replaced by the coreferential name 'Trully':

(1) Cicero was a Roman orator.
(2) Tully was a Roman orator.

On TDR, an utterance of (1) expresses the same proposition as an utterance of (2). Now, consider the following utterances:

(3) Sarah believes that Cicero was a Roman orator.
(4) Sarah believes that Tully was a Roman orator.

Sentence (1) is a syntactic constituent of (3) and (2) is a constituent of (4). According to TDR, the singular terms 'Cicero' and 'Tully' contribute nothing other than their respective referent to the propositions respectively expressed by an utterance of (3) and by an utterance of (4). Now, since 'Cicero' and 'Tully' have the same referent, the proposition expressed by an utterance of (3) is no other than the proposition expressed by an utterance of (4).

In 'On Sense and Reference', Frege (1970) advances the following argument against TDR:

(*a*) If an utterance of (1) and an utterance of (2) did express the same proposition, then belief-ascriptions (3) and (4) ought always to have the same truth–value (for a given assignment of referent to 'Cicero' and 'Tully').
(*b*) (3) and (4) may have different truth-values.
(*c*) Conclusion: an utterance of (1) and an utterance of (2) do not express the same proposition.

Frege concludes that the contribution of a singular term to the proposition expressed by the utterance of a sentence containing it is its sense (*Sinn*) or the mode of presentation of its referent.

Salmon's response

In his book, *Frege's Puzzle*, Salmon (1986), who argues for TDR, rejects premise (*b*) of the Fregean argument. In favour of this premise, common sense has it that Sarah, a perfectly rational person, could well assent to the proposition expressed by an utterance of (1) or she could express that proposition by uttering (1). Concomitantly, she might rationally dissent from the proposition expressed by an utterance of (2) and rationally utter the negation of the proposition expressed by an utterance of (2).

In such a situation, common sense would unhesitatingly say that this is a case in which an utterance of (3) is true and an utterance of (4) is false. Salmon claims that, if the proposition expressed by an utterance of (3) is true, then so is the proposition expressed by an utterance of (4). In his view, even though Sarah might give all conceivable evidence that, although she holds true the proposition expressed by an utterance of (1), and does not hold true the proposition expressed by an utterance of (2), nonetheless utterances of (3) and (4) have the same truth-value. So if (3) expresses a truth, then so must (4). On Salmon's view then, given that Sarah holds true the proposition expressed by an utterance of (1), she is committed to holding true the proposition expressed by an utterance of (2). If she does not, then she must be confused, not quite rational.

Salmon's strategy is twofold. On the one hand, he holds that the logical form of the proposition *explicitly* expressed by either belief-ascription (3) or (4) is:

$(\exists m)$ BEL (Sarah, p, m)

where 'BEL' is a three place predicate, 'p' is the Russellian singular proposition (expressible by an utterance of either (1) or (2), and 'm' is the mode of presentation of the singular proposition (which is no other than the mode of presentation of the singular term). So the proposition explicitly expressed by an utterance of either (3) or (4) contains an existential quantification over modes of presentation of Cicero. It does not refer to any mode of presentation of Cicero.

On the other hand, Salmon holds that, in addition to the unique proposition explicitly expressed by an utterance of either (3) or (4), reference is made to particular modes of presentation of Cicero in what is communicated by utterances of (3) and (4). What is communicated by an utterance is richer than the proposition the utterance explicitly expresses for it includes what is *implicitly* communicated. What is communicated by an utterance of (3) and what is communicated by an utterance of (4) contain a reference to a particular mode of presentation of Cicero.[7] But the mode of presentation of Cicero referred to by what is communicated by (3) differs from the mode of presentation of Cicero referred to by what is communicated by (4).

So an utterance of (3) and an utterance of (4) explicitly express the same true (let us suppose) general proposition, which contains an existential quantification over modes of presentation of Cicero. In addition, an utterance of (3) conveys implicitly the true proposition that Sarah recognises the singular proposition under the mode of presentation of Cicero associated with 'Cicero' and an utterance of (4) implicitly conveys the false proposition that Sarah recognises the same singular proposition under the mode of presentation associated with 'Tully'.

Salmon's account is strongly revisionistic: if he is right, then ordinary usage errs in supposing that an utterance of (3) can express a true proposition, and an utterance of (4) a false proposition. Suppose Sarah gives all conceivable evidence that she holds false the proposition expressed by an utterance of (2). This would suggest that an utterance of (4) is false. Then (5) would seem to express a true proposition:

(5) Sarah does not believe that Tully was a Roman orator.

Intuitively, (5) seems to deny what (4) asserts. Salmon assumes a distinction between the proposition explicitly expressed by an utterance of (4) and what is overall communicated by an utterance of (4). He seems therefore to face the following dilemma. He may suppose that what is denied by an utterance of (5) is either the proposition explicitly expressed by an utterance of (4) or the overall content communicated by it.

Suppose he chooses the first possibility. On his view, the proposition explicitly expressed by an utterance of either (3) or (4) is the true general proposition to the effect that there exists a mode of presentation of Cicero under which Sarah believes that Cicero is a Roman orator. Suppose, then, that he takes the view that the proposition explicitly expressed by an utterance of (5) is false, not true. Remember: Salmon had the resources to argue that although an utterance of (4) is true [if an utterance of (3) is], still, an utterance of (4) may misleadingly convey the false information that Sarah believes that Cicero was a Roman orator under the mode of presentation of Cicero associated with 'Tully'. However, it is hard to see how Salmon could explain how by expressing a literally false proposition, an utterance of (5) may simultaneously communicate overall what we take to be true information.

Suppose Salmon chooses the second possibility. He may assume that what is denied by (5) is not the proposition explicitly expressed by an utterance of (4) but the overall content communicated by it. The hard problem then faced by Salmon would be to explain how something which was merely communicated by, and was not part of the proposition explicitly expressed by, an utterance of (4), becomes, by virtue of the presence of the negation in (5), part of the proposition explicitly expressed by an utterance of (5), i.e. part of the truth-conditions of an utterance of (5).[8]

Fodor's Account

In SAIB, Fodor (1990b) considers a variant of Frege's argument against TDR:

(*e*) The content expressed by a complex expression is a function of the contents of its constituents.

(*f*) The content of an expression is its denotation.
(*g*) Two expressions with the same content can always be substituted one for the other *salva veritate* without altering the truth-value of the more complex expression.
(*h*) As examples (3) and (4) show, two expressions with the same denotation are not always substitutable one for the other in a complex expression without altering the truth-value of the complex expression.
(*i*) Frege concludes that the culprit is assumption *f* and that the content of an expression does not reduce to its denotation.

Because he wants at all costs to avoid Content Holism, Fodor's goal is to protect *f* from Frege's argument, i.e., to show that the non-substitutability *salva veritate* of coreferential expressions in belief contexts is compatible with the assumption that they have the same content. Since the context in which the substitution is performed is a belief context, Fodor's defence of the denotational theory depends on a theory of belief-attributions. Before arguing for his theory, Fodor refines his theory of belief-*states* as presented in his book, *Psychosemantics* (1987). *Believing* is now treated as a *four* place relation between (*a*) a person, (*b*) a proposition (or content), (*c*) a vehicle (or mental symbol), and (*d*) a functional role.

The propositional content is what in other contexts Fodor and other philosophers call *broad* content. The vehicle is a symbol token; it has therefore syntactic and semantic properties. Two vehicle tokens belong to distinct types either if they are syntactically different or if they express different contents. Two vehicle tokens belonging to two distinct types can—like symbols of public languages—express one and the same proposition. Fodor's main innovation resides in his insertion of the functional role as one of the parameters of the belief relation. The functional role (of the vehicle) is its inferential role in a belief system: 'its causal role in (certain) mental processes' (SAIB: 168). Two vehicle tokens belonging to one and the same type will have distinct functional roles if they belong to different belief systems. Conversely (*cf.* SAIB, note 8: 176), two vehicle tokens belonging to two distinct types (with distinct syntactic or semantic properties) may have one and the same functional role: they may contribute to the tokening of one and the same belief within two distinct belief systems.

Since believing is a four-place relation, in Fodor's view a person may entertain two different belief-states towards one and the same propositional content. The difference in belief-states will come from a difference between either the vehicles or the functional roles. The Fodorian view does not have Salmon's revisionist implications.

From Fodor's standpoint, one can distinguish more or less specific belief-attributions according to whether they aspire to capture, besides the content, the vehicle of the belief. A vehicle-specific attribution will by assumption be more specific than an attribution which merely captures the content. According to Fodor, the norms guiding an ascriber in his or her specification of the vehicle are less constraining than the norms guiding the specification of content (in the sense that the former are subject to more pragmatic considerations than the latter):

> the embedded formula must express *the very proposition* that the 'beliefs' predicate attributes. I think, however, that it is otherwise with the specification of the vehicle: here everything is slippery and pragmatic. Roughly, what's required is a degree of isomorphism to the vehicle that is appropriate to the purposes at hand; and there isn't any purpose-independent specification of how much isomorphism is enough (SAIB: 171).

Let us represent the less specific attribution (the merely content-specific attribution) by means of a formula such as (A1):

$$(\exists x)\ (\exists y)\ ([x = \text{a vehicle}\ \&\ y = \text{a role}]\ \&\ [\text{BEL}\ \{\text{Sarah, P, x, y}\}])$$

where 'P' refers to the content expressible by an utterance of (1): 'Cicero was a Roman orator'. Such a formula, which refers to the propositional content, contains an existential generalisation over both vehicles and functional roles, neither of which is specified. Such a formula bears affinity to what is traditionally called a *de re* attribution; it captures so-called broad content.

We can represent a more specific attribution which also aims at specifying the vehicle by (A2):

$$(\exists y)\ ([y = \text{a role}]\ \&\ [\text{BEL}\ \{\text{Sarah, P, c, y}\}])$$

where 'c' refers to the vehicle (or mental symbol) in Sarah's head—a symbol 'isomorphic' to the English sentence (1): 'Cicero was a Roman orator'. Such a formula, which refers both to the content and to the vehicle, still generalises over functional roles. It is closer than (A1) to what is traditionally called a *de dicto* attribution and it delivers more specific information about the belief-state than (A1).

Suppose an utterance like (3) or (4) is taken to be as little specific as possible—specifying no more than the (broad) content of Sarah's belief on the model of the double existential quantification over vehicles and functional roles (contained in [A1]). On this interpretation, an utterance of (3) and an utterance of (4) contain no means of referring to the vehicle. The referential expressions have their ordinary references: they are mutually substitutable *salva veritate* in (3) and (4). And it is possible to hold true both an utterance of (3) and an utterance of (4). Let us now interpret an utterance of (3) as a more specific attribution aimed at specifying the content and the vehicle of Sarah's belief-state. In this case, we could represent respectively (3) and (4) on the mode of the simple existential quantification (over functional roles) contained in (A2):

(3′) (∃ y) ([y = a role] & [BEL {Sarah, P, c, y}])

where 'c' refers to the vehicle in Sarah's head 'isomorphic' to the English sentence (1): 'Cicero was a Roman orator'.

(4′) (∃ y) ([y = a role] & [BEL {Sarah, P, t, y}])

where 't' refers to the vehicle in Sarah's head 'isomorphic' to the English sentence (2): 'Tully was a Roman orator'. On this more specific interpretation, in spite of the fact that 'Cicero' and 'Tully' are coreferential or that an utterance of (1) and an utterance of (2) have the same truth-value, (3') would be a true formula and (4') would be a false formula.

Fodor's Dilemma

It follows that Fodor's theory is not open to the objections I voiced earlier against Salmon's. However, its strength might well turn out to be a weakness from within the general perspective of Fodor on scientific psychology as encapsulated by the three assumptions recapitulated above. Consider the fact that Fodor allows two vehicle-specific attributions to be representable by two logical formulae with distinct truth-values. He thereby concedes that the semantics of vehicle-specific belief-attributions is not strictly compositional: the content of an attribution—its truth-value—is not just a function of the contents (the denotations) of its constituents. It also depends on a non-semantic aspect of one or more of its constituents—non-semantic at least by Fodor's standards according to which the entire content of a belief-state resides in the proposition independent both of the syntactic properties of the vehicle and of its functional role. The difference between the truth-value of (3') and (4') can be attributed to a difference between the syntactic properties of the mental symbols (or the vehicles) in Sarah's head respectively 'isomorphic' to 'Cicero' and to 'Tully'. From Fodor's standpoint on scientific psychology, one can assume that the explanatory success of a belief-ascription depends jointly on the ability of the ascription to refer to the intentional content and to the syntactic properties of the vehicle whose semantic property is the content referred to in the ascription, since according to the Formality Condition, the causal properties of a mental symbol are its syntactic properties.

Now, consider a *ceteris paribus* commonsense psychological generalisation such as 'If a person *X* forms the perceptual belief that a 400-kg bull is galloping towards her, other things being equal, she will acquire the desire (or the intention) to distance herself from the bull's trajectory.' The antecedent of such a generalisation contains a belief-ascription which, from Fodor's point of view, is analysable as a relation between a person *X*, proposition *P* (that a 400-kg bull is galloping towards the person *X*), a vehicle and a functional role. The consequent contains a desire-attribution analysable as a relation between the person referred to in the antecedent, the proposition *Q* (that *X* locate herself at a respectable distance from the trajectory of the bull referred to in the antecedent), a vehicle and a functional role.

Unlike vehicle-specific attributions such as (3′) and (4′) whose truth-value and explanatory role depend—besides reference to content—upon the syntactic properties of the vehicle, the explanatory success of the generalisation may depend upon the *functional role* of the attributed belief. The generalisation will successfully explain the behaviour of a person who prefers to maximise the distance between herself and the 400-kg bull galloping towards her. But a person may believe that she is on the trajectory of a 400-kg bull without believing herself to be that person. In other words, a person may believe *of* (or *about*) herself that she is on the trajectory of a bull without believing that she is such a person. Imagine a complicated system of mirrors by means of which a person sees herself on the trajectory of a 400-kg galloping bull without realising that she is that person.

Alternatively, a person may believe that a 400-kg bull is galloping towards her and nonetheless prefer to remain on this trajectory because her belief system includes the belief that the soul of a person trampled to death by a bull will go straight to heaven and, other things being equal, she prefers for her soul to go straight to heaven. A person with such a belief and preference may have the belief that a 400-kg bull is galloping towards her. She may have, according to Fodor, a belief with the same intentional content as the belief of a person who does not believe that the soul of a person whose body has been trampled to death by a bull will go straight to heaven. The vehicle by means of which *X* (who believes that the soul of a person whose body has been trampled to death by a bull will go straight to heaven) believes that a 400-kg bull is galloping towards her may be a mental symbol token of the same type as the vehicle by means of which *Y* (who does not believe that the soul of a person whose body has been trampled to death by a bull will go straight to heaven) believes that a 400-kg bull is galloping towards her. The difference between the belief-states of *X* and *Y* when they both believe that a 400-kg bull is galloping towards them comes from the insertion of two vehicle tokens of the same type within two distinct belief systems.

In these conditions, Fodor's theory of belief-ascriptions seems to face the following dilemma: either Fodor takes the functional role to be an ingredient of the causal role of the vehicle; or he takes it to be an ingredient of the content of the vehicle. In the first alternative Fodor assumes that the inferential potential of a belief (the belief that a 400-kg bull is galloping towards the person having this belief) contributes to the causal role of the belief. In this case, he must concede that the causal role of a mental representation (a vehicle) does not reduce, contrary to the Formality Condition, to purely syntactic properties. Perhaps Fodor could distinguish between a narrow causal role (possessed by a symbol by virtue of its syntactic properties) and a broad causal role (attributed to the functional role of the symbol). The latter, not the former, might depend upon the doxastic environment of the vehicle. But this would imply a broadening (or a liberalisation) of the Formality Condition. In the second alternative Fodor denies that the functional role of a symbol has causal efficacy and he takes the functional role to be an ingredient of the content of the mental symbol. In this case, he must give up a purely denotational (atomistic) theory of content and revise his rejection of Content Holism. In my own view, what Fodor ought to do is both.[9]

Notes

1. This assumption is more controversial than it may sound at first since some philosophers (such as Churchland 1989) deny it.
2. Schiffer (1992: 502–3).
3. See Fodor 1990a. Intentional Realism too pushes him towards Content Atomism and away from Content Holism.
4. See Fodor 1987 (chs. 1 and 3).
5. The locus classicus is Fodor 1981b. See also Fodor 1987 (ch. 1).
6. Cf. Jacob 1991a.
7. Salmon assumes that the content implicitly communicated by an utterance is generated via Gricean mechanisms.
8. The last two paragraphs were prompted by remarks by Recanati. I have criticised Salmon in detail in Jacob 1991b.
9. I am grateful to the organisers and participants of the NISTADS Workshop on Rationality in Cognitive Science and Social Science.

References

CHURCHLAND, P.M. 1989. *A Neurocomputational Perspective*, Cambridge, Mass.: MIT Press.

FODOR, J.A. 1975. *The Language of Thought*, New York: Crowell.

———. 1981a. *Representations*, Cambridge, Mass.: MIT Press.

———. 1981b. Methodological Solipsism Considered as a Research Strategy in Cognitive Psychology. In J.A. Fodor 1981a.

FODOR, J.A. 1987. *Psychosemantics*, Cambridge, Mass.: MIT Press.

———. 1990a. *A Theory of Content and Other Essays*, Cambridge, Mass.: MIT Press.

———. 1990b. Substitution Arguments and the Individuation of Beliefs. In J.A. Fodor 1990a.

FREGE, G. 1892. On Sense and Reference. In P. Geach and M. Black (eds.), *Translation from the Philosophical Writings of Gottlob Frege*, Oxford: Blackwell, 1970.

JACOB, P. 1991a. Are Mental Properties Causally Efficacious? *Grazer Philosophische Studien* 39, 51–73.

———. 1991b. Semantics and Psychology: The Semantics of Belief-Ascriptions. In N. Cooper and P. Engel (eds.), *New Inquiries into Meaning and Truth*, Hertfordshire: Harvester.

KAPLAN, D. 1979. Dthat. In P.A. French, T.E. Uehling and H.K. Wettstein (eds.), *Contemporary Perspectives in the Philosophy of Language*, Minneapolis: University of Minnesota Press.

SALMON, N. 1986. *Frege's Puzzle*, Cambridge, Mass.: MIT Press.

SCHIFFER, S. 1978. The Basis of Reference. *Erkenntnis*, 13, 171–206.

———. 1992. Belief Ascriptions. *The Journal of Philosophy*, lxxxix, 10, 499–521.

STICH, S. 1983. *From Folk Psychology to Cognitive Science*, Cambridge, Mass.: MIT Press.

Action Sentence and Action for Welfare

PARTHASARATHI BANERJEE

Introduction

There is currently a great deal of debate regarding whether the injunctions on actions are the same as injunctions on knowledge (or reason). On one side are proponents who hold that the injunctions on knowledge are the same as injunctions on actions undertaken by an actor. The other side of the debate holds that these two types are not the same. Proponents of the view that the two injunctions are identical do not propose a singular theory. In the Indian theoretical discourses there were strong and very subtle arguments favouring Jnan Vidhi (injunctions on knowledge), especially as propounded by the Bhartrmitra and his school, as also by some advaita (the nearest equivalent of which would be monist) philosophers. In contemporary Anglo-American philosophy too, although it is significantly different from Indian philosophy in terms of both purpose and content, there are positions that defend reason (or impartial reason) alone. For example, in the Anglo-American debate attention is devoted to the question of rational choice prior to the action proper; questions of morality or justice are considered either as an input to or as an outcome of such a choice scheme.

Rationality acts as a qualifier of choice, although it is often defined in terms of choice; choice, too, if not defined in terms of rational choice, would mostly derive from a psychology of maximisation, which again may be described by some in terms of rational choice. The debate thus centres around a construct of contradiction between reason and matters of fact on the one hand

and passion, interest and action on the other. Even when a causal (or intentional) explanation of action is offered, the engagement of reason with intention remains unexplored. It is thus difficult for the proponents of the causal view to explain intending, or to do away with modes of descriptions, or to justify how propositional pride may provide reason for both action and emotions, etc. There seems to be an uneasy relation between reason(ing), intention (intending) and act (action).

The near exclusion of reason or cognition from (human) action dissociates description (especially the scientific) from action-sentences. It also dissociates the moral from desire or interest (if not wholly, at least partially—even while using rational restraints). This distinction appears most acute in the separation of morality (into the public sphere) from desire (into the private sphere). In order to overcome this dissociation, one approach has been to break up the public sphere into a series of individuated events, or to reduce the private (name) into a tag (or a quantified common name, etc.). Inductive causal laws are then replaced by rules, norms. The demand of reason then appears as being normative. Such an approach would, however, require a teleology that would in turn demand a presiding sovereign, a personal abiding interest and a belief. Injunctions offer the scope. Speech, injunction and verbal cognition, we would argue, give us the scope to understand and explain cognition, desire and action in a causal framework, such that in order to act for one's welfare one remains unencumbered by the rational–moral or reason–passion dichotomy. If a sentence is understood in a certain way, which we also call rational, injunctions, desire, and action follow. This prescription is not merely for acting but for arriving at the fruit of action; it is not meant for arriving at a rational–valid cognition. Injunctions are also derivable from a discourse, from observing a discourse-driven action.

It is not the purpose of this article to show that these strands of opinions and theorisation, as mostly available in Anglo-American literature do indeed share with Indian literature a common guiding principle such as replacing injunctions on actions by injunctions on knowledge. But assuming that this is so, this article defends a view on injunctions on action that considers welfare to be the fruit of action. The points that will be raised are from certain Indian tracts, although when necessary, references to other Indian views that are contrary to our argument will also be included.

I shall argue that words cause pravrtti; the nouns, verb root, cases, inflexions etc., all are employed to initiate a person. Because of this, the cause of the cognition that causes pravrtti (the disposition to act) is owing to the word-universal. However not all the words (or parts of a sentence) cause pravrtti (the cognition of meaning over the associated property of being action)—with the same directness. Some terms (injunctions) cause cognitions of meaning having content over karyatvanvita through directly causing feelings; while some others (the arthavad) cause feelings that only indirectly generate meanings over karyatvanvita. We can therefore infer injunctions even from descriptions, or narrations. Such imaginative inferences, that initiate pravrtti indirectly, complete a sentence. The narrations and descriptions are therefore not reduced to simple descriptions of facts. Cognitions from words *are* possible through the completion of a sentence. This completion entails causing the disposition to act in discourses or descriptions where there is no explicit use of injunctions.

Rational Action and Complete Sentence

Every action is followed by a change; therefore, an action is a cause. However, we define action as a particular type of cause. Action is distinct and different from a relation in such a way that whatever appears as the non-associated cause of relation is to be understood as an action. A typical example of an action is contraction, or motion etc. For our purposes, it is not necessary to enter about whether an action can only be inferred or can be perceived directly. It is sufficient to briefly explain this notion of causality. Whatever always appears before the fruit of action is called a cause. The fruit of action (or act, or action) is defined as the negation of a prior-absence. An example of prior-absence is that absence implied in the usage: 'here (should be a) pot'. A pot is the negation of this absence, and is therefore an act. In this essay, I often use the term 'action' in the sense conveyed by the use of pot here. A pot or a mat is a karya, and an action, as defined first, is a karma.

A pot or a mat appears to be associated with one type of cause. For example, a mat remains associated with its cause, the thread. This thread is called an associative cause. The form (rupa) of pot is associated with the pot; therefore the pot is the associative cause of the form 'pot'. We may now consider the association of threads in a mat. This association of threads is a cause of the mat and is called the non-associated cause, which can be defined as a cause that remains associated on the same entity as the fruit-of-action or the cause. The intermeshing of threads is thus a non-associated cause for the mat or, the form (rupa) 'thread' is the non-associated cause of the form 'mat'. In the first example this cause remains associated with the fruit of action, viz., the mat. In the second example of form, this cause remains associated with the cause (and not with the fruit). Briefly, the relational cause that appears along with an associated cause is called a non-associated cause. While an associative cause is generally material, a non-associated cause, is either a quality (or property) when applied to materials, or is a quality, or is an action when applied to quality. A third type of causality, viz. the contingent causality, is also proposed by a group of theorists. However, as will be explicated in this essay, the associative cause is nothing but the material cause, and the non-associated cause is nothing but a component of the contingent cause.

Let us now look at the change, which I will refer to as parinama, that follows or remains associated with an action. A person can act in three ways: physically, through speaking, and mentally. Though these actions and their consequent parinamas are distinct, they are not absolutely inidentical (identity = abheda; inidentity = bheda or absolute non-identity). These actions are therefore neither absolutely identical nor absolutely inidentical. A distinct physical action is sort of behaviouristic. There are actions of speaking (but not the same as 'speech act') and actions of cognising in mental affairs (called 'manan'). One can get a better idea about the identicality and inidenticality of actions through the relations that the parinamas have between themselves.

According to the Vyasbhasya of Patanjal Sutra (3/13), there are three types of parinamas,[1] namely, dharma–parinama (change in mental states, properties or qualities), laksana–parinama (closer to semantic, or suggestive changes in meaning) and avastha–parinama (configurational or situational change). Here dharmi (the bearer of dharma) has as its locus the dharma (the locus of changes in states, the distinct etc.; otherwise also called asraya, the residence). Changes in dharma are therefore located in the dharmi (for example, mental metamorphoses are located in the states).

Changes in laksana are located in the dharmi and changes in avastha are located in the laksana. A parinama is the disappearance of one state in the locus followed by the appearance of another in the same locus.

Avastha–parinama is a reflection of physical action. No physical action can be undertaken without a corresponding change in its locus, the laksana; nor can physical action be undertaken without the employment of a sentence (as spoken, if not to a listener, at least to oneself). A word is not the bearer of laksana; only a sentence bears it. Hence there cannot be any pure behaviouristic (or motoric) action. Such an action is only possible through a machine. Similarly nothing can be spoken without a change in its own dharmi, or without a corresponding change in the mental state. Nor can there be any dharma–parinama, i.e. changes in dharmi (the bearer) without a respective change in the dharma. The dharmi, the mind-substrate is 'karyattva' (the property of being action), on which appears the karya (action), as exemplified in the appearance (bhava) of a mental state. Karyattva is therefore the uninterrupted real, on which the parinama appear and disappear.[2]

A sentence (unless it is the chanting of a snake-charmer) is therefore made complete or 'one' in the dharma by the employment of a dharmi and, if necessary, by the employment of laksana as well. A complete sentence has four parts (or is fourfold). These are vidhi, arthavad, mantra and namdheya; when put together these constitute one sentence. A vidhi is injunctional, optative. Arthavad refers to description; mantra, according to Yaska[3] refers to that which concerns itself with mental activity (like reasoning, for example). A mantra is somewhat comparable to a rule. Namdheya refers to names that are uttered. A sentence that refers to one complete act or one complete cognition, is partly injunctive; it is partly descriptive, in that it indirectly refers to the injunctive; it is partly mental in that it reminds one either of injunctions and what is desired, or of those materials (or their qualities) which are being acted upon; and partly it distinguishes and differentiates through naming.[4] The point to emphasise is that one complete sentence, refers to one complete cognition. The property of such a complete sentence, called 'ekavakyata', distinguishes one cognition by way of delineating the imperatives attached to an action (and also the act), the karan (the cause) and the karya (the act). Such a sentence is therefore at least initially a dharmi defined on and associated with the dharma.

It may be observed that this one sentence binds together all the three modes of actions. The pivotal role is played by the vidhi, which is usually expressed through an inflexion like a grammatical instrument (called vidhipratyaya), such that the action is never reduced to performing the verbal root alone. A vidhi is therefore an injunction on the mind defined on the mind-substrate, the dharma; and is the residence or locus not only of mental acts, but also of sentential and physical acts. A mantra unites or conjoins (synthetically or syntactically the term used is anvaya). It does so in three ways, such as by conjoining the vidhi action with the substrate karyattva, or with the sentence (which appears through laksana), or with the physical or dispositional (which appears through laksana) or with the physical, dispositional (which acts on substance, real). That is why Yaska describes three types of mantras.[5] The first relates action to the deep philosophical or spiritual positions; the second expresses itself and the third describes the other, the material (on which action is being performed). Arthavad, as has been noted, is not empowered to indicate a complete meaning (a cognition with vidhi) of the action by itself alone. It follows a vidhi, by praising and describing it. This is so because it describes a siddha (an accomplished matter or a real) and therefore cannot prescribe by itself a performance or a complete cognition. This siddha is a particular. The absence of an arthavad (and often a mantra) is likely to render the action, as prescribed through the vidhi, into an imaginary act. An action without the associate material or real (devata) is unknown, hence a vidhi is to be followed by arthavad (and mantra). Thus an arthavad sometimes describes a quality (predicating on). It would otherwise describe or distinguish a particular. It may be reasoned, then, that an arthavad resides in the vakya i.e. the sentential part and is characterised by laksana. A proposition in science, for example, is an arthavad. As a follower of the vidhi, such a descriptive sentence can generate a performance on the physical–dispositional, on the material–real, or can associate itself with a sense or mode in the mental. Again, namdheya, residing in the sentential, associates itself with the meaning of the verb root of the vidhi and qualifies that meaning by way of designating or naming.

A vidhi, however, is not always spelt out. When heard, it gives rise to immediate cognition. This is called sruti and immediate cognition, since it unfailingly follows sruti, is a pramana (a validation). Arthavad, mantra and namdheya do not give rise to such an immediate cognition but attain to it only secondarily or indirectly. When the vidhi of a particular performance is not clearly spelt out, it is inferred, imagined and cognised through both arthavad and mantra. These two therefore appear as pramana only while in association with the heard or inferred (and cognised) vidhi. A vidhi would thus have an expectancy (in order to achieve one complete cognition or sentence) for arthavad, mantra and sometimes for the namdheya as well. This brings out the notion of a complete action or performance. A complete action thus requires the fourfold cognition of vidhi, viz. first that which speaks about the action (what is it? how does it begin); second, that which spells out the adhikara (who can or would perform it? an adhikara spells out the relation between the action and its fruit); third, that which speaks about the form of the total action, about the relations or benefits that the secondary actions have with or form the principal action; and the fourth vidhi speaks of the sequence that the actions should follow in the performance of the total action. Therefore, if only one vidhi were known (or told) beforehand, a performance, to be complete, would have to infer, imagine or reason out the three other vidhis.

This was a fourfold structure. It has been seen that both vidhi–arthavad–mantra–namdheya and karyattva–mental–sentential–physical and/or dispositional are also fourfold. The causal relation has been discussed as well. It is perhaps necessary to point out that here, a cause is more extensional (vyapta) and less direct or perceivable (suksma) then the caused; though another scheme, not necessitating extensionality may also be conceived. In this scheme, a movement away from the denoted, as though through the designated and the modal etc. would be considered not only as a movement towards the cause (through the caused and cause, act and the action, event and cause) but also as part of a complete cognition, sentence or performance. Such a sentence therefore contains not only the determinate (empirical or real) but also both the layers of causes and their relations. For example, a completion is achieved by going through a sequence from acts or events, through to their determination, to action, and finally to the injunctions. Thus, if we have to define rational, it would be in terms of a sentence, cognition or action which satisfies the criteria just outlined. This way of defining the rational, it may be noted, has injunction as both the actuator (or initiator, pravartaka) and the principal. Rationality resides in this oneness, in the unity of the four through their threefold parinamas (or at least threefold anvayas), through the causality to the final fruit or end of action; and also in their being four at the three levels of determinacy. In case the vidhi is not specified, it may arise in the cognition through inference or expectation from the descriptions given.

There remains, however, the question of whether the vidhis are heard (sruti) and therefore whether they are directly cognised.[6] While we cannot go into this question here, it may be noted that regarding a dharma, that knowledge of the vidhis unless specified, can be obtained through inference rather than direct perception or cognition. Therefore, when a vidhi has not been specified but only inferred, precedence or primacy should be given to another vidhi which is specified in the performance of an action, in case there is a doubt.

It may be noted that rational, as defined above, refers to the content or object (visaya) of the cognition. An actuator (pravartaka) is derived from the notion pravrtti (the disposition or will to act; such a state of mind is very different from attitude or habit). The cause of pravrtti is desire (or wish, called cikirsa) which in turn is caused by knowledge (or cognition). There are many views regarding the content of this cognition. The first criterion of rationality demands that the content of this sort of a cognition has to be such that it may cause a cikirsa which in turn would cause the pravrtti. To delineate a few positions on the nature of the content: one position maintains that it is the sense of the word–sound (sabda-bhavana); another maintains that it is the property of the goal-fulfilling, will-fulfilling actions (istasadhanata; 'ista' is difficult to translate); yet another maintains that it is the cognitive designator (jnapak) of the istasadhanata. In contrast, it seems reasonable to argue that this content (visaya) of cognition is nothing but karyata. This karyata, with the above-mentioned ideas on vidhi, is the cognition such that '(that) is the action (what ought to be done)'. It is therefore indicative of the property of 'ought' performances, called kartavyata. We argue that the cognition of 'that is the action' appears to be associated with the karyatva, the dharma.

However, the 'that' is actually not present in the first appearance of the cognition. Action does not appear as delimited by 'that'. The object of cognition and cognition, being similar, appear delimited by 'that' only when the cogniser employs the ego (abhiman) through an adjunct instrument. The very first appearance of cognition on and in the karyatva, is closely related to the notion of pratibha (prophetic, poetic, pure aesthetic). Such a cognition, not yet formulated as a sentence (or words, such as a vidhi), has also not yet associated itself (the cognition) with the self-referring speaker (or the cognising person). It is only after this that the cogniser, through the conjunct employment of its own self-referring, arrives at the cognition that 'that action can be (or ought to be) done by me'. This latter cognition may then be taken up by the cogniser, as though for the purpose of comparison with the cogniser's earlier state of desire fulfillment, at the level of cikirsa. Thus the first appearance of 'rational' is, say, in the form of '(that) is the action'. With the appearance in cognition of the ego, and then the 'that', there follows the appearance of the actor (kartrtva)–enjoyer (bhoktrtva), e.g., the cognition that 'that can be done by me', 'that (fruit) can be enjoyed by me'. Rational can therefore be defined as this karyata or kartavyata, which the content of cognition causes through cikirsa, the actuator, etc.[7]

Cognition and Injunction on Action

The competing cognitivists who argue that injunctions on action are indeed injunctions on knowledge (for some it is belief), would now argue that the meaning of a vidhi lies in actuating (or prescribing); in 'pravartana' and not in the good or in the absence of non-good. The vidhi-designator (vidhi-pratyaya) does not signify pravrtti. The vidhi-designator, it is argued, signifies the fact of 'actuating owing to the pravrtti state of mind' only as a class and it cannot signify the particular causality of the pravrtti state of mind. These cognitivists maintain that the vidhi-designator would enable an action to be performed but need not necessarily relate to the good (artha). The example given by them is that of the four uses of vidhi, viz., order, pray, request, advice. In the first two of these four cases, the person receiving an order or the person who is praying would not, while acting, pay attention to the content of cognition (or as a result of that, the pravrtti) of the person who has ordered, or prayed etc. Such a cognition is necessary only in the third and fourth uses of vidhi, viz., request and advice. Here, the pravrttis are different, but pravartana resides in all the four uses of vidhi. Pravartana being the common factor, a vidhi-designator would signify merely actuating.[8]

While agreeing to the fact that injunctions exist, this argument does not permit the actions under injunction to bear any specific fruit (phala), or good. According to proponents of this view, a sentence is pivoted around the verb, and since the other terms of the sentence are relative to it through the karaka (a grammatical device, the near-equivalent of which would be case), the sentence acquires unity through the verb, the fulfilling action (sadhan). Primacy of such an action, or the verb, is due not only to the nature of words but also to the nature of the (determinate) material or its constituent causes, the guna. The fulfilling action takes up, for example, the performance (which is to be acted because of the vidhi), with the help of what is fulfilled (siddha). This argument of the competing theorists refers to a position described in Jaimini Sutra 6/1/1, in which the constituent cause (guna) or the material which exists (as fulfilled, the siddha) enters the performance (in the action) as its constituent (guna). Satisfaction (or heavenly pleasure, a pleasure which is not limited by pain) delimited in material or space; or justice, morality delimited in material or space, etc. enter the performance as input guna. The vidhi-action, according to these theorists, therefore, cannot bear these inputs, as its fruit (or good). The input here is the fulfilled (or accomplished, siddha). Further, if the verb is considered to be the inflexion (akhyata—that in which the action or the verb is principal) alone and not the entire word for performance as the goal or end, even then this inflexion cannot be associated with the fruit of the performance. The end or goal is the performing itself.

It is further argued by these theorists that even if vidhi on action and its specific fruit-bearing nature is accepted, these actions and fruits would only entice the performer towards knowledge (or rational, absolute). The example adduced takes the following form: if some Devadatta is asked to visit a far-off town and is expected to reach it via a number of villages on his way so that he can stay and enjoy the fruits of those villages as well—then though Devadatta enjoys the fruit-bearing villages on his way as per the injunction, this enjoyment as an itinerant (performer) only enables him to obtain the final fruit of getting to the town (the rational, absolute knowledge). This, it is argued, is as per the ekavakyata. Ṣuch injunctions, according to this line of argument, by allowing the passion of the actor to be in action while performing only, enables the actor to overcome the passion. Such a dispassionate (overcoming of passion through allowing passion to merely perform) actor becomes eligible for a rational or valid/true knowledge. Again knowledge, if it has to follow the injunctions on knowledge, would accure to the performer not as a result or outcome of the end (phala) of the performance, since injunctions cannot bear a fruit, but only because a cognition (as though infallible) is similar to or dependent on the cognisable object. A cognition unrelated to the cognisable object is false. Hence knowledge appears not as a fruit of the injunction-following action, but as the result of the postulate that the cognition was true.

Though this theory is seductive, it is unacceptable. This view of cognition suffers from a naturalism because if the differences in pleasures or pains that are evidenced are not due to differences in individual actions, then they can only be accepted as 'natural' or *a priori*. Individual action, the fruits of which can be visualised (drsta), would then be undertaken through procedures of agreements in presence or in absence (that is, whenever there is such an action, it is followed by such a fruit, or whenever such an action is absent, no such fruit follows). In such a case, the desirous person would not be in need of injunctions. An action can be guided by inference. Such actions would nevertheless follow inferred norms, rules or positive laws. It is interesting to note that norms/rules in the domain of economics–commerce can be specifically separated from rules which are in the form of injunctions and as are framed/ordained by the sovereign authority (as per the arthasastra). At the level of rules/norms and the respective theories, this differentiation is inevitable and necessary.

In a certain sense this difference is due to the dissimilarity that exists between the individual and the sovereign authority. However, actions bearing visible fruits (drstartha) or those bearing fruits that are not visible (adrstartha) can both exist only in the domain of arthasastra, i.e., the domain of stability and welfare (abhyudaya) under a sovereign penal authority. Dharma is concerned with this invisible (adrstartha). The distant (temporally) fruit of, say, economic actions is not invisible, and therefore cannot concern itself with dharma. It can be argued (as has been in the previous section) that dharma, as a complete sentence or as a complete action, is not merely cognitive or spiritual but is also fruit-bearing. Therefore such rational actions, as dharma (or adrsta) are performed not only for the good, welfare (abhyudaya) of the subject but can also be undertaken ordinarily or frequently under the power of injunctions of the sovereign penal authority (danda). Dharma, as has been argued, is the complete sentence, pivoted around the vidhi, and its meaning—the good (artha)—can be obtained through the karyatva. Such is the invisibility (adrsta), that it does not require a separate validation, or a separate valid–cognition (pramananantar). Vidhi actions are therefore not necessarily related to temporally distant fruit.

Moreover, according to the competing cognitivists, the knowledge-injunctions would be as prescriptive as the action-injunctions. We would argue that the validation of a vidhi is in the accomplishment (siddhi) of the action-injunction. A vidhi satisfies only its own content (i.e., in its accomplishment). If we consider that there is a vidhi on the knowledge of the ultimate (the ultimate rational, say) then as per the cognitivists' argument, the knowledge can be accomplished. Yet, we would argue, that due to the sameness (or similarity) of the content of knowledge and the object of knowledge, this accomplished knowledge would remain invalidated, thus denying the opponents' claim. Again, it is not acceptable that the same sentence signifies the following two (because a sentence by definition is the signifier of the unique)—viz., first, a sentential significance owing to the injunction applied through another sentence signifying a knowledge-injunction; and second, the significance owing to the similarity between the content of knowledge and the object of the knowledge. For example, in the cognition, owing to an injunction such as 'and that (I) cognise', what can be validated is the object qualified by the 'and' (that) form and not of the 'and' formed (or with the 'and' content) object. These two cognitions cannot be the outcome of the same sentence. Therefore, even if we accept vidhi on the ultimate rational knowledge as is argued by the opponents, since the cognition can only be validated as qualified by the 'and', in this case it cannot because of the similarity of the cognition to the object of knowledge. This latter cognition is due to the imposition of say, 'and that', 'similarity'. Consequently, our argument is that in order to achieve ultimate knowledge, we do not need injunctions on knowledge; and even if we accept injunctions on knowledge, the knowledge attained cannot be validated. Moreover, if validation of the cognition of a material (or real) can be attained through transposing another form (formal), it is not necessary to have a knowledge-injunction for material validation.

The desire for fruit is also shown in the common desire for (the class of) fruit. The common would terminate in a determinate ('and that'). A common desire can cause pravrtti (disposition), but pravrtti requires the knowledge of the determinate means. Desire (of, say, pleasure) is thus indispensable for any specific performance. The desire is the same whether it applies to the pravrtti for visible fruit or whether it applies to invisible fruit. Action under injunction and obtaining its fruit are much too difficult in this ephemeral world. The fact that the invisible fruits or good (adrstartha) can be obtained in an unknown and indefinite future, is due not to the power of injunction but to the power of the material (or real). An appearance of a good must follow the laws and the power of matter. Because of this temporal difference, the visible would certainly appear stronger than the invisible, if the two were in opposition. In fact, actions under the two modes for attaining the same fruit are possible simultaneously.

If a vidhi performance cannot bear the pleasurable fruit (especially that endowed with uninterrupted pleasure alone—an aesthetic, not limitable by material or space, the svarga), then it would not be undertaken by any person. Every action is painful and it is directly cognised as such by the undertaker. A direct cognition need not resort to an inference, and therefore the (painful) performance cannot attach (anvaya) itself with the performance (the meaning of verb-root). The direct cognition of the pain of the performance, however, appears on the same word for performance, i.e., on the meaning of the verb-root (and therefore has a stronger cognitive validation). This word-meaning on the pain of performance has to attach itself meaningfully (in order to attain the good) to the other word in the sentence, viz., the 'to-be-accomplished' (siddha). This to-be-accomplished fruit appears as uninterrupted and not-experienced pleasure, notwithstanding the fact that a sentential cognition is of secondary validation compared to the cognition appearing off a word. The painful action under injunction is performed only because it will bear the fruit as pleasure.

A person resorts to volition in performance only because it brings happiness and pleasure—the good. The life-long performances are meant for this common (or universal, samanya) happiness–pleasure–good. This common good is available to a subject in the many determinate qualified states of oneself being in pleasure or oneself having achieved good. For the cognitivists, this often relates to the questions of part–whole or to the archimedean point (as in the example of the visit to the town via the fruit-laden villages). We would argue on the contrary, that an injunction sentence is self-contained and has a complete meaning (svarasya), while a sentence in ordinary discourse does not. Sentences of the latter type depend for their meaning on a different kind of validation (pramananantar) which is not required by vidhi sentences. The speaker in an ordinary discourse arrives at the meaning through a different validation, and then uses sentences to convey the meaning. Vidhi sentences do not signify meaning arrived at through a different validation; here the meaning is acquired by dint of sentential-instruments (linga). The laksana used in such a sentence is not meant for overcoming an alternative validation. As per sentential or word–sound (used here in the sense of an ordinary discourse), the implication that Devadatta receives/enjoys the fruit-laden villages is valid. However, only when we pay attention to the speaker's intent or desire do we cognise that Devadatta receives these only as an itinerant and that he is to enjoy thereafter the town (to be visited). This latter cognition (illocutionary) is related to another different validation—only following which the ekavakyata (or here, the discourse) is accepted. This ekavakyata is not sentential (or does not preside over ordinary discourse), but is dependent on another validation. If the meaning were sentential (i.e., according to the sabdasvabhav or nature of the word–sound), its meaning would have been unvarying and unrelated to the existence or non-existence of another validation. Discourse is delimited by illocution. Similarly, sentences that prescribe earning wealth, for example, if considered in isolation will have only materially implied validation (vastuvrttyanusar) unless they appear as part of an injunction (while the meaning appears sententially). Therefore, both discourse and material implication are delimiting, and in such cases actions pertaining to them would only be delimited by obtaining the (visible) fruit. A vidhi sentence coalesces through one sentence, and there too a fruit is obtainable. It may be noted that none of the instances of good just cited have been defined as the indefinable good. Ultimate knowledge alone is indefinable. The visible and the invisible good are the convolution (vivarta) of that same ultimate, related through the relation of identity/inidentity (called tadatmya). Therefore, the same person can acquire eligibility (adhikara, the relation of action to the fruit) to gain the ultimate knowledge, the invisible good and the visible good.[9]

Injunction: From Desire to Scientific Description

What is the relevance of such vidhis in the contemporary world? While it is not possible to go into this question in detail here, a few points of relevance can be indicated. Examples may be drawn from constitutional law, group/administrative/associational laws/directives, or even from the formation of contracts etc., where such injunctions often form the visible core. From another point of view, and especially given the existence of party (organised group), large organisations, and inheritance of property, even norms might appear to be injunctional. Moreover, if we examine ordinary discourse and especially the learning process of a child (for the formation of cognition), the abundant presence of vidhi is amazing. The pre-eminence of injunction in the description of science, in meaning and human actions cannot be doubted. It is not necessary to look for empirical abundance here. The question is whether the cognition or the meaning of a sentence can do away with injunctions; or whether there can be sentences as descriptions on accomplished (siddha) terms. One point of view stresses that there are both to-be-accomplished and accomplished sentences and therefore knowledge is ultimately separable from the property of action. Another point of view attempts to reduce the action–property into cognition thereby recognising cognition alone. The attempt in this paper has been to vindicate action, and its property of fruit-bearing and therefore, the argument here is that sentential meanings are arrived at through action–property, action–injunctions. Sentences in ordinary discourse use action–injunctions and the cognition of that, either by the participant in a discourse or by the observer of the discourse, leads to action. Consider the formation of cognition in a child, as the child watches a discourse take place between two adults, one of whom is the speaker and the other the listener. The child sees directly that the speaker initiates the listener into an act through the use of speech. Hence the speaker is the initiator (prayojak) and the listener is the initiated (prayojya).

Now the question is whether it is the gestures or the speech symbols, or whether it is the (perceptual) feeling of the speech on the property of being action that initiates the listener adult. The first view, as held by the vantage (prasthan) of Nyaya (as pertaining to the study of the means to valid knowledge), admits a memory-actuator to the remembered (smarak-smarya) relation on the word or term to the word-meaning (pada-padartha). A word is a symbol that can actuate its memory on to memory. The cognition cannot therefore be validated at the memory-actuated word-meaning stage, but through such associations on the word-meaning as are delimited on the property-of-being acts (karyata). This is close to Bhatta's position (discussed briefly in the last part of the previous section). The other, more commonly followed view (held by the Prabhakaras) on the perceptual–felt necessity of associations upon the hearing of word-sound, argues that the memory of a word-meaning arises upon hearing the word(sound). The word (sound) triggers off an association, and evokes the word-meaning from the memory. A word as felt is valid, and there is in the word an instrumental power (karaka sakti) to evoke the feeling for association. Such association (anvaya) takes place over the domain of the property-of-being action (karyatva). The knowledge of the word 'pot' therefore causes the cognition of the word(sound) 'pot' associated with the property-of- being action (karyatvanvita). Why should the karyatvanvita cognition of the word(sound) pot be preferred over the simpler cognition of the meaning of the word-sound pot? It can be argued that a child forms cognition by watching the adult actions in the same manner. This cognition is that the class of word(sounds) (padatva) is the delimiter of cause and that on the karyatvanvita appears the class of cognition of the word(sound) (savdabuddhitva) as the delimiter of the property of acts (karyatvanvita). Such a cognition appears initially to the child where the class (samanya) of word(sound) causes the cognition of the word-meaning associated with the property-of-being action; and only thereafter the cognition of the word-meaning pot is attributed by the child to the determinate word 'pot'.[9]

The child watches and feels that following the utterance by the initiator adult of such a sentence as 'bring the pot' there is a disposition or desire (pravrtti) on the part of the initiated adult, and therefore the child wants to know the reason for the utterance (he infers that there is a reason or necessity for the utterance of the sentence). By using the principle of agreement in presence or in absence, the child knows that the pravrtti of the initiated adult follows the utterance of the sentence by the initiator adult. That is, the child infers that the uttered sound must be causing something unique towards satisfying the specific necessity of the initiated adult. It is also true that the child has already validated the knowledge of breast-feeding (the property-of-being act). The child knows that he had his own pravrtti towards breast-feeding and that he was satisfied through an act—that is, nothing other than the knowledge of the property-of-being act is the cause of pravrtti. Though the child has not yet acquired the cognition of sentential meaning from a sentential utterance, he nevertheless knows about the appearance of cognition directly through his senses and from inferences. Therefore, he infers that the pravrtti of the initiated adult is related to the (implicational) power of the word-sound (causal knowledge of the cognition of property-of-being act in the word-sound), in the class of sentences. Following this generalised cognition, the child recognises through the process of eliminative thinking, the specific word-meanings (associated with action) of the particular sentence. Since the general causal relation (of the cognition of word-sound) appears as part of or related to karyatvanvita, the specific terms also appear as part of the same karyatvanvita. Here the karyatva is the property, instantiated in the karya (act) which is an element of the locus of the property.

However, as has been outlined in the last section, the other view holds that even without considering specific words as the bearers of the loci of (or over) karyatva, i.e., as part of karyatvanvita, these specific words, if put together with words which are action-designating, would cause sentential cognition over karyatvanvita. In this view of the competing theorists, any word that is object-wise syntactic (or associative) can cause the cognition of such a word-meaning. These terms, which designate the property-of-being acts, when put together, cause through a special character, a cognition of meaning which is content-wise karyatvanvita. Therefore, this view questions the necessity of accepting a word as causing cognition of meaning that is over the karyatvanvita. If the word is an action-designating one, the other words that appear with it would cause cognition of meaning over karyatvanvita. Again, even in the absence of a vidhi-denoting (pratyaya) inflexion, a sentence would still possess a cognition of syntax or association.

This criticism does not deny the fact that the cognition of the initiated adult has a content of associated (or syntactic) acts, the cause of which is the sentence of the initiator adult, and the child infers that the cognition in the initiated adult has been caused by the sentence. However, the competing theorists virtually deny the content of associated acts (karyanvita) in the cognition causing the disposition to act (pravrtti) by first defining this cognition as having a content of syntax or association, which then together with the action-designating terms only evokes a meaning-content as associated acts (karyanvita). But this is to unnecessarily complicate the issue. The point is that the child infers about the nature or type of cognition that has initiated, i.e., that has given rise to the pravrtti of the initiated adult. Now pravrtti is similar or has a similar measure to its cause, the knowledge. In fact, we may admit as a second criterion of rationality the similarity prevailing between act or the disposition to act, and the knowledge (the cause) of it. Therefore, the child, being rational (as evidenced by his satisfaction following breast-feeding), would infer the initiated adult as rational—as a person not having two cognitions (the anvitartha and the karyanvita content) of the same sentence. The cognition of the initiated adult would therefore have a content of associated actions. The competing theorists deny the word-universal (pada-samanya), the word-cognition with a generative character of meanings over the associative property-of-being action (karyatvanvita). This can be refuted by the assertion that the argument offered (anyalabhyatva) is inappropriate. We need to look into the initial (adya) cognition of the child alone. The child does not have in his cognition of injunction the designation of the property of act, but initially, it appears to him to be a designation of act.

What are present are the cognition of direct perception, and the cognition of inference. Therefore the inferred knowledge acts as a cause. Having his earlier knowledge about breast-feeding, the child cognises as the generator of pravrtti a content of meaning as is over karyatvanvita and not one that is syntactic over pure associativity (anvitartha). However, the opponents argue that the words must cause the feeling of associativity, since even the karyatvanvita is associative; but it must be pointed out that these two are different because cognition differs according to its object. In a cognition, its object (or content, visaya) is an adjective. Adjectival differences cause different cognitions. Again, the noun, the adjective and their relation are different from the qualified determinate (object, visista). Hence, because of the first reason, the two cognitions of meaning-contents that are of karyatvanvita and of anvita are different. Owing to the second reason, the cognitions of word-sounds that are qualified by the properties of karyatvanvita and the properties of anvita are different. As an example it may be recalled that the two cognitions—of the floor and of the floor having a pot on it—are different.

The meaning associated over the property-of-being action (karyatvanvita) at its first appearance is simply an action-feeling (bhavana). The onlooker of an action-generating discourse, by way of employing his ego, cognises the sameness of meaning as 'that action'. 'That action' in the onlooker's cognition is about the other person's (the initiated) ability to perform. It is indeed a confirmation of all the qualifications (or adjectives) of the ability to conduct the performance which the initiated adult performs. Again, since the delimited property of the delimiter of causes (karanatavacchedaka dharmavacchinna) is the direct instrumentality for an accomplishment, and since such a delimited property is present in the 'that action' cognition (in fact here the delimiter of causes is the property of being such a cognition), therefore the cognition of 'that action' (which the able initiated can perform and is performing) is the direct means of pravrtti, or the desire (towards action) of the onlooking child. These qualities or adjectives are the initiators of the 'can be done (or ought to be done)' cognition. Only after this would the learning child search for self-referencing of the presence of such qualities, following which a cognition such as 'that action can be done by me' would arise. 'That action' is a cognition drawn inferentially about the initiated and 'that action can be done by me' is the cognition, arrived at through the employment of an assisting cause (by associating the search of self-reference) on the earlier inference, and this is the immediate assisting cause of desire, pravrtti. Because of the absence of this cause an action is not always undertaken, even while there is in the subject a cognition of 'that action'. Weakness of will or the notion of pragmatism that the competing theorists resort to, are due to their admission that the cognition 'that action can be done by me' arises first and immediately, in the onlooker, as an inference along with the admission that this cognition is the direct instrumentality for the pravrtti. Our argument, it may be observed, allows the child even a deferred action following the arousal of desire—deferred since the appearance of the cognition of that action.

Moreover, the location of 'my ability to perform the action' should not be possible in an action, yet to be performed in the future. The ability of the initiated is available to the child. The competing theorists, however, deny it. Their argument is that in the scheme of 'action is caused by desire, which is caused by the cognition' cognition and desire (or desire and action) are not similar (or do not have a commensurate measure); rather, they have similar (or commensurate) qualities (or adjectives) or these are resident in the same qualities. It is argued by them that on the property of being happy, there is the cognition of the happy person (a name-tag, individuated; as it appears before the child about himself having been happy following breast-feeding, as a fact) as known (out of an accomplished action), and there appears in the child on the same property of being happy the desire (out of the known cognition of happy person) for the future, unaccomplished or non-appeared happy person. This desire cannot be for the known happy person, but only for the non-appeared happy person. It needs to be noted that in this argument, the property of being happy is the qualifier, the adjective, and the known happy person is the noun in the cognition; the non-appeared happy person is the noun in the desire. In this causal transition, i.e., from the present to the future, there are two nouns on the same quality.

I believe that the desire due to cognition has similarity (measure) with its cause, or action due to desire has similarity with its cause. This follows my second condition of rationality. In the first case, the cognition of the accomplished happy-person is in the past and the desire of non-appeared happy person is in the future. But in the second case, both the desire and the action are in the future. I would argue that the noun 'happy person' remains the same over causation, that is the object (or cognition) remains the same. The noun here is of a special type being used adjectivally. For example, while a pot in front is directly perceived (therefore cognised), a pot-universal (beyond space and beyond time) is perceived indirectly (alaukika) through an indirect nearness (alaukika pratyasatti) to (or with) the form (rupa) whose designation is universal. It is said that a cognition of a self-referring noun appears as though an adjective, and being of such property (or attribute) is a relation between the pot-universal and the eye of the beholder. In fact it is not just a pot-universal, but the designator of such an attribute that is adjectival such that the latter is similar to the universal. This special relation or nearness (called samanyalaksanapratyasatti) with the pot-universal while directly perceiving the pot in front, also entails having the same relation with the pot itself (not the universal pot) in the event of a cognition of a 'pot on the floor'. Similar to the cognition of a 'pot on the floor' is the cognition of 'happy person on me (my being at various states or times)'. The child therefore has this special relation (samanyalaksanapratyasatti) with the happy person (which is here a universal happy person), while having a direct cognition of his having been the accomplished happy person, or his having the cognition of his future, non-appeared happy person. It is feasible now to have a cognition of a future object, the non-appeared happy person or the non-appeared ability to perform the (future) action. According to the competing theorists, the property of being happy remains the same, that is, the adjective remains same or similar and this property of being happy actually acts as the designator of the sameness of the cognising person. My argument, however, shows that in contrast to the adjective remaining the same, it is the special noun acting adjectivally that remains the same. Happiness is not an adjective, but it is a changed state (parinama) which action can bring about. The cognition of oneself being happy person, is a special cognition of samanya (the near-equivalent universal has been used earlier) whose adjectival employment qualifying one's personhood (for example over time) gives rise to the cognition of the qualified self; and while action can bring a change in this qualified self, it cannot change the samanya happy person.

It may be concluded therefore that words cause pravrtti; nouns, verb-root, cases, inflexions etc., are all employed to initiate a person. Because of this, the cause of the cognition that causes pravrtti (the disposition to act) is related to the word-universal. However not all the words (or parts of a sentence) cause pravrtti (the cognition of meaning over the associated property-of-being action) in the same direct manner. Some terms (injunctions) cause cognitions of meaning having content over karyatvanvita through causing feelings directly, while some others (the arthavad) cause feelings that only indirectly generate the meaning over karyatvanvita. Therefore, injunctions can be inferred even from descriptions or narrations. Such imaginative inferences that initiate pravrtti indirectly, complete a sentence. The narrations and descriptions are thus not reduced to simple description of facts. Cognitions from words are possible through the completion of a sentence. This completion entails causing the disposition to act in discourses or descriptions where there is no explicit use of injunctions.

Notes

1. See Yogendranatha (1953a), p. 319.
2. See Mbh, *Anusasanaprakaranaparva*, 6.
3. See Yaska, *Naighantukakanda*, 1/1/1.
4. See Medhatithi on Manu, 2/6.
5. See Yaska, *Daivatakanda*, 3/6
6. See Tilak (1978).
7. See Yogendranatha (1981), p. 116; and other chapters.
8. See Yogendranatha (1953b), p. 108.
9. See Shankarabhasya and Bhamati, 1/1/3.

References

Indian Sources

JHA, MM. GANGANATHA. 1964. *Purva-Mimamsa in its Sources*. Varanasi: Banaras Hindu University.

JHA, V.N. 1987. *The Philosophy of Injunctions*. Delhi: Pratibha Prakasan.

Mahabharata.

Manusmriti: Medhatithibhasya, especially chapters 1, 2 & 12.

MISRA, MANDANA. *Vidhi Viveka* (Vol. 1), 1982. Edited by S. Bhattacharya. Calcutta: Sanskrit College. (in Bengali)

TARKALAMKAR, JAGADIS. 1980. *Savdasaktiprakasika*. Edited and translated by Madhusudan Nyayacarya. Calcutta: Sanskrit College.

TILAK, BAL GANGADHAR. 1978. *Gitarahasya* tr. into Bengali by Jyotirindranatha Thakur, Calcutta: Harap.

UDAYANACARYA. *Nyayakusumanjali* 1973. Edited and translated by Pt. Durganatha Jha. Varanasi: Varanaseya Sanskrit University. (in Hindi)

YASKA. *Niruktam (Durgacaryavrttisahitam)*.

MM. YOGENDRANATHA. 'Tarka-Samkhya-Vedantatirtha'. 1953a. *Our Heritage* July–December (in Bengali).

———. 1953b. 'Different Views of Ancient Mimamsa'. *Our Heritage* January–June (in Bengali).

———. 1981. *Vakyartha Nirupaner Darshanik Paddhati*. Calcutta: Sanskrit Book Depot. (in Bengali)

English-language Sources

BENTHAM, J. 1973. *Principles of Morals and Legislation*. New York: Anchor Books.

DAVIDSON, D. 1980. *Essays on Actions & Events*. Oxford: Clarendon.

DONNELLAN, K.S. 1970. 'Proper Names and Identifying Descriptions'. *Synthese* 21.

GAUTHIER, D. 1986. *Morals by Agreement*. Oxford: Clarendon.

HARE, R.M. 1952. *The Language of Morals*. Oxford: Oxford University Press.

HARSAYI, J.C. 1982. 'Morality and the Theory of Rational Behaviour' in A. Sen and B. Williams (eds.). *Utilitarianism and Beyond*. Cambridge: Cambridge University Press.

HART. H.L. 1983. *Essays in Jurisprudence & Philosophy*. Oxford: Clarendon.

———. 1982. *Essays on Bentham*. Oxford: Oxford University Press.

KAPLAN, D. 1978. 'Dthat' in P. Cole (ed.) *Syntax and Semantics*. New York: Academic Press.

KRONFELD, A. 1990. *Reference and Computation*. Cambridge: Cambridge University Press.

MILL, J.S. 1987. *The Logic of the Moral Sciences*. London: Duckworth.

MOORE, G.E. 1968. *Principia Ethica*. Cambridge: Cambridge University Press.

RAWLS, J. 1973. *A Theory of Justice*. Oxford: Oxford University Press (paperback).

RUSSELL, B. 1953. *Mysticism and Logic*. Harmondsworth: Penguin.

SCHELLING, T. 1960. *The Strategy of Conflict*. Cambridge, Mass.: Harvard University Press.

SEARLE, J.R. (ed.). 1971. *The Philosophy of Language*. Oxford: Oxford University Press.

SELBY-BIGGE L.A. (ed.). 1975. *Enquiries Concerning Human Understanding and Concerning the Principles of Morals for Hume*. Oxford: Clarendon.

What 'Rational' Could Mean in the Human Sciences

ALLAN GIBBARD

The term 'rational' figures in a wide variety of inquiries into human beings and their activities. This immediately raises the question of what the term means which is also the main concern of this article. I do not claim to account for all uses of the term, but I shall sketch and comment on a theory that addresses an especially puzzling cluster of uses.

Specifically, I shall ask about the use of the term 'rational' in scientific theories. If the term 'rational' does find a home in scientific theory, it will be in the human sciences as broadly understood: in the sciences of cognition and sociability, and in the sciences of human-like systems, natural and artificial. Instead of treating specific examples of the human sciences, I shall explore two broad kinds of meaning that the term 'rational' might be given: *normative* and *descriptive*. The positivistic tradition in the philosophy of science holds that these two kinds of meaning are distinct, and that normative terms cannot properly appear in scientific theories. I shall sketch all account of normative concepts that endorse these claims, and briefly explore two difficulties for the account.

Normative Concepts

The term 'rational' applies both to *people* and to their *doings*: we speak of people as being rational or irrational, as also of acts and beliefs. In this I shall concentrate on acts and beliefs. My question, I should stress, will not be what it *is* for an act or belief to be rational, but what it *means* to call it rational. That was the topic of my book *Wise Choices, Apt Feelings* (1990). I shall outline and be sketching the theory contained in that book and comment on it.

'Rational' as applied to acts figures in a cluster of terms that can seem especially puzzling and problematic. A few of these terms are 'a reason to' (as in 'The fact that one would enjoy doing something is a reason to do it'), 'ought' in a non-moral sense (as in 'We ought to get out of the building before the fire spreads dangerously'), 'ought' in a moral sense (as in 'You ought not to lie, even when it would be to your advantage to do so'), and 'it makes most sense to' (as in 'It makes most sense to postpone the decision until we have more information'). These terms are often grouped together as 'normative'.

During the first half of this century, it was specifically *moral* language, not normative language in general, that chiefly figured in philosophical controversy about meanings. Analytical moral philosophy during that period was caught up in debating whether there was something peculiar about moral concepts—and if so, what explained this peculiarity. G.E. Moore (1903) argued that *good* is a 'simple, non-natural quality'. The term 'good', he insisted, is not definable in the 'natural' terms that figure in the empirical sciences—psychology and sociology included. The emotivists of the 1930s agreed that moral terms are not definable in natural terms, but rejected Moore's talk of 'non-natural' properties. It is a mistake, they argued, to try and explain moral language as attributing properties at all; moral language does something else. Ayer (1936) held that moral language is peculiar because it serves not to state facts but to express feelings. Stevenson (1937, 1944) held that moral language demands that the audience join the speaker in certain feelings. Hare (1952) rejected emotivism, but kept its *anti-descriptivism*: its insistence that moral language does something other than straightforwardly describe things, attribute properties to them.

My own view, like Hare's, is anti-descriptivistic but opposed to emotivism. Moreover, it broadens the scope of the inquiry: There indeed *is*, I maintain, something special about moral language, but it is not a peculiarity of moral language alone, but of normative language in general. Both for normative language in general and moral language in particular, I reject naturalism, non-naturalism, and emotivism. All three traditions, though, have features that could figure in an improved view of the workings of normative language.

It might be advisable to start, not with a *theory* of normative language, but with a characterisation that, if not theory-free, is, like the following, light on theory. Terms like 'rational', 'ought' in a non-moral sense, and 'a reason to' have seemed to many philosophers somehow special and problematic. They also seem interdefinable: the 'rational' thing to do or what one 'ought' to do is what one has most reason to do, the thing supported by the preponderance of reasons. Going the other way, it can be said: A *reason* to do something is something it is rational to treat as weighing in favour of doing it. Other terms that are traditionally considered normative seem definable using terms like 'ought', 'rational', or 'a reason to'. Morally 'wrong' acts, for instance, are—roughly, and in a narrow sense of the term—acts of a kind one is normally to blame for if one has performed one of them. In turn, one is *to blame* for an act if it is rational for people to have feelings of blame toward one for doing it: feelings of guilt on the part of the agent, and feelings of resentment on the part of others. Thus, normative language is language that shares whatever special feature it is that, in the bulk of their uses, makes these terms and terms they serve to define somehow special.

This characterisation is theory-light, in that it leaves it fully open what feature it might be that makes language normative. Only one kind of theory is precluded by the definition—though it is an important kind of theory. What is excluded is a theory that explains away the impression that terms in this cluster are special. And this, indeed, is as it should be: someone who picks out some terms and not others as 'normative' is taking it that these terms have some distinguishing characteristic.

What, then, might explain the impression left by Moore and his successors that there *is* such a thing as normativity, that these terms have something special and problematic about them, that they somehow differ in kind from terms that figure in the empirical sciences? Long after Moore first wrote about moral language, A.C. Ewing offered a theory that can be considered to be a theory of normative concepts in general. Ewing's theory, like Moore's, was non-naturalistic, but the two differed on what the fundamental non-natural concept is. Moore (1903) thought *good* to be a simple concept, in terms of which other moral concepts could be defined. Later, he felt forced to recognise another simple concept, the moral *ought*. Ewing, on the other hand, proposed that both of these could be defined in terms of a single primitive concept—along with terms that could figure in empirical science. Both, he thought, could be defined in terms of a primitive, non-moral *ought*. (In what follows I shall use plain 'ought' for Ewing's primitive 'ought', and use 'ought_m' if I need to speak of the moral 'ought'.) A state of affairs is *good* if one ought to desire it to obtain. One ought_m to do something if, and only if, everyone ought to reprehend one's not doing it. One might, then, foist on Ewing the following theory of what it is that makes normative terms normative: that this primitive 'ought' figures in their definition.

One might identify this primitive *ought*—the *ought* that makes concepts normative—with the *ought* of rationality. The *rational* act, let's say, is the one a person ought (in Ewing's primitive sense) to perform. The *rational* thing to believe is what one *ought* to believe. The terms will differ in that 'rational' permits and 'ought' requires: more than one alternative may be rational; an act is rational if it's not the case that one ought not to do it.

A Theory of Normative Meanings

My own proposal is to do to Ewing's primitive 'ought' what the emotivists did to Moore's primitive 'good'. I will claim—along with the Ewing I've constructed—that there is indeed something special about this primitive concept of *ought*; it is the conceptual atom that makes the concepts it figures in normative. I will deny, though, that this primitive *ought* picks out a non-natural property. (Indeed, I will deny that there is any such thing as a non-natural property.) Normative language serves not strictly to describe things or to attribute properties to them. It does something else: it serves to *express* a special kind of state of mind. This state of mind, I propose, is something other than belief in a matter of plain fact.

What kind of state of mind, then, could it be? I propose that normative statements express the speaker's *acceptance of norms*. The state *accepting a norm*, is one peculiar to human beings. It would have to figure in any adequate human psychology. By a *norm* I can mean simply an imperative, like 'Look before you leap'. The question of substance, then, is not how to characterise a 'norm', but how to characterise the state *accepting* a norm. I characterise it by placing it within a speculative psychology: if any human psychic state works the way this psychology pictures 'accepting' such-and-such a norm, then that is the state that will count as *accepting* the norm.

My psychological speculation centres around human evolution: why, in a highly social, linguistic species like ours, might we expect such a state to evolve—or more precisely, the *capacity* for such a state, for a state that looks like human acceptance of a norm? This question could also be posed in the following way: I identify 'accepting a norm that says to do *X*' with thinking that one *ought* to do *X*. Why might we expect the capacity for a state like *thinking that one ought* to do something to have evolved in a linguistic, highly social species like ours?

My speculative answer appeals to Thomas Schelling's famous discussion of bargaining and coordination (1960), and to John Maynard Smith's application of similar game-theoretic concepts to the evolution of behaviour (1982). Schelling explains how, in bargaining among instrumentally rational agents, wasteful conflict is avoided if expectations are coordinated. Imagine a situation where we are haggling over the price of some object. If there is a price that, by common knowledge, we both expect to settle on in the end, and for each of us a sale at this price is better than no sale at all, then each of us will hold out for this price rather than agree to something less advantageous. Each, moreover, will correctly expect the other to hold out for this price. Each, then, will expect that this particular price is the best he can manage, and an agreement will be arrived at quickly. If, in contrast, the buyer expects a low price and the seller a high one, then each will hold out for a long time for the price he expects. Time and effort will be wasted in bargaining.

John Maynard Smith showed how like considerations applied to the genetic evolution of propensities for behaviour. The genes will tend to settle on 'conventions' such as territoriality and pecking orders. My speculation is that as our own species evolved, social life became too intricate for such rigid animal coordinating devices—devices like territoriality and pecking orders—to work well. Emerging capacities for language, though, allowed for a more refined and flexible kind of coordination.

Suppose there is a psychic state *S* that works as follows: let us describe state *S* as 'accepting' norm *N*. (*N* might be, say, 'Let the hunter whose spear killed the animal decide how the meat will be divided.') First, being in state *S* motivates one toward certain kinds of actions (say, letting the hunter whose spear killed the animal divide its meat, and insisting on that role oneself if it's one's own spear that killed the animal). I will call this kind of motivation *normative governance*. Second, being in state *S* tends one toward uttering certain kinds of words (words I will informally gloss, in our example, as meaning 'the hunter whose spear killed the game ought to allocate the meat'). Third, those to whom such utterances are directed tend to comc to share the state *S* that evokes the utterance. If all these things obtain, then the exchange of such utterances—call this *normative discussion*—will tend, though a process of mutual influence, toward everyone's accepting like norms. Then, because of normative governance, people who accept like norms will tend toward coordinated actions. The mechanism I have sketched will tend toward coordinated actions.

This, of course, can only be a vastly oversimplified sketch of a model. A good account of the relevant genetically coded propensities that our species has evolved would have to be vastly more complex and refined than anything I have said, here or elsewhere. I am suggesting, though, that the kind of story I have been telling might serve as a first approximation to what we humans are like.

Even as a crude first approximation, the model suggests features of human life that should figure in an adequate human science: it stressed the importance of *speech* to our species, speech that isn't at all confined to straightforward description. It suggests why some kinds of beliefs—the ones whose voicings one would translate in normative terms—tend to motivate their holders in special ways. On this kind of picture, one would expect human beings, often to be engaged, in effect, in renegotiating their relations, but speaking not in bald terms of self-interest, even when their moral appeals are transparently self-serving.

My strategy in explaining normative concepts has been the following. I started with a view I foisted on Ewing: What makes a concept normative is that a primitive *ought* of rationality figures in its analysis. I sought then to explain this primitive 'ought' by saying what state of mind it serves to express. To identify this state of mind, I placed it in a speculative human psychology: if anything, it is the state of mind that plays the role in our psychic lives that this psychological speculation depicts. The speculation pictured the workings of psychic mechanisms whose biological function is to coordinate actions. These mechanisms make for normative consensus through discussion and mutual influence, and for coordinated action through normative judgments governing us to act accordingly. To explain what the term 'rational' means to explain the kind of state of mind the term serves to express.

Rationality in Human Sciences

The term 'rational' can figure in theories in diverse branches of the human sciences: microeconomics, decision and game theory, the sociology of tradition and modernisation, psychiatry, and artificial intelligence—to name a few. These theories, of course, can be disputed, and if they make claims as to what constitutes an act's or a belief's being rational, these claims can be disputed. What am I saying that such disputes consist in?

Disputes over a characterisation can, of course, be empty: the term 'rational' can be introduced into a theory with its meaning stipulated, so that it would make no more sense to challenge its proffered characterisation than it would to challenge a mathematician's definition of a term like 'well ordered'. The term can be given a non-normative sense suitable for framing causal/explanatory hypotheses. If these hypotheses are disputed, the questions in play may then be non-normative scientific questions about how to explain phenomena. They may not concern rationality in the normative sense.

Still, even with sheer stipulations, one can ask whether the meanings stipulated serve to address the questions that vex us—questions that may come to us as questions of what is 'rational' in an ordinary, unstipulated sense. Some talk of rationality in the human sciences, moreover, seems to range beyond stipulation, and then disputes about its characteristics may have substance and be normative disputes. Economists dispute what constitutes rationality for economic agents; decision theorists and game theorists dispute whether classic decision-theoretic axioms adequately characterise rationality.[1] One school of psychologists proposes that we are equipped—perhaps genetically—with certain mental 'heuristics' for dealing with complex aspects of reality, *heuristics* in the sense that they are tractable and work pretty well, but don't come with guarantees: they fall short, in some circumstances, of perfect rationality.[2] Clinical psychologists may distinguish 'rational' from 'irrational' beliefs and motivations. In all these cases, it can be questioned whether the acts, beliefs, or patterns of preference the theorist is *calling* 'rational' really *are* rational, and the ones he is calling 'irrational' really are irrational.

This raises the puzzle my theory is meant to address: What is in dispute if such questions are raised? The theory I have sketched offers an answer. If we disagree about what kinds of acts are rational, we disagree as to *what norms to accept* as governing action. If we disagree about which beliefs or preferences are rational, we disagree as to *what norms to accept* as governing belief or preference. This answer may not apply straightforwardly to every such dispute, but it offers a style of hypothesis that could be elaborated in different ways in different contexts.

Disputes that can be explained in this style are the ones I will now class as 'normative'. I have moved from an initial, theory-light characterisation of 'normative' terms to a theory of what is special and problematic about a central range of uses of such terms as 'rational', 'ought', and 'a reason to'. They are terms used not to state causal/explanatory facts, but to express a speaker's acceptance of norms. Disputes about what is rational in this sense can hinge on questions as to what norms to accept, distinguish, then, between things the term 'rational' might be used to mean in a purely causal/explanatory theory from genuinely normative uses of the term. It is the genuinely normative uses my theory picks out.

Nevertheless, my theory does tie in with a piece of causal/explanatory speculation. In the course of developing my account of what the term 'rational' means in a genuinely normative sense, I sketched a picture of how normative language and normative discussion might work in human thought, discussion, and motivation. It is a picture of why normative thought and speech—thought and speech that could have been framed in terms of the concept *rational*—play a crucial part in human affairs. We are biologically adapted, I speculated, to think and discuss in such terms, and to be guided, in part, in our acts, beliefs, and feelings by the conclusions we reach.

The model, for all its defects, has its virtues. It helps explain why normative convictions seem stubbornly individual in a way.

We would not have been adapted simply to soak up the normative judgments of those around us, since such a genetic design would make the organism far too pliable for its genes' own 'good'. The model also helps explain why normative convictions are strongly social. The mechanisms that produce them have it among their biological functions to coordinate, by tending people toward consensus. Selection pressures toward coordination were not the only pressures that shaped the genetic design of these mechanisms, but they form an important part of the story. They are shaped partly by the 'advantages' (to an individual's genes) of coordinated action: thus they are shaped to achieve normative consensus through mutual influence, and to motivate people according to the consensus.

This is not to say that the term 'rational', in a genuinely normative sense, could be given a straight definition in causal/explanatory terms. We could explain the social dynamics of norm acceptance and normative discussion without ever saying what makes a normative judgment correct, or what makes an act rational or not. It is normative *judgments* or *convictions* that figure in the positive model that I am sketching: the model is of individuals' *thinking* some sorts of human doings—some rational and others not—influencing each other in these thoughts, and so at times achieving consensus on normative matters. It depicts what it is for people to *make assertions* and *come to conclusions* as to what sorts of acts, beliefs and feelings are rational. In offering a causal/explanatory account of judgments and assertions about rationality, one does not oneself make assertions about rationality.

The model does suggest, though, why no one—scientist or otherwise—could satisfactorily lead a human kind of life without having normative convictions and pondering and discussing normative questions. It suggests why normative judgments are an indispensable part of human life. I am not proposing, therefore, that a human scientist should be devoid of convictions as to what kinds of acts are rational and what kinds are not. I am suggesting that these normative convictions will not be voiced in the course of giving a causal/explanatory account of human normative thought and discussion.

Because even during working hours the scientist looking for causal explanations must make normative judgments and heed them. (S)he must be guided by judgments of what theories and hypotheses it is rational to accept given the evidence, and what lines of inquiry and experiment it is rational to pursue. Norms apply both to acts and to beliefs, and scientists in their work need norms for both. They need to settle practical normative questions of how to proceed in their investigations: what experiments and investigations is it rational to pursue? They also must settle what to believe on the basis of their evidence: what, given the evidence, is it rational to believe? Science follows norms for how to investigate a question and what it is rational to believe on the basis of investigation. This goes for all of science, not just for human sciences.

I am distinguishing, then, normative questions from causal/explanatory questions. As mentioned earlier, according to my theory, normative questions are not purely factual; they are questions of what norms to accept. Purely causal/explanatory questions are purely factual, although they raise normative questions about how to investigate them and answer them. Among the factual questions a social scientist can investigate are questions about peoples' normative judgments and assertions, and the social dynamics of such judgments and assertions.

Descriptive Concepts and Supervenience

The position I have sketched can seem fantastic. Won't normative claims *have* to appear in good causal/explanatory theories of humanity? Will an adequate theory of the human mind not picture it as rational in certain ways? Will an adequate theory of social phenomena not picture various beliefs, goals, and stratagems as rational? And will the term 'rational' in any plausible causal/explanatory theory of human doings not take its ordinary, normative sense? I have agreed that in a purely causal/explanatory theory, the term 'rational' may well occur, but I claim that if it does, it will be used in a non-normative sense. Can I reasonably claim such a bifurcation of senses for the term?

I not only insist on the bifurcation, but I also have more to say on the matter. Descriptive and normative senses of 'rational' may well tie together intimately. The key lies in Moore's talk of *supervenience*, and the explanations an expressivist can offer for the kind of supervenience that Moore noted. We must return, then, to Moore and the metaethics of this century—with an eye to how its lessons might apply to normative and causal/explanatory concepts of rationality and of content. Moore's own treatment was confusing and confused[3] but I think his doctrine can be restated so as to be intelligible.

Moore stoutly denied that the meaning of 'good' can be given in natural, scientific terms. Still, he claimed that goodness *supervenes* on natural properties: there couldn't be two things that were the same in all their natural properties, one of which was good and the other not. Supervenience has often been discussed in more recent literature, but we should note some features of supervenience as Moore conceived it that recent discussions often miss. The supervenience of *good* on the natural is both necessary and *a priori*. It is necessary in the sense that not only are there *in fact* no two things alike in all their natural properties but differing in goodness, there are two possible things alike in their natural properties but different in goodness—and this is necessary, not contingent. Moore's supervenience is *a priori* in that we don't need empirical investigation to discover it. A defence of our claim to know that *good* supervenes could omit all appeal to the facts of experience.

Moore distinguished *good* from what he called *the* good. Interpreting this distinction is problematic, but he may have meant something like this: the meaning of the term 'good' cannot be put in natural, scientific terms; the concept *good* is not a natural concept. Nevertheless, there is a natural property that all and only good things have. This natural property—call it *N*—may be intractably complex, but nevertheless it's a property.

To see how this works, imagine that Helen is a clear-headed hedonist. She thinks this *N* to be the property of being pleasurable; she thinks that all and only good things are pleasurable.[4] She doesn't think that 'good' *means* pleasurable. To see this, imagine a value pluralist Ken. Ken thinks that some kinds of knowledge are good even if they aren't pleasurable. If Helen thought that 'good' *meant* pleasurable, she'd have to think that Ken was confused in his understanding of language. Helen, being clean-headed doesn't think this; she thinks that Ken has a substantively mistaken moral opinion. He gives correct expression to his incorrect moral views—so Helen thinks. Good, then, she will think, doesn't *mean* pleasurable; the statement 'All and only pleasurable things are good' is synthetic rather than analytic. Still, she will insist, it is a necessary truth, knowable *a priori*. It isn't something that would fail to be so in some possible circumstances, and isn't something we learn from experience.

Moore himself denied hedonism. He may, though, have thought there to be another natural property—far more complex and reflecting principles of 'organic unity'—that all and only good things have. That all and only good things have this property, he would have to say, is a synthetic *a priori* necessity.

Expressivists, of course, reject all talk of non-natural properties. Still, they claim to explain the features of normative concepts that led Moore to his strange doctrines. When Helen expounds her hedonism, the expressivist says she is expressing her acceptance of norms for valuing states of affairs. 'Value a state of affairs,' she is saying, 'to whatever degree that state of affairs is pleasurable.' This applies, she tells us, to any possible circumstance, and is to be maintained come what empirical news may—in these senses, she's putting this imperative forth as necessary and *a priori*. She is saying, 'Come whatever empirical news might, for any possible state of affairs in any possible circumstance, value it is to the degree that state of affairs is pleasurable.' To say this—to express this norm—is to treat goodness as supervening on pleasurableness.[5]

Goodness, according to the expressivist, is not a natural property, it is not in the strict sense a *property at all*. Pleasurableness, though, is a natural property, and goodness, Helen maintains, supervenes on pleasurableness. Claims of goodness are not translatable into the causal/explanatory language of pleasure; they express the speaker's acceptance of norms. Goodness, though, supervenes on the natural: there is a synthetic *a priori* necessity of the form. 'A state of affairs is good to the degree that it is *N*,' where *N* is a descriptive predicate signifying a natural magnitude.

All this suggests a stance that an expressivist about rationality could take toward the cognitive and the social sciences. The term 'rational' in a normative sense, we can say, will not appear in purely causal/explanatory theories about human beings. Whether an act or belief is rational, though, *supervenes* on its causal/explanatory properties. There is something to be said of the form 'An act is rational if and only if it is *N*,' where *N* is a causal/explanatory predicate, a predicate that could appear in a purely causal explanation in the human sciences, 'All and only acts that are *N* are rational' is a synthetic *a priori* necessity. We may not agree on what this predicate *N* is, and indeed most of us will be undecided individually as to what property fits the bill. What we will agree on is that there *is* such a property.

The term 'rational', then, could be given a causal/explanatory sense in a scientific theory that fits in with its normative sense. We won't after all, be completely undecided as to what properties go to make an act or a belief rational, and we won't be completely at odds on the issue. Truisms will abound as to what is rational and what is not. A causal/explanatory statement and a normative statement may be *truistically equivalent*: truisms may entail that one of them holds if and only if the other does. A scientist, then, may appropriate the term 'rational' for causal/explanatory purpose and yet end up saying things that are truistically equivalent to what (s)he would be saying, in the same words, if (s)he meant the word 'rational' normatively.

Indeed since in speech we don't make distinctions that don't matter for our purposes, there may well be no fact of the matter whether she is using the term normatively or causally. There may even be no need to clear the matter up, since (s)he would gladly assert what (s)he does in both senses. Whether (s)he is using the term in a normative sense, a causal/explanatory sense, or a sense that somehow combines the two may be indeterminate, and left so. You and I may agree normatively; we may agree as to what natural property *constitutes* an act's or judgment's being rational. Or we may agree closely enough for purposes at hand. If so, we can speak together in a scientific vein of the causes of rational and irrational human doings and their effects, in a way that carelessly but harmlessly mixes normative and causal/explanatory senses.

With a term like 'rational', though, such talk will often be contentious. Another scientist whose normative judgments opposed ours might agree with the purely causal, scientific component of our account, and yet disagree about whether it was an account of *rationality* in a normative sense or of something else. When such disputes arise, their clarification will demand great care in sorting out normative claims from causal/explanatory claims.

Meaning Ascriptions as Normative

As I did in my book (1990), I've been treating meanings as parts of the natural world, open in principle to empirical, scientific study, and so candidates to figure in causal explanations of human phenomena. Rationality in a normative sense, in contrast, would not figure directly in causal explanations: in nature, I maintained, there is no such thing as a normative property. My book, then, made two contrasting claims, one for meanings and judgments and the other for normative rationality. The realm of meaning is a part of the causal order. Judgments, with their content, are natural human states, and this applies equally to non-normative and normative judgments. When we report a person's judgments or when we say what he means by his words, then we are, in a strict sense, attributing a *property* to him, a property that will play a causal role. When, in contrast, we call something 'rational' and so *express* a normative judgment, we are not straightforwardly attributing a property. Judgments of rationality are in the causal order; their correctness or incorrectness is not.

A major strain in recent philosophy of language and philosophy of mind, though, has insisted otherwise: It has proclaimed that 'meaning is normative'. Less crudely, might be taken to mean the claim that ascriptions of meaning are normative statements: when I say what a person *meant* by an utterance, I am making a normative claim.[6]

This raises problems for the theory of my book in which I classified statements that ascribe meanings as causal/explanatory and hence descriptive, and normative statements as non-descriptive. If statements that ascribe meanings are themselves normative statements, these classifications can't both be right.

Suppose, then, I accept that 'meaning is normative'. Could I still deny that normative statements are descriptive? Could I go on excluding the rationality or irrationality of human doings from the causal world? Could I rule out strictly causal explanations of human doings in normative terms? If I do, I seem to face two problems: first, it may seem fantastic to suppose that the human sciences could proceed without talk of meanings, judgments, attitudes and the like—which I have excluded from the realm of strict causal explanation.[7] Second, and perhaps worse, my theory has by its own lights now become non-factual. The theory, after all, is of what normative terms mean. If claims of meaning are normative and if normative statements are not purely descriptive, then my theory is itself not purely descriptive.

Could I, then, accept that 'meaning is normative', and still maintain a sharp fact–norm distinction? I am claiming that there is a sharp line to be drawn between descriptions of natural fact, on the one hand, and normative statements on the other. Normative statements are not descriptive statements of fact at all and indeed the simplest of normative statements might most clearly be expressed by imperatives. Can I maintain all this if it follows that those very claims are non-descriptive?

In what follows, I will take a gingerly first look at this question. What might a theory be like that treats meanings as something beyond pure matters of fact? Could my theory intelligibly renounce all claim to be purely descriptive and scientific? I will propose that it could: such a claim might not be self-undermining. In calling normative claims non-descriptive, I am after all not denying them a serious use. It is part of my theory that normative judgments and assertions—judgments and assertions that go beyond sheer description—are a crucial, inescapable part of any life that is fully human. Perhaps in saying what terms mean, in classifying them as descriptive or non-descriptive, I am not simply offering a proto-scientific description of the human world. Instead, I am saying how to *use* these terms. In ascribing a judgment to a person, descriptive or normative, I might be saying how to *take* the person, what *stance* to take toward the person.[8]

How, then, shall we interpret a claim that 'meaning is normative'? A variety of interpretations are suggested in the literature, but the one I will try to develop takes as its slogan '*Means* implies *ought*'. Statements ascribing meanings, the claim would be, have normative entailments: they entail things about what people have *reason* to do or are *committed* to doing.

Saul Kripke's famous exposition of Wittgenstein introduces a claim in this vicinity: Kripke's Wittgenstein considers a strange kind of meaning sceptic who questions what I have meant, up to now, by '+'. This sceptic challenges me as to whether my previous use of '+' was such that if I had been asked 'What is 58 + 67?' it would have been correct for me to answer '125'. If I meant *plus* by '+' (and meant the standard things by the rest of my language), then I *ought* to have answered '125'. In this sense, the claim that I meant *plus* by '+' has at least one normative implication: an implication as to how I *ought* to have answered.[9]

Kripke's exposition has evoked immense discussion. But I won't dwell on whether I have rightly interpreted Kripke's talk of 'normativity' or whether his claims, so interpreted, can withstand examination.[10] Instead, I will explore a few aspects of how such a claim might work out.

First, is it plausible that an adequate cognitive science and adequate social sciences might dispense with all notions of meaning and judgments with content? That would also mean dispensing with all notions of desires, aims and goals, since those too have content: if I want to be alone, the content of my desire is *that I be alone*. Now claims indeed abound in the current philosophical literature that intentional notions like these have no place in science: that they belong to another kind of discourse, and that a mature science could stick to notions of syntax or neurology.[11] I myself am sceptical of these suggestions. It is not that I think that scientific treatments will leave talk of meaning or conceptual content unchanged, or that these will appear as items in the kind of mental analogue of Newtonian physics that Hume may have had in mind. The human brain, however, is the most complex kind of object in the universe—so far as we know. Descriptions of what it does cannot be couched in basic physical, chemical or neurological vocabulary, or they would be intractable as substitutes for talk of judgments, aims and the like. Even if we could give a full theory of the brain in such terms, we would need theories couched in higher-level terms to see the forest for these twigs.

We need, then, a scientific picture that integrates human thought and judgment with a conception of ourselves as parts of a physical universe. The organising principle for studying any living being is Darwinian: all life is a product of natural selection and natural selection mimics intentional design. Organs will have functions, the things the organ is designed to do. With the brain the design in question will be cybernetic and cybernetic design involves states that are judgment-like and states that are goal-like. In the brain we can expect to find mechanisms whose biological functions are to code *ways things are*, and there will be states whose function is to code *goals* for the organism to attain. The programme that seeks to define representational content in terms of ideas like these has been dubbed *teleosemantics*.[12]

Why then, even suspect that ascriptions of content might be non-descriptive? One answer is that teleosemantics raises puzzles. How are we to specify what a state represents—especially when it is the product of complex experience and complex social interactions? There is no state whose biological function is to represent, say, electrons, or even a wheel. Human genetic evolution, after all, came before electrons were conceived of and before wheels even existed. To carry out the teleosemantic programme, we would have to say how a mechanism can be said to represent kinds of states of affairs that weren't around when selection pressures shaped genetic plans for our cognitive faculties. Attempts to carry out such a philosophical programme may lead to problems of indeterminacy, problems analogous to those explored by Quine in a different naturalistic framework.[13]

How these problems work out in the end remains to be seen, but here is one possibility: certain kinds of indeterminacy may not matter for explanatory purposes. For purposes of accepting or rejecting what someone asserts, though, these indeterminacies will matter a great deal if on one interpretation, what the person says is acceptable, whereas on another, which is equally good explanatorily, what he says is unacceptable.

A normative account of the content of meaning ascriptions might remove this indeterminacy—in a sense. In ascribing content to a person's judgments and assertions, let's try saying, we are settling under what conditions *to accept* what he judges or what he says. Considerations of teleosemantics may leave it indeterminate what content to attribute, and for explanatory purposes it may not matter. The purposes of deciding whether or not to agree with the person, though, one must decide. (We may, to be sure, decide to regard the judgment as indeterminate in content, and so neither to agree nor to disagree. The claim that the content of the judgment is indeterminate will still be a normative claim: it tells one, under the circumstances, neither to agree nor to disagree.)

Supervenience and Meanings of 'Meaning'

I have been heading in two directions: On the one hand, I have been claiming that adequate human sciences will need causal/explanatory concepts that look much like our old concepts of meaning. On the other hand, I have been exploring the claim that the concept of meaning is normative. How could I have it both ways—that the concept of meaning is causal/explanatory and that the concept of meaning is normative—and still insist that normative concepts cannot be causal/explanatory?

My suggestion is that we need two distinct concepts of meaning, factual and normative. When we press our ordinary concept of meaning, we'll find we need to develop it both in a factual and in a normative direction. For causal/explanatory purposes, we will need to develop it teleosemantically. For purposes of accepting or rejecting things people say, for agreeing or disagreeing, we will need to develop it normatively.

Still, we can expect a close tie between these two lines of development: a tie of supervenience. It seems plausible that what a speaker means in the normative sense will supervene on what he means in a causal/explanatory sense. In other words, our acceptance or rejection of things a speaker says should correspond to our acceptance or rejection of what he *means*—with 'means' taken in a sense that best helps explain human doings. Take the sense of 'means' that turns out to be most fruitful in explaining human doings—explaining them causally. Label this causal/explanatory sense means_{ce}. Then the best norm for accepting or rejecting things people say might read like this: 'Accept what a person says if and only if you ought to accept what it means_{ce}.' Suppose Ursule says: 'La neige est blanche' and 'La neige est blanche' in her utterance means_{ce} that snow is white. Then the norm as applied to this case will say, 'Accept what Ursule said if and only if you ought to accept that snow is white.' To say things like this—and to put them forth as *a priori* necessities—is to say that meaning in the normative sense *supervenes* on meaning_{ce}.

What I have put forward is, of course, is not a finished proposal but a suggestion for a line of thought. It would need much working out before it could be ready for scrutiny and evaluation. Here, though, let me break off with just one more thought.

The term 'rational', can be given both non-factual, normative senses and factual, causal/explanatory senses. The two will always be distinct but for many purposes we will be able to conflate them harmlessly. Often it won't matter whether the things we say with the term 'rational' are taken factually or normatively. Rationality, though, is contentious. What causal/explanatory property rationality in the normative sense supervenes on is highly contestable. Often, then—even if a causal/explanatory sense is stipulated for the term—things said using the term will be contentious because they suggest a corresponding normative claim which is contentious. We will find we cannot afford to be careless about which kind of sense we are using.

Meaning is less contentious normatively and so when we use the term, we may find we can be careless about whether we are speaking normatively or factually. We might be able to be far more careless on this score than we can when we use the term 'rational'. That the term 'rational' can be used both in normative and in causal/explanatory ways will bear heavily on the requirements of clear thinking in the human sciences. That the term 'meaning' can be used in both these ways may be more a fine point in the theories of meaning and normative concepts. The logic is the same in both cases but the scope of truistic equivalence may differ greatly.

Notes

1. For discussions of these disputes, see Fishburn (1981), Broome (1991).
2. See, for instance, Nisbett and Ross (1980).
3. Moore (1903). Frankena (1939) is the classic article on Moore's 'Naturalistic Fallacy'; Moore (1942, p. 27) speaks of 'confusions and invalid arguments' in *Principia Ethica* (1903).
4. More precisely, she thinks that the *intrinsic value* of a state of affairs—the value the state of affairs has in itself equals the net amount of pleasure the state of affairs contains.
5. Blackburn (1984, pp. 182–87).
6. See Kripke (1982, esp. p. 37); Wittgenstein (1953); McDowell (1984, esp. p. 336).
7. Some philosophers do, in various ways, deny that concepts of intentional states will figure in straight, causal, scientific explanations. See, for instance, Churchland (1986), Davidson (1971) and Stich (1983). I do not accept any of these views, although a discussion of where I differ and why is beyond the scope of this article.
8. Dennett (1987) speaks of an 'intentional stance'.
9. Kripke (1982, esp. p. 37).
10. For a review of some of the discussion evoked by Kripke, see Boghossian (1989).
11. See Stich (1983); Churchland (1986).
12. See Gibbard (1982; 1990, ch. 6); Millikan (1986); Dretske (1986).
13. For a discussion of problems encountered when trying to construct a teleosemantic concept of representation, see Dretske (1986).

References

AYER, A.J. 1936. *Language, Truth and Logic*. London: Victor Gollancz.

BLACKBURN, SIMON. 1984. *Spreading the Word: Groundings in the Philosophy of Language*. Oxford: Clarendon Press.

BOGHOSSIAN, PAUL. 1989. 'The Rule-Following Considerations'. *Mind* 98: 507–49.

BROOME, JOHN. 1991. *Weighing Goods: Equality, Uncertainty, and Time*. Oxford: Basil Blackwell.

CHURCHLAND, PATRICIA. 1986. *Neurophilosophy. Toward a Unified Theory of Mind/Brain*. Cambridge, MA: MIT Press.

DAVIDSON, DONALD. 1971. 'Psychology as Philosophy' in Davidson *Essays on Actions and Events*, 229–39.

———. 1980. *Essays on Actions and Events*. Oxford: Clarendon Press.

DENNETT, DANIEL. 1987. *The Intentional Stance*. Cambridge, MA: MIT Press.

DRETSKE, FRED. 1986. 'Misrepresentation'. In Radu J. Bogdan (ed.), *Belief: Form, Content, and Function*. Oxford: Clarendon Press.

EWING, A.C. 1939. 'A Suggested Non-Naturalistic Analysis of Good'. *Mind* 48: 1–22.

FISHBURN, PETER. 1981. 'Subjective Expected Utility. A Review of Normative Theories'. *Theory and Decision* 13: 139–99.

FRANKENA, WILLIAM. 1939. 'The Naturalistic Fallacy'. *Mind* 48: 464–77.

GIBBARD, ALLAN. 1982. 'Human Evolution and the Sense of Justice'. *Midwest Studies in Philosophy* 7: 31–46.

———. 1990. *Wise Choices, Apt Feelings: A Theory of Normative Judgment*. Oxford: Oxford University Press.

HARE, R.M. 1952. *The Language of Morals*. Oxford: Oxford University Press.

KRIPKE, SAUL. 1982. *Wittgenstein on Rules and Private Language*. Cambridge, MA: Harvard University Press.

MAYNARD SMITH, JOHN. 1974. 'The Theory of Games and the Evolution of Animal Conflicts'. *Journal of Theoretical Biology* 47: 209–21.

———. 1976. 'Evolution and the Theory of Games'. *American Scientist* 64: 41–45.

———. 1978. 'The Evolution of Behavior'. *Scientific American* 239: 176–92.

———. 1982. *Evolution and the Theory of Games*. Cambridge, England: Cambridge University Press.

MCDOWELL, JOHN. 1984. 'Wittgenstein on Following a Rule'. *Synthese* 58: 325–63.

MILLIKAN, RUTH. 1986. 'Thoughts without Laws: Cognitive Science with Content'. *Philosophical Review* 95: 47–80.

MOORE, G.E. 1903. *Principia Ethica*. Cambridge, England: Cambridge University Press.

———. 1942. 'An Autobiography'. In Paul Schilpp (ed.), *The Philosophy of G.E. Moore*. Lasalle, Illinois: Open Court.

NISBETT, RICHARD and ROSS, LEE. 1980. *Human Inference: Strategies and Shortcomings of Social Judgment*. Englewood Cliffs, N.J.: Prentice Hall.

QUINE, W.V.O. 1960. *Word and Object*. Cambridge, MA: MIT Press.

SCHELLING, THOMAS. 1960. *The Strategy of Conflict*. Cambridge, Ma: Harvard University Press.

STEVENSON, CHARLES. 1937. 'The Emotive Theory of Ethical Terms'. *Mind* 46: 14–31.

———. 1944. *Ethics and Language*. New Haven: Yale University Press.

STICH, STEPHEN P. 1983. *From Folk Psychology to Cognitive Science: The Case Against Belief*. Cambridge, Ma.: Bradford/MIT.

WITTGENSTEIN, LUDWIG. 1953. *Philosophische Untersuchungen/Philosophical Investigations*. New York: MacMillan.

Meanings and Rationalities in Social Choice Theory

SERGE-CHRISTOPHE KOLM

Introduction

In the grand tradition of the critical philosophy of Descartes, Kant, Hegel and Popper, Arrow's (1963) work *Social Choice and Individual Values* (SCIV) tried to make sense of utilitarianism after ordinalism and of Pareto–Lange–Bergson–Samuelson's Social Welfare Functions, and brilliantly refuted that the social good could be defined in this manner.[1] This *critique of economic or utilitarian reason* in social choice rested on the interpretation that these theories provide a solution for several sets of individuals' preferences (indeed all of them) with meaningful comparisons across them. This interpretation is certainly valid for utilitarianism, yet Bergson, following Little, rejected it (and Samuelson joined them in pushing it aside, out of the field of Welfare Economics). However, we will see that these criticisms of Arrow's way of posing the problem are self-defeating. Indeed, if one rejects the plurality of the sets of individuals' preferences, or the way in which Arrow uses it to impose 'conditions' on social choice, then by the same token one must reject the idea that the social good can meaningfully be defined by the maximisation of something.

Numerous works by other scholars have proved Arrow's conclusion under somewhat more general conditions, thus undermining hopes to salvage Social Welfare Functions or to rescue some sort of utilitarianism. We call this literature *social choice theory in the strict sense* (SCT). These logically enticing pieces constitute a domain with an extraordinary outlook, indeed, a surrealistic one (it is reviewed in particular by Kelly 1978 and Sen 1986). It is not so much that each failure constitutes a success—since not solving the problem extends impossibility. Rather, what is most impressive is the equanimity with which each specific assumption is accepted, as if it were self-evident and necessary. This is, of course, the very opposite of what Arrow does. Arrow's own work strikes one by its broad view, its philosophical culture and its deep and pervasive concern for meaning and significance.

By contrast, the typical SCT study rightfully bows to SCIV, but it uses it as tonic and alibi for jumping boldly into exciting technicalities with no discussion of the actual meaning of what it is doing, and without a word for such basic questions as:

- What does it mean that each individual has several preferences?
- What does it mean that society has several sets of possibilities?
- What does it mean that 'society' has preferences or tastes or a consistent mind?
- Why should the social optimum be defined by the maximisation of something (or by a 'social ordering')?
- What does 'rational choice' *normally* mean and what are the many consequences of this meaning for the problem at hand?
- Why require answers to apparently absurd questions, such as the best social choice on impossible sets of possibilities with impossible individual preferences (e.g. unanimously suicidal ones), and why then regret the absence of an answer rather than find it reassuring?
- Why is the social optimum *necessarily* defined in terms of individual preferences, and what are the consequences of this necessity?
- Who holds the opinion that the social choice should satisfy certain conditions?
- Where are the variables that can express the liberties and equalities that are the material of all the individualistic social ethics—since they manifest agency and non-arbitrariness?

Every work in this literature will be addressed by all or most of our questions, either directly or after obvious restatement. Yet, in order to focus the referencing, and because it is the initial and the most, indeed the only meaning-minded and philosophical work, my references will be essentially taken from SCIV (last edition, 1963), which I shall thus henceforth include in SCT.

In SCT's presentations of its own topic and objectives, one can distinguish two different pictures. On the one hand, SCT aims at answering the grand question of what is the overall social optimum, the best possible state of society. This question is brought about by the history of thought, since both utilitarianism and the theory of the Social Welfare Function can answer it. Yet, on the other hand, SCT is sometimes presented as trying to determine the good collective choice for specific questions and situations (particularly, for example, in choosing a voting procedure). However, the two cannot be separated since, if one knows the overall social optimum, one knows the best outcome of each specific question; and if one knows the best procedure in each specific case, one has the overall optimum in lumping them together. Yet, SCIV does not consider the sequential and decentralised collective decision-making that would make explicit the relation between both aspects. Social Choice Theory that is not 'in the strict sense' has turned into varied studies of the various processes of collective decision-making. Yet, SCIV and the (strict) SCT that follows it closely aim primarily at the grand problem of the social optimum—as the motivating history of thought did—although we will see that the noted ambiguity is at the heart of at least some of the problems of this theory.

What a Question!

Consider that society has to choose one option out of 50 available ones. Then ask the question: What should society's seventeenth best choice be, if individuals' preferences were the opposite of what they are, and if the set of available options were a completely different one? Who would ask such a strange, useless and seemingly silly question? The answer is: 'Social Choice Theory' in the strict sense.

Then SCT complains that it cannot find an acceptable answer to questions of this type—not just to only one, but to the myriads of such forged, non-posed, imaginary questions. This is the 'impossibility' result. Note that the question regarding preferences, contrary to the actual ones, implies not only that each individual always prefers to have less of each consumption good, and that his family dies under torture, but also that he himself instantly dies—since actual living individuals want to survive—so that we know the corresponding best social choice if the stock of nuclear weapons is still available. Yet, more generally, should one expect it to be possible to provide rational answers to absurd questions?

The Problem of Social Choice Theory

Of course, if there is a social ethical problem, the question would be: What should the unique social choice out of the *actual* possibilities (defined as mutually exclusive) be, given the *actual* individual preferences and other relevant items? Now, SCT chooses to answer this question by imposing *a priori* a particular and particularly devious way (inspired in part by utilitarianism and its epigones).

Indeed, it first seeks a *social ordering* of all the alternatives. By definition, such an ordering enables one to find a 'socially best' state out of *any* possibility set, as a maximal element of the ordering on that set. The actual possibility set is just one particular case. Moreover, the very consideration of an ordering requires that it be meaningful to consider the choice on several possibility sets, since the ordering is defined as a set of relations among such choices—indeed, this definition compares choices among *all pairs* of alternatives.[2] Note that in order to take as favourable a view of SCT as possible, we avoid speaking of 'social preferences', whereas the SCT literature is replete with such expressions as 'society prefers' or even 'collective tastes' (SCIV, p. 3) which suggest an embodiment of the group that is hard to make sense of and is hardly consistent with SCT's overall individualistic outlook. (This *per se* weakens the reason for the consistency conditions—transitivity or its equivalents—that gives these social choices the structure of an ordering.) SCIV itself is indeed very clear and explicit that the social ordering it considers means 'nothing more' than social choices: 'I regard social values as meaning *nothing more than social choices*', 'I locate (social values) in the *actions* taken by society' (p. 106); '"collective rationality" (is) that social choices . . . were so interrelated as to satisfy the definition of an ordering' (p. 107).

Second, SCT wants the social choice to be justified only by the *individuals' preferences*, meaning their values or tastes whatever their nature or reasons, and represented by *orderings* of the same alternatives. A set of such individual preference orderings, one for each individual, is called a *profile* in SCT literature.

Third, SCT wants to define a social ordering for each profile of individuals' preference orderings.[3] This defines a function that I shall call the *social ordering function* (since the term 'social welfare function' has been used in several different senses). Applying the social ordering function to the *actual* profile of individuals' preferences yields the social ordering to be used in order to obtain the social choice from the actual possibility set.

Fourth, SCT sets a number of seemingly reasonable *conditions* on the social ordering function, and it shows that no such function can satisfy them (Arrow 1963).

The Problems with Social Choice Theory

This theory has been abundantly and dramatically interpreted (impossibility of democracy, etc.), and also technically refined. Yet, it raises the following interrelated questions.

Multiplicity and Counterfactuality Problems

Once we accept SCT's outlook, framework and structure, the biggest question of all is probably the 'multiplicity' or 'plurality' one. That is, what is the *meaning* of considering *several* preference profiles and *several* possibility sets—indeed, SCT considers all imaginable ones (this is 'unrestricted' or 'universal' domain for the profiles and it is implied by the ordering for the possibility sets)? The individuals' preferences and the possibility set are what they are. Why require an answer to unraised, fanciful questions, why demand a solution to non-posed, irrelevant problems? If one cannot find an answer that is required to answer also all these false, absent questions and non-problems, why should one deplore this 'impossibility'? It says nothing on the possibility of solving the actual social ethical problem, except that the particular way chosen by SCT is not a fruitful one. The rest is impossibility of the impossible, and there is nothing wrong with this—indeed, it is rather reassuring. Should not the social choice abide by a principle of *independence of irrelevant social choices*? Yet, it is apparent that this question requires more subtle discussion.

Rationality Problems

One argument that can be proposed in defense of SCT comes from a demand of *rationality* in the ordinary sense of *for a reason*. That is, if I choose this *because of* that, there must exist another, hypothetical case, where I would have chosen something else. Yet, the various cases would now be exclusive ones, and this raises a problem of meaning for SCT's 'conditions' that compare them, and that must be shown to be necessary. Furthermore, at least as puzzling is the fact that this very requirement of rationality can be shown to undermine the possibility that a social ordering existed and that several of SCT's 'conditions' on the social ordering function be satisfied.[4]

Individual Preferences Problem

SCT's 'conditions' imply that its social ordering—hence its social choice for a given possibility set—is concerned with the world only by the intermediary of individuals' preferences. Now, individuals commonly manifest poor moral judgement. However, there is a notable logical and epistemic reason in favour of this structure, which is that only individuals' opinions can exist in the overall society. Yet, this very reason implies that SCT's 'conditions' on the social ordering function represent values of the very individuals it considers; as a result, these conditions become trivial (either trivially satisfied or *a priori* impossible).[5]

Endogenous Values Problems

Indeed, the very formation of these values must be considered, and this undermines the very logic of SCT, as is shown for instance by the 'paradox of the democratic dictator' who agrees with the good social ordering because it is good, and *ipso facto* turns out to be a dictator in SCT's sense, hence rendering this ordering no good.

Problems of Liberty and Equality

While SCT has, in a sense, too many variables from its pluralities, counterfactualities, 'universal domain' and assumption of a social ordering justified by choices, on other grounds it bypasses certain variables that are necessary for solving its problem. Indeed, it does not explicitly consider the variables that can relevantly express the two values that all individualistic social ethics and all theories of justice use to solve the problem: liberty and equality. It can straightforwardly express the value of freedom by replacing, in the definition of 'social states', the outcomes of actions and processes by means of these free actions and interactions. Yet, it does not do so. The discussion of equality also requires other explicit variables (were it only to be able to define the basic social ethical principle of 'equal treatment of equals').

It should be noted that the initial work, SCIV, is aware of the questions of endogenous preferences (pp. 6, 7) and of the intrinsic value of social decision processes (pp. 89, 90), although its treatment of the latter is not the simple, straightforward consideration of the value of liberty that I have pointed out.

Requirements of Social Choice Theory

Although I have not yet presented the 'conditions' and the proofs of SCT, we nevertheless need to know some of their conceptual requirements in order to see what actual meanings can or cannot be attributed to this theory. To this end, I propose that SCT needs the following things, which may be necessary either for the definitions, for the meaningfulness of the concepts, for the proofs and working out of the theory, or for several of these aspects. For individual preferences and possibility sets there are many statements, because various parts and *variants* of SCT require one or the other, and various interpretations or problems of SCT raise difficulties for one or the other.

There are individuals (at least three).
There are alternative social states (at least three).

Individual Preferences:

(*i*) Plurality: each individual 'can' have several alternative orderings of the social states.
(*ii*) 'Universal (or unrestricted) domain': each individual 'can' have all logically possible alternative orderings of the social states, and all preference profiles are possible.

Possibility Sets:

(*i*) Plurality: there 'can be' several alternative possibility sets for social choice, each of which is a subset of the alternatives.
(*ii*) These possibility sets are not all disjointed; some have common elements.
(*iii*) All the pairs of alternatives are possible alternative possibility sets.
(*iv*) 'Universal domain for social choice': all the subsets of alternatives are possible alternative possibility sets.

For possibility sets, condition *iii* is necessary for making sense of a social ordering (i.e. for expressing the transitivity conditions). It is *a priori* problematic (is society's choice ever restricted to two alternatives?). This condition is implied by condition *iv* and it does not hold if condition *i* (or condition *ii*) does not hold. Conditions *i* and *ii* together may provide necessary conditions for an ordering.[6] One of these conditions that will be used several times in this article is the so-called 'Chernoff condition', according to which, if the choices correspond to an ordering (i.e., one that does not depend on the possibility set), it is not possible that the choices from two intersecting possibility sets both belong to both sets and are distinct.

We denote as *proper* a function the domain of which has more than one element. The logic of SCT rests on two proper functions endowed with particular conditions: the *social ordering function* and the *social choice function* (that defines the possibly chosen social states in each possibility set, as a function of this set). The properness of the social ordering function is the plurality of preference profiles for which the social ordering (and the social choice) are defined. It is necessary for the definition or application of most 'conditions' on this function (all but the unanimity-Pareto condition, discussed in the next section). The properness of the social choice function is necessary in order to make sense of using a social ordering. The conditions on it are those that define an ordering (transitivity or equivalent conditions).

I shall now investigate the different meanings of SCT's plurality of preference profiles and of possibility sets.

SCT's Pluralities are not Successive Social Choices of the Same Individuals

The definition or the application of all SCT's conditions on the social ordering function except the unanimity-Pareto one are replete with expressions such as: Individual *i*'s preference ordering or tastes or values *change* in such a way, *become* that, incur an *alteration* or a *variation*, take a *new* form different from what they *formerly* were (SCIV p. 25), etc. SCT's proofs often mention that the possibility set of the social choice *becomes* this and that. This tends to suggest a time or sequential interpretation of these pluralities.

Indeed, SCIV begins with the example of successive choices by the same set of individuals and varying possibility sets taken from the same set of alternatives at various dates—with steady, dated individual preferences: 'It is expected that choices of this type have to be made *repeatedly*, but sometimes not all of the three alternatives will be available' (p. 2). The aim is to show an intransitivity in the social choice implemented by voting. The result is that this procedure does not provide a dated social ordering that would be the same at all dates (as the dated individual preferences are). Yet, why don't the individuals initially make the social choice for all the present and future dates at once, as a contingent choice if the future possibility sets are not known with certainty? Each individual's preferences in fact bear on the states at all dates at once. The outcome would be a unique, intertemporal social choice, with which no social ordering and no intransitivity could be defined.

Indeed, SCIV also states: 'Each alternative is a complete description of the state of every individual throughout the future' (p. 87). These variables are presented as an alternative to variables that are government's instrumental variables but, presumably and classically, the latter are valued by the individuals indirectly, through their influence on the former variables.

Considering social states that include variables at all (or all future) dates is standard economic theory. More importantly, it corresponds to the actual question. Then, at once, social states (and hence the objects of individual preferences), and possibility sets (and hence the problems of choice they pose), are intertemporal.

An intertemporal interpretation of SCT's plurality of preference profiles and possibility sets can only be made through a model with temporal independence of the individual preferences and social possibilities, repetition of the dated alternative states, yet changes in dated one-period preferences and possibility sets (uncorrelated changes, for defining social orderings and social ordering functions). The actual world is not like this. It is, indeed, very different. Such a theory would describe *only* counterfactuals. Actually, two variables at two different dates can only be two different variables (hence possibility sets at two different dates can have no common element). Intertemporal trade-offs in the choice of the (intertemporal) social state are essential questions for both individual and social choice (saving, investment, etc.). Indeed, the very set of individuals changes in time as some arrive and others leave, contrary to what is assumed by the working of SCT (the question of population change is discussed in the following paragraphs).

SCT's Pluralities are not Different Populations

Another interpretation of SCT's pluralities of preference profiles and possibility sets is that each profile or possibility set corresponds to a different population. These populations would not interfere with one another, and SCT would be the theory of the social choices for which they would provide instances. This view does not exactly fit with the letter of SCT, that commonly speaks of different preference orderings for the same individual. As an instance among many, SCIV states: 'Several different ordering relations are being considered for the same individual' (p. 19). Yet, this interpretation is consistent with 'Such a social welfare (ordering) function would be universal in the sense that it would be applicable to any community' (p. 24). I.M.D. Little (1951) explicitly notes (without being rebuked) that 'changes in tastes' can be a substitution of individuals. This interpretation was indeed considered favourably by Arrow as well as by others. Indeed, in all SCT the expression '*individuals*' preferences' frequently slips into '*individual* preferences'. Arrow advocates not to 'take very seriously the exact words' used in 'explicit definitions' (p. 107n).

Unfortunately, before any other consideration, this interpretation is precluded by one of the 'conditions' SCT requires from the social ordering function. This condition is 'non-dictatorship' or the absence of a 'dictator'. A dictator is defined as an individual such that the social ordering coincides with his own individual preference ordering for all preference profiles. Then, if the set of individuals is different for each profile, there cannot be a dictator, since the social ordering function is defined for at least two profiles (indeed, it is defined for all profiles—'unrestricted domain'). As a consequence, the non-dictatorship condition is always trivially satisfied and does not make sense.

Note also that according to SCT the alternative populations should all contain the same number of individuals. This is implied by the working of the theory (definitions of a number of 'conditions' and proofs). It is also implied by the SCT's requirement that the various social orderings (one for each profile) order the same social states. Indeed, an important characteristic of a social state is the set of individual bundles of consumption goods and leisure—one bundle for each individual—and a characteristic of this set is the number of bundles it contains.

Furthermore, a population is characterised not only by its preference profile, but also by the capacity of its members to produce and consume, individually and collectively. Hence, for each profile there is a correspondingly unique possibility set, and the corresponding social ordering cannot be defined from alternative social choices with different possibility sets.

It is hardly worth adding that different populations cannot have the same social states, and certainly not the same Pareto-efficient social states, as the definitions and proofs of the theory require it, because of individuals' consumption choices and skills and of society's possibility sets—even apart from the number and size question. For example, Germany and a Chinese province of the same size cannot have a common Pareto-efficient set of individual diets; the allocation of Picasso's drawings is a feature of a social state only in the population that includes Picasso; populations are even accompanied by more or less specific types of resources provided by nature or by history.

SCT's Pluralities Cannot be Social Choices in Different Fields

A society may make different social choices for different issues—such as one for distribution, one for certain public goods, etc. Yet, this cannot be what SCT means by plurality since the social states ordered by the orderings in the different fields are different, and the individual preferences on the one hand and the possibility sets on the other are interdependent across the fields (what they are in one field depends on the social choice for the other issues).

SCT Cannot be a Theory of Constitutions

SCT is often presented and defended as a theory of constitutions. This assertion aims at justifying the plurality of profiles of individual preferences, and 'constitution' denotes the proper social ordering function (Kemp and Asimakopulos 1952; SCIV 1963; Samuelson, 1967, etc.).

Indeed, a constitution is a set of rules that transforms various sets of individual preferences into social choices. However, the foregoing discussion shows that this interpretation does not hold: SCT cannot be considered as describing constitutions, nor can constitutions provide meaning to SCT because:

1. The application of a constitution does not provide a social ordering of the alternatives, nor does it impose that the determination of the social choice be made by way of a social ordering that is then applied to the possibility set in order to determine the choice.
2. Indeed, the application of a constitution by judges, voters, governments and others commonly yields social choices that are inconsistent with the hypothesis of an underlying steady, possibility-independent social ordering, as well as with other 'conditions' required by SCT. Yet, this result comes from

considerations of fairness, equity, or rationality (in the normal sense of 'for a reason'), as will be shown later on in this article.

3. A constitution is applied in several successive choices, and we have seen that this cannot be what SCT means. To begin with, successive choices are on different social states (for the future at each date).
4. The various applications of a constitution are often for different fields, issues and questions, and we have seen that this cannot be what a proper social ordering function means. Again, the social states would differ in different social choices.
5. The successive applications of a constitution usually imply different sets of individuals, since the population changes (more or less) as time elapses. The number of individuals changes, and after a sufficient interval, the whole population is different, and the individuals who remain change as they age. This provides several additional reasons why these successive choices cannot provide the plurality of preference profiles and of possibility sets that SCT requires: variation in the number of individuals leads to disjoint possibility sets because the number of individual consumption–leisure bundles varies, and because the individuals' productive and consumptive capacities vary (because the individuals change or age); actual one-to-one relation between preference profiles and possibility sets; and triviality of the non-dictatorship relation when the whole population changes (unless we consider hereditary dictatorship).

If SCT's social ordering function were a constitution, it would be applicable to all countries and times, by virtue of its definition and the hypothesis of 'unrestricted domain'. SCIV explicitly draws the conclusion that: 'Such a social [ordering] function would be universal in the sense that it would be applicable to any community' (p. 24). This has the following two consequences.

6. This cannot provide meaning to SCT's plurality of preference profiles since these specific applications would be with different populations (as has just been explained).
7. This universality is against the classical political meaning of a constitution. The term and concept of a 'constitution' has been given meaning in political theory and philosophy. It is

not a virgin term that one can use *ad libitum*. An aspect of this meaning is that the form of a constitution is and should be *specific to the people* to which it applies, and eventually transformed and amended through history to accommodate changes. This is indeed the very theme of Montesquieu's influential work *L'Esprit des Lois*. The classical theory of democracy assigns a specific place to a constitution: the Rights of Man are universal, the constitution is people-specific, the laws are topic-specific, public actions are event-specific. These characteristics are indeed respected by a constitution democratically chosen by a society with specific characters, uses, conceptions and problems. This is even more the case with constitutions that are constituted by tradition and by the very content of a specific history (the classical English view). It would also hold for constitutions resulting from an explicit or implicit contract (Buchanan's (1954) theory would be a case in point). And Rousseauan constitutions were the most careful of all in adapting to the characters and customs of the populations. On the contrary, SCT's social ordering function would be a universal, unique constitution, valid for all time, all places and all people. Translated into SCT's language, variables and concepts, the specificity of the constitution implies that the search for the good constitution should consider less than universal (unrestricted) domain for the preference profiles—since people are here characterised by their preferences. This means that, for a given people in a given historical period, certain values are impossible or are more or less likely or more or less shared (hence, probabilistic extensions of SCT considering the likelihood of profiles are interesting). Since interpersonal distribution is a major issue this would in particular describe the spread of various senses of justice, of solidarity and of altruism. Yet other very relevant aspects of peoples, such as the cohesiveness or divisiveness of the society, the propensity to trust and cooperation or to diffidence, etc., are not so straightforwardly modelisable by SCT.

A Priori Uncertainties do not Justify SCT's Pluralities

The literature of SCT casually but frequently justifies the plurality of preference profiles by the fact that individuals' preferences are not known in advance (e.g. SCIV, pp. 24, 28, 49, 59, 60, 61, 62, 74, 89). The same remark can be applied to the plurality of possibility sets that is required to define a social ordering (indeed, SCIV, p. 12, defines any ordering as 'before knowing the [possibility] set').

However, what we want is to determine the social state that must be implemented. SCT goes about it by finding a maximal element of the social ordering on the possibility set. It needs to know both these factors for determining this optimum. Therefore, when it has the social ordering function it seeks to define, it will also need to know the actual individual preferences in order to obtain the social ordering from this function. Further, SCT also has to know the possibility set in order to define the optimal social choice as a maximal element of the social ordering on it.

But since, with SCT's method, one has to know the individual preferences and the set of possibles in order to perform what is actually demanded—the determination of the chosen social state—we can as well start directly with the knowledge of these two factors and try to define this choice directly from them. The specifics of SCT then appear to be unnecessary, particularly its two intermediary concepts of a social ordering and of a social ordering function. These concepts, with their 'conditions' on the function, may rule out a solution: this is of no avail since they are unnecessary if they are justified by this uncertainty.

This reasoning can also be applied to the first half of SCT's method: the determination of the social ordering function. Assuming we want a social ordering, *a priori* uncertainty cannot justify determining it by the intermediary of a proper social ordering function, since deriving the ordering from this function requires that we know the actual preference profile. But if we have this information, we can directly derive the social ordering from it, without imposing on ourselves the obligation of defining a social ordering for each (or several) preference profile. Hence, this reason does not justify a need for a proper social ordering function, nor does it not justify that we mourn the absence of a good one.

Yet, it is common for public actions to be implemented without a perfect knowledge of individual preferences and other relevant factors. This, however, requires a full-fledged theory of social ethical choice in uncertainty, which SCT is not. I have presented this theory elsewhere.[7] However, even with the most favourable theory, the basic problems would remain. The *ex ante* choice of instrumental variables is unique and therefore does not provide a basis for making sense of an *ex ante* social ordering. Several possible *ex post* possibility sets are considered *ex ante* but there is no choice on *each* of these sets, so that the conditions for an *ex post* social ordering are not satisfied either. Furthermore, the set of these plausible *ex post* sets certainly does not contain all the pairs of social states, which is necessary for defining the *ex post* social ordering. Similarly, one *ex ante* contemplates several preference profiles, so that SCT conditions that require this plurality can be stated. Yet, not all profiles are *ex ante de facto* plausible (no 'universal domain'). Furthermore, since *ex post* preference profiles and possibility sets are unique, the meaningfulness of the numerous SCT conditions that are 'transversal' is in serious doubt: This affects both conditions on the social ordering function and the meaningful existence of this ordering.

Rational Moral Counterfactuality

Can SCT be justified by rationality? Is SCT rational? 'Rational' in its normal and common sense means 'for a reason'. It means a capacity to begin to provide a serious answer to the question 'why?' (a complete answer is endless and hence impossible). This question applies to a *choice*. It can be the choice of an action, of a belief, of a hypothesis, of an expression (that is a particular case of action), of a judgment, etc.[8]

The contrary of *rational* is *arbitrary*, and a synonym of rational is *justified*. Hence, rational means non-arbitrary or justified. Rationality means non-arbitrariness or justification. Irrational means arbitrary, non-justified, unjustified. Irrationality means arbitrariness, lack of justification.

An answer to the question 'why' would contain two elements, one general and one specific. In the expression 'I choose y because x is the case', x is the specific and the general is the functional relation from x to y. For such an expression to count as an answer there must exist another specific x' different from x and such that, if it were the case instead of x, the same functional relation would apply and would lead to a choice y' different from y. This implies that this function must have a *domain* that contains at least *two* elements (x and x'), only then can we call it a *proper* function. Further, it must have a *range* containing at least two elements (y and y')—we can then call it an *actual* function; and that the actual specific belongs to this domain (x here). The elements of the domain and of the range other than the actual specific x and the actually chosen y are *counterfactuals*: they are imaginary. To consider them is a conceptual experiment (they may however actually be the case in other, comparable questions). Therefore, rationality requires that one consider *at least one counterfactual specific* and *one counterfactual choice*.

A particular case of rationality is that of laws or rules that are intended to apply to *several* actual instances, indeed to a set of instances that are not pointed out individually but are characterised by certain common features that define the domain of applicability. These rational laws or rules can be found in all domains of rationality, for choosing actions or evaluations, particularly ethical ones (such as keeping one's promises, not lying, distributing income in a certain way, etc.), or for choosing a belief based on induction or logic. They are formally akin to certain other answers to the question 'why', namely causal scientific laws, that are not *per se* instances of rationality because their object is not human choice (the falling of a stone is not a human action) although they provide reasons to believe in predictions (*e.g.*, the law of gravitation).

Laws or rules often satisfy requirement of rationality by a pair of conditions that implies it: 'universal domain' and 'sensitivity'. 'Universal domain' means that they apply to the entire domain on which they have been *a priori* defined, and 'sensitivity' means that there exist, in this domain, two specifics that yield different conclusions. This implies that for the actual choice, there exists a counterfactual (that may be similar to another factual) that gives another conclusion. For scientific causal laws, 'universal domain' reflects Aristotle's contention that 'science is only general'. Now, these two conditions are similar to two conditions stated by SCT for the social ordering function defined on the preference profiles, the 'unrestricted domain' of profiles and the 'non-imposition' of the ordering of each pair of alternatives. And the conditions for the social choice to be considered as representing a social ordering also require many possibility sets—at least all sets of pairs of alternatives for necessary and sufficient conditions.

However, we have seen that SCT cannot be the description of a set of *several actual* social choices: it can describe neither choices at different dates, nor choices by different populations, nor choices in different fields. Hence, neither the social ordering function nor the use of a social ordering can constitute a rule or a law. Note that, by comparison, if a scientific 'law' were to apply to only one case, it would turn out to be either false or unfalsifiable (i.e., 'metaphysical' in Popper's sense). Unique events are explained by several laws, each of which applies to several instances.

Yet, ethical stances can be checked on counterfactuals and can be given generality by conceptual experiments. This is part of Plato's dialectical method as described in *The Republic*, as it is of Rawls' (1971) 'reflective equilibrium'. The imagined cases of application are then appraised by our more or less primary ethical judgments ('intuition' is a common term here, although a misused one).

The unique counterfactual that is necessary for an answer to the question 'why' to make sense, is a minimal requirement that may contain only little information. For instance, the case where 'social indifference' among all alternatives results from every individual's such indifference, can be taken as counterfactual for the case of any other social ordering resulting from any other preference profile. Further, for the possibility set, any other set of alternatives that is disjoint from it leads to another choice by definitional necessity! Considering other counterfactuals is thus informative. However, what counterfactuals should we consider? If requirements concerning counterfactuals that could not exist lead to impossibility, they erase the possibly relevant information one can derive from others. But how do we determine what feature of reality we should imagine away in order to define what 'could' or 'could not' exist? Expressions such as 'possible', 'plausible' or 'realistic' are ambiguous. Strictly speaking, only the actual is 'possible'. Yet the laws of physics enable one to answer questions that are wildly counterfactual. We have already seen that SCT constrains itself to provide answers to strange, useless and silly questions. Yet, consider the following question: what would the trajectory of the Eiffel Tower be if it were a satellite of Jupiter, this taking place two centuries ago, and if in addition it were painted red? This question is certainly no less strange, useless and silly! Nevertheless, the law of universal gravitation answers it. There is a difference, however. Both from the point of view of scientific knowledge (for refutation or corroboration), and that of usefulness (for application or explanation), a scientific law requires of itself only that it answers actual cases. Its answering counterfactuals is a bonus, and to pose questions about impossible cases is metaphysics, mythology, poetry, science-fiction, etc. rather than science.

By contrast, since SCT actually applies to only one case, it has to consider counterfactuals (at least one) in giving a reason for its choice. In addition, all laws have limits of validity. Even Newton's law fails at speeds close to the speed of light. We may admit that the laws of zoology should apply to the counterfactual unicorn (after all, it would just mean adding a horn to a horse). But should we also require that they also apply to mermaids, centaurs, sphinxes, forked-footed devils, winged angels, Anubis, Ganesh and chimeras? Is the Kantian maxim 'not to lie in any circumstance' at all possible? Should we require the social ordering function to grind out a social ordering for a case where every individual would prefer that he and his family perish by torture, or where each prefers to have less of every good in all situations? It may make sense for the social choice function to satisfy the requirements SCT demands for a number of possibility sets. But is it sensible to consider social choice among pairs, indeed among all pairs, as necessary for meaningfully defining the ordering? Even if it were, could the question, 'Why should society choose state *a* if it has to choose between states *a* and *c*?' be validly answered by: 'Because it would choose *b* in the pair (*a, b*) and *c* in the pair (*b, c*)'? Remember that only one choice situation can hold, and that these choices are not those of a single individual (this would be A. Bergson's (1966) position, which is on opposition to SCT's position). Acyclicity among longer chains would be still more difficult to make intuitive sense of.

Transversal Conditions

Consider a function $y = f(x)$: $X \rightarrow Y$ with $x \in X$ and $y \in Y$ (for any variables x and y). We consider *conditions* on this function, but we distinguish several kinds of conditions.

A *point condition* is expressed with reference to only *one* x and *one* y = f(x), although its domain of validity can be any subset of X (including X itself).

A *transversal condition* is expressed with reference to several *different* x's, Say $x_1, \ldots x_m$ with $m > 1$, and the corresponding $y_i = f(x_i)$. It is indeed a condition on the function $f^{(m)}: X^m \rightarrow Y^m$. Its domain of validity is a certain set $Z \subseteq X^m$. The integer *m* is called its *order* (a transversal condition of order 1 is a point condition). A transversal condition is said to be of *full order* if its definition requires all the x's of X (examples are conditions of the form 'there exists an x—or m x's—such that . . .').

Point conditions imply certain transversal conditions, but not all transversal conditions can be deduced from point conditions. Transversal conditions that cannot be deduced from point conditions are said to be *irreducibly transversal conditions* (implying that their order is larger than one).

Assume that the x's represent mutually exclusive existants, i.e., if one were actually the case, no other one could be. This can result from their mere definition. Then, point conditions can make sense, but irreducibly transversal conditions cannot, since several different x's cannot exist together.

If the mutual exclusion results from the *definition* of the variable x, the consideration of several different x's within the same condition goes against this very definition: it is thus a *counterconceptual*. If the x's considered in an irreducibly transversal condition, or at least one of them, is a counterfactual, the condition is in a sense counterfactual in the second degree.

For example, the structure of a causal law is a point condition from the definition. That is, if $x \in X$ denotes the cause (where the x's are defined as mutually exclusive), $y \in Y$ denotes the effect, and the function f represents the causal law, with y = f(x), a causal structure can be of the form $f(x) \in Y(x) \subset Y$, but not irreducibly of the form $\{f(x_i)\} \in \bar{Y}(\{x_i\}) \subset Y^m$ where $m > 1$ is the number of i's. Irreducibility means that $\bar{Y}(\{x_i\})$ is *not* a Cartesian product of the form $\bar{Y}(\{x_i\}) = \Pi\, Y(x_i)$ with $Y(x_i) \subset Y$ for each i. Hence, a hypothesis on this causal law can be a point condition but not a transversal condition.

All of SCT's conditions, except one, are of the irreducibly transversal type. We have already noted that SCT's use of certain counterfactuals is justifiable (at least one is necessary for rationality). This, however, provides no justification for requiring counterconceptuals.

The only point condition used by SCT is the Pareto or unanimity condition on the social ordering function (if we do not count as a 'condition' the fact that the social choice should be possible). All the other conditions are irreducibly transversal. For the social ordering function, the x's are preference profiles, y is the social ordering and f is the social ordering function. For the existence of a social ordering, the x's are possibility sets, y is the chosen alternative and the function f is the 'choice function'. Let s_j denote alternative social states.

The *unanimity-Pareto* condition states that if s_1 is not ranked below s_2 in all individual orderings and is ranked above s_2 in at least one, the social ordering ranks it above s_2 (with possible variants). Only one preference profile is considered: this is a point condition.

The *monotonicity* condition states that if the ranking of an alternative worsens in no individual ordering and improves in at least one, and nothing else changes, the ranking of this alternative does not worsen in the social ordering. This is an irreducibly transversal condition of the second order.

The *sensitivity* condition ('non-imposition') states that for each pair of alternatives, there exist two preference profiles that yield social orderings that order them differently. This is an irreducibly transversal condition of the full order.

The *non-dictatorship* condition states that there does not exist an individual *i* such that the social ordering coincides with this individual's ordering for all preference profiles. This is an irreducibly transversal condition of the full order.

The *independence of irrelevant alternatives* condition states that the social choice in a possibility set depends only on individuals' orderings of these possible alternatives. Along with other SCT assumptions, it also means that the social ordering of any subset of the alternatives depends only on the individuals' orderings of *these* alternatives. This also occurs if, and only *if* it holds for every pair of alternatives. In all these formulations, this condition implies the absence of a change when a certain other change occurs: the social choice among certain alternatives, or their social ordering, does not change when the individual orderings of other alternatives, or comparisons of other alternatives with the considered ones, change. Hence this condition is irreducibly transversal of the second order.[9]

The conditions for the *existence of an ordering* are also transversal. The conditions described in the section entitled 'Requirements of Social Choice Theory' demand at least two different possibility sets for their definitions. A condition of transitivity requires three different possibility sets of two elements each for its definition. The full conditions of transitivity that define an ordering require the consideration of all the two-elements possibility sets.

Our analysis sheds light on the debate that opposed Arrow on the one hand and Little (1951) followed by Bergson (1966) on the other. The latter's basic objection to SCT is that transversal conditions on the social ordering function do not make sense. Arrow's answer (SCIV, p. 105) states the issue very clearly: SCT's 'conditions are of two kinds: those that relate to the social choice function produced by any given set of individual orderings, and those that relate to the way social choices vary with respect to changes in the individual orderings. It is the latter that are rejected by Bergson and Little. According to Little

> If tastes change, we may expect a new ordering of all conceivable states; but we do not require that the difference between the new and the old ordering should bear any particular relation to the changes in tastes that have occurred. We have, so to speak, a new world and a new order; and we do not demand correspondence between the change in the world and the change in the order.

SCIV states that the independence of irrelevant alternatives condition is the only transversal one while, as we have seen, the unanimity-Pareto condition is the only condition that is not irreducibly transversal. The independence of irrelevant alternatives is the only condition about which SCIV expresses some doubt as a general and necessary rule—a very sound evaluation of this condition and one shared by all philosophers of social choice, although SCIV sees this condition to be somewhat more general than others do. Bergson notes the transversality of both this condition and monotonicity, while Little had mentioned monotonicity. (Little is confused about the independence of irrelevant alternatives for the reason noted in footnote 9).

Nevertheless, Little and Bergson's attack is globally inconsistent. The conditions on the social ordering function, once this device is accepted, do have the intuitive appeal SCIV says they have. Little and Bergson argued against the meaningfulness of transversal conditions on the ground that the implied comparison is counterfactual. It is indeed counterfactual, but the very plurality of preference profiles that this comparison requires is itself counterfactual. Little and Bergson should therefore have rejected this plurality. But then, they should have rejected *justifying* the social ordering by the preference profile, since this implies at least one counterfactual. Consequently, they should have rejected all conditions referring to this relation from the profile(s) to the social ordering(s) (including unanimity-Pareto), and they should indeed have rejected a classical social welfare function with this meaning.

Little and Bergson also accepted the existence of a social ordering, and hence its meaningfulness, although this meaning requires counterfactual possibility sets and transversal conditions on the social choice function. Yet, they had a good reason for this: they see the social ordering as representing an individual's ethical opinion rather than as a condition on social choices, as SCT does. However, if this individual is rational in the normal sense, he also needs at least one counterfactual preference profile in order to justify his Bergson social welfare function. What Little and Bergson failed to do is to distinguish among the various levels and types of counterfactuality; to point out the epistemic necessity of counterfactuality and of conditions based on it; and to derive from the former consideration what counterfactuality is required and hence what type of conditions is legitimate.

Notes

1. This essay is a follow-up to my three previous pieces on the meaning of this theory of social choice: 'Choix collectif et optimum social' (*Revue d'Economie Politique,* 1980), *Le sens du choix social* (chapter 16 of my book *Philosophie de l'Economie*, 1986), and in my book *L'Homme Pluridimensionnel* (chapter 6, 1987). It is an excerpt of 3 more complete analysis of this question, 'What Sense Social Choice?'

2. With the 'transitivity' condition if a is chosen in the pair (a,b) and b is chosen in the pair (b,c), then a is chosen in the pair (a,c) ; plus the classical consideration of possible ties.
3. With m alternatives and n individuals there are $(m!)^n$ profiles of only strict preferences.
4. See 'What Sense Social Choice?', Section III, p. 16.
5. This is explained in 'What Sense Social Choice?' Section III, p. 17.
6. Let X denote the set of alternative 'social states', $S \subseteq X$ a possibility set, $C(S) \subseteq S$ the set of possibly chosen social states if the possibility set is S, or 'choice set' (SCIV), S_i several particular S's indexed by the integer i. Then, C(S) can be the maximal elements on S of an ordering of X (independent of S) only if, for all collection (S_i) with several i's, the following two conditions are satisfied:

$$C(\cup S_i) \supseteq \cap C(S_i) \supseteq [\{x_i 38 : \{x_i\} \subseteq \cap S_i \text{ and } x_i \in C(Sd2i) \ \forall_i].$$

   The first condition is 'the choice set of the union contains the intersection of the choice sets'. The second one states that choices from different possibilities sets that belong to all these possibility sets are choices for all these possibility sets. These conditions for all (S_i) practically imply an ordering on X.
7. S.-Ch. Kolm, 'What Sense Social Choice', Section II.
8. Hence the expression *rational behaviour a priori* is nonsense because behaviour does not necessarily result from a choice, since it can be instinctive, impulsive, compulsive, reflex, akrasic, etc. By contrast, *rational conduct* makes sense, and *rational action* is acceptable.
9. In SCIV, the *Literary presentation* and the *examples* given of this condition (rather than its definition itself) refer to the *disappearance of certain alternatives*, hence the changes in the possibility set, rather than to individual preferences concerning impossible alternatives (p. 26). This has misled, e.g. Little (*op. cit.*). On this interpretation, this condition would only be that the social ordering does not depend on the possibility set. It then becomes a condition on the meaningfulness of the considered possibility-independent social ordering rather than on the social ordering *function* of the preference profiles *per se*.

References

ARROW, K. 1963. *Social Choice and Individual Values*. 2nd edition, New York: John Wiley.

BERGSON, A. 1938. A reformulation of certain aspects of welfare economics. *Quarterly Journal of Economics*, 52: 310–34.

———. 1966. *Essays in Normative Economics*. Cambridge: Harvard University Press.

BUCHANAN, J. 1954. Individual Choice in Voting and the Market. *Journal of Political Economy*, 62: August: 334–43.

CHAPMAN, B. 1980. *Individual Rights and Collective Rationality*, Mimeo. London: Westminster Institute for Ethics and for Human Values.

GIBBARD, A. 1973. Manipulation of Voting Schemes: A General Result. *Econometrica*, 41: 587–601.

HOOK, S. (ed.) 1967. *Human Values and Economic Policy,* New York: New York University Press.

KELLY, J. 1978. *Arrow Impossibility Theorems,* New York: Academic Press.

KEMP, M.C. and A. ASIMAKOPULOS. 1952. A Note on Social Welfare Functions and Cardinal Utility. *Canadian Journal of Economics and Political Science*, 18: 195–200.

KOLM, S.-CH. 1980. Choix Collectif et Optimum Social. *Revue d'Economie Politique.*

———. 1985. *Le Contrat Social Liberal.* Paris: Presses Universitaires de France.

———. 1986, *Philosophie de l'Economie*. Paris: ed. du Seuil.

———. 1987. *L'Homme Pluridimensionel,* Paris: ed. Albin Michel.

———. 1994, 'What Sense Social Choice?' in *The Social Optimum* , Cambridge, Mass: MIT Press.

LITTLE, I.M.D. 1951. Social Choice and Individual Values. *Journal of Political Economy,* 60.

PARETO, V. 1913. Il massimo di utilitá per una collettivita. *Giornale delle economisti.* See also *Mind and Society*.

PLATO. *The Republic.* Any edition.

RAWLS, J. 1971. *A Theory of Justice*. Oxford: Oxford University Press.

SEN, A. 1986. Social Choice Theory. In K. Arrow and M. Intriligator (eds.), *Handbook of Mathematical Economics*. Amsterdam: North-Holland.

The Coherence of Rights

SATISH K. JAIN

Introduction

My main concern here is with the idea of coherence of rights-assignments and its formalisations. A rights-assignment is called coherent if *and only if* the exercise of rights by itself never leads to an empty choice set, no matter which profile of individual orderings and which non-empty finite subset of alternatives are considered. In other words, a rights-assignment is coherent if and only if the Gibbard paradox (Gibbard 1974: 388–92) is not possible for any profile of individual orderings and for any non-empty finite subset of alternatives.

I shall discuss two of the formalisations in the social choice literature relating to the coherence of rights-assignments and show that neither of the two formalisations is equivalent to coherence of rights-assignments, if by coherence of rights-assignments we mean the absence of the Gibbard paradox for all profiles of individual orderings and for all non-empty finite subsets of alternatives. Let $D = (D_1, \ldots, D_n)$ be a rights-assignment, where D_i is the set of ordered pairs assigned to individual *i* (individual *i*'s protected sphere), and $(R_1, \ldots, R_n)$ a profile of individual orderings. One formalisation states that D is coherent if and only if there exists an ordering-extension of each and every $D_i \cap R_i$.[1] I shall show that while the existence of an ordering-extension of each and every $D_1 \cap R_i$ is sufficient to ensure the coherence of D, it is not necessary (Theorem 1).

The second formalisation states that D is coherent if *and only if* there exists no critical loop in D. A critical loop in D is a sequence of at least two ordered pairs (x_μ, y_μ), where μ ranges from 1 to t, such that (*a*) each (x_μ, y_μ) belongs to some $D_{i\,\mu}$;

(*b*) the second element of each ordered pair is identical to the first element of the succeeding ordered pair, if a succeeding ordered pair exists; (*c*) the second element of the last ordered pair is identical to the first element of the first ordered pair; and (*d*) the set $\{i_1, \ldots, i_t\}$ is not a singleton.[2] I shall show that, like the first formalisation, the absence of a critical loop in D is sufficient to ensure the coherence of D but is not necessary (Theorem 2). It is generally believed that these two formalisations are equivalent.[3] I shall show that this belief is not correct. Indeed, the two are logically independent (Theorem 3).

Fortunately, slight modifications in the two formalisations make them equivalent to coherence, and consequently to each other. We show in Theorems 4 and 5 that (*a*) a rights-assignment is coherent if *and only if* there exists an ordering-extension of each and every $D_i \cap P(R_i)$; (*b*) a rights-assignment D is coherent if *and only if* there is no modified critical loop in D. A modified critical loop is a sequence of ordered pairs (x_μ, y_μ) which, in addition to satisfying the conditions mentioned in the definition of a critical loop, satisfies the restriction that $x_1, \ldots, x_t$ involved in the loop are all distinct.

I shall also discuss the important special case of symmetric rights-assignments. For symmetric assignments it is true that D is coherent if and only if there exists an ordering-extension of each and every $D_i \cap R_i$ (Theorem 6).

The analysis of coherence of rights-assignments has some important implications for the way the idea of a liberal individual is formalised. Consider the following definition of a liberal individual: Let Y be the set of all ordering-extensions of each and every $D_i \cap R_i$. An individual *j* is defined to be liberal if *and only if* $R^*_j = R_j \cap R^j$ for some $R^j \in Y$, where R^*_j is the relation that individual *j* wishes to be counted in social choice.[4]

There are difficulties associated with this formulation. It does not ensure that R^*_j is a sub-relation of R_j. For instance, if individual *j* is assigned the ordered pair (x, y) but not (y, x), then if individual *j* happens to be indifferent between x and y, every ordering-extension belonging to Y would have x preferred to y and consequently we would have x preferred to y in terms of R^*_j as well. Thus, R^*_j may not even faithfully reflect the preferences of individual *j* over his own protected sphere.[5]

However, the most important difficulty with this formulation arises because the set of ordering-extensions Y may be empty even though the rights-assignment is coherent, as has been shown in Theorem 1. Consequently, it may be impossible for anyone to be liberal in the sense of this formulation, notwithstanding some individuals' desire to respect other individuals' rights.

Using the result of Theorem 4, I will provide a formalisation of the idea of a liberal individual which is free from the difficulties just cited. This formulation, in addition, will have the property that, whenever the profile of individual orderings is such that the weak Pareto-criterion and the coherent libertarian condition do not conflict, only ordering-extensions which preserve individuals' unanimous strict preferences are used. Consequently, it can be shown that if there exists at least one liberal individual in the society, then there exists a collective choice rule with unrestricted domain, satisfying the coherent libertarian requirement, conditional weak Pareto-criterion and satisfying the property that, whenever the profile of individual orderings is such that the weak Pareto-criterion and the coherent libertarian requirement do not conflict, the chosen elements are Pareto-optimal in the sense of the weak Pareto- criterion (Theorem 7).

Definitions and Assumptions

The set of social alternatives are denoted by X. It is assumed that X is finite and $\# X = m \geqslant 3$. The finite set of individuals constituting the society is denoted by N and it is assumed that $\# N = n \geqslant 2$.

Let R be a binary relation on a set X. The asymmetric and symmetric parts of R, to be denoted by P(R) and I(R) respectively, are defined as follows: $\forall\ x, y \in X : [\{(x, y) \in P(R) \text{ if } (x, y) \in R \ \&\ (y, x) \notin R\} \ \&\ \{(x, y) \in I(R) \text{ if } (x, y) \in R \ \&\ (y, x) \in R\}]$. A binary relation R on a set X is (*a*) reflexive if $\forall\ x \in X : (x, x) \in R$; (*b*) connected if $\forall$ distinct $x, y \in X : [(x, y) \in R \vee (y, x) \in R]$; (*c*) transitive if $\forall\ x, y, z \in X : [(x, y) \in R \ \&\ (y, z) \in R \rightarrow (x, z) \in R]$; (*d*) an ordering if it is reflexive, connected and transitive; (*e*) consistent if $\forall\ x_1, \ldots, x_t \in X, (t \geqslant 2) : [(x_1, x_2) \in P(R) \ \&\ (x_k, x_{k+1}) \in R, k = 2, \ldots, t-1, \rightarrow (x_t, x_1) \notin R]$; (*f*) acyclic if $\forall\ x_1, \ldots, x_t \in X, (t \geqslant 2) : [(x_k, x_{k+1}) \in P(R), k = 1, \ldots, t-1, \rightarrow (x_1, x_t) \in R]$.

$(x, y) \in R$ will at times be written as xRy. $P(R)$ and $I(R)$ might sometimes be written simply as P and I respectively.

It will be assumed that each individual $i \in N$ has an ordering R_i over X. The set of all orderings of set X will be denoted by T. The n-fold Cartesian product of T will be written as T^n. Profiles of individual orderings will be written on the pattern $a = (R^a{}_1, \ldots, R^a{}_n) \in T^n$, or simply as $a = (R_1, \ldots, R_n) \in T^n$.

Let R_1 and R_2 be two binary relations on a set X. R_2 is said to be an extension of R_1 (or equivalently R_1 is said to be a sub-relation of R_2) if (*a*) $R_1 \subset R_2$ and (*b*) $P(R_1) \subset P(R_2)$. If R_2 is an extension of R_1 and R_2 is an ordering, we say that R_2 is an ordering-extension of R_1.

Remark 1: In an important contribution Suzumura (1983) has proved the following theorem: A binary relation R has an ordering-extension if R is consistent.[6]

We denote by K the set of all non-empty subsets of X; $K = 2^X - \{\emptyset\}$. A choice function C defined over K is a function which, for every $S \in K$, assigns a unique non-empty subset $C(S)$ of S. A collective choice rule (CCR)f defined over $W \subset T^n$ is a function which, for every profile of individual orderings $a = (R^a{}_1, \ldots, R^a{}_n) \in W$, determines a unique choice function C^a over K; $C^a = f(a)$.

Unrestricted Domain (U): A CCR is said to have unrestricted domain if domain $W = T^n$.

For every $a \in T^n$ and every $S \in K$, we define:

$$C^a{}_p(S) = [x \in S \mid \sim \{ \exists y \in S: (y, x) \in \bigcap_{i=1}^{n} P(R^a{}_i)\}];$$

$$C^a{}_{\bar{p}}(S) = [x \in S \mid \sim \{ \exists y \in S: (y, x) \in P(\bigcap_{i=1}^{n} R^a{}_i)\}]$$

Weak Pareto-criterion (P): A CCR with unrestricted domain satisfies the weak Pareto-criterion if $\forall a \in T^n$ and $\forall S \in K : C^a(S) \subset C^a{}_p(S)$.

Strict Pareto-criterion ($\bar{P}$): A CCR with unrestricted domain satisfies the strict Pareto-criterion if $\forall a \in T^n$ and $\forall S \in K : C^a(S) \subset C^a{}_{\bar{p}}(S)$.

The twofold Cartesian product X^2 is the set of all ordered pairs of X. H denotes the set of all subsets of X^2; $H = 2^{X^2}$.

Let $D_i \in H$ $(i \in N)$ denote the set of all ordered pairs assigned to individual *i*, i.e. the protected sphere of individual *i*. $D = (D_1, \ldots, D_n) \in H^n$ is called the rights-assignment for the society.

For any set X, we define $\Delta_x = [(x, x) | x \in X]$.

Coherence of Rights-Assignments

Let $D = (D_1, \ldots, D_n)$ be a rights-assignment for the society. For all $a \in T^n$ and for all $S \in K$ we define:

$C^a_D(S) = [x \in S | \sim \{\exists y \in S : \exists i \in N : (y, x) \in D_i \cap P(R^a_i)\}]$

A rights-assignment $D = (D_1, \ldots, D_n)$ is called coherent if $\forall\, a \in T^n$ and $\forall\, S \in K : C^a_D(S) \neq \emptyset$. In other words, a rights-assignment D is defined to be coherent if the exercise of rights by itself never results in an empty choice set no matter which non-empty subset of alternatives and which profile of individual orderings are considered.

In the social choice literature, the idea of coherence of rights-assignments has been formalised in two different ways. In what follows, it will be shown that neither of the two formalisations is equivalent to the coherence of rights-assignments as defined above.

E-coherence: $D = (D_1, \ldots, D_n)$ is defined to be E- coherent if for every $(R_1, \ldots, R_n) \in T^n$ there exists an ordering-extension R of each and every $D_i \cap R_i$, $i = 1, \ldots, n$.

L-coherence: Let $D = (D_1, \ldots, D_n)$ be a rights-assignment. A critical loop in D is a sequence of ordered pairs $< (x_\mu, y_\mu) >^t_{\mu=1}$ ($t \geq 2$) such that

(*i*) $(x_\mu, y_\mu) \in \bigcup_{i=1}^{n} D_i$, $\forall\, \mu \in (1, 2, \ldots, t)$

(*ii*) there exists no $i^* \in N$ such that $(x_\mu, y_\mu) \in D_{i^*}$, $\forall\, \mu \in (1, 2, \ldots, t)$

(*iii*) $x_1 = y_t$ and $x_\mu = y_{\mu-1}$, $\forall\, \mu \in (2, \ldots, t)$.

D is defined to be L-coherent if there exists no critical loop in D.

Theorem 1:

(*a*) Coherence of rights-assignment $D = (D_1, \ldots, D_n)$ does not imply E-coherence of D.

(*b*) E-coherence of D implies coherence of D.

Proof:

(*a*): Proof consists of an example.

Let $X = \{x, y, z\}$, $N = (1, 2)$, $D = (D_1, D_2)$

$D_1 = X^2 - [(y, x)]$
$D_2 = \Delta x = [(x, x), (y, y), (z, z)]$.

The Coherence of D is obvious.

Consider the following profile of individual orderings:
(*i*) $x\ I_1\ y\ I_1\ z$
(*ii*) $x\ P_2\ y\ P_2\ z$
$D_1 \cap R_1 = D_1$
$D_2 \cap R_2 = \Delta_x$
As D_1 is not consistent [we have $(x,y) \in P(D_1)$, $(y, z) \in D_1$, $(z, x) \in D_1$], there does not exist any R which is an ordering-extension of each and every $D_i \cap R_i$.

(*b*): Suppose D is incoherent. Then there exists a profile a = $(R^a_1, \ldots, R^a_n) \in T^n$ and an $S \in K$ such that $C^a_D(S)$ is empty. Choose S such that S is the smallest set or one of the smallest sets belonging to K for which $C^a_D(.)$ is empty. Let $S = \{x_1, \ldots, x_t\}$, where $x_1, \ldots, x_t$ are all distinct. From the definition of $C^a_D(S)$, it follows that $t \geqslant 2$.

As S is one of the smallest sets belonging to K for which $C^a_D(.)$ is empty, it follows that there exists a one-to-one correspondence θ: $S \rightarrow S$ such that $\forall\ x_k \in S : [\ \exists\ i_k \in N : \{\theta\ (x_k), x_k\} \in D_{i^k} \cap P(R^a_{i^k})]$
Define:
$y_1 = x_1$
$y_{k+1} = \theta\ (y_k),\ k = 1, \ldots, t-1.$
We conclude:
(*i*) $y_k \neq y_{k+1}$, $k = 1, \ldots, t-1$, and $y_t \neq \theta\ (y_t)$, as each $P\ (R^a_i)$ is asymmetric;
(*ii*) $y_1, \ldots, y_t$ are all distinct and $\theta\ (y_t) = y_1$, as S is one of the smallest sets belonging to K for which $C^a_D(.)$ is empty.
Thus we have:
$\exists\ j_1, \ldots, j_t \in N$:
$(y_1, y_t) \in D_{j^t} \cap P(R^a_{j^t})$
$(y_t, y_{t-1}) \in D_{j^t-1} \cap P(R_{j^{t-1}}^a)$
:
$(y_2, y_1) \in D_{j1} \cap P(R^a_{j1})$.

Suppose there exists an ordering R which is an extension of each and every $D_i \cap R^a_i$, $i = 1, \ldots, n$.
For any $i \in N$ and any $x, y \in X$ we have:

$(x, y) \in D_i \cap P(R^a_i) \rightarrow (x, y) \in D_i \cap R^a_i$
$(x, y) \in D_i \cap P(R^a_i) \rightarrow (x, y) \in P(R^a_i)$
$\rightarrow (y, x) \notin R^a_i$
$\rightarrow (y, x) \notin D_i \cap R^a_i$

Therefore:
$(x, y) \in D_i \cap P(R^a_i) \rightarrow (x, y) \in P(D_i \cap R^a_i)$
$\rightarrow (x, y) \in P(R)$

Consequently we must have:
$(y_1, y_t) \in P(R), (y_t, y_{t-1}) \in P(R) \dots\dots\dots\dots (y_2, y_1) \in P(R)$.
However, since this violates transitivity or asymmetry of P(R), there cannot possibly exist an ordering-extension of each and every $D_i \cap R^a_i$, i = 1, . . ., n.

This establishes the theorem.

Theorem 2:
(*a*) Coherence of rights-assignment $D = (D_1, \dots, D_n)$ does not imply L-coherence of D.
(*b*) L-coherence of D implies coherence of D.

Proof:
(*a*): Proof consists of an example.
Let X = (x, y, z), N = (1, 2), $D = (D_1, D_2)$
$D_1 = \Delta_x \cup [(x, y), (y, x)]$
$D_2 = \Delta_x \cup [(y, z), (z, y)]$
The coherence of D is obvious.

Consider the following sequence of ordered pairs $< (x_\mu, y_\mu) >^4_{\mu=1}$:
(*i*) $(x_1, y_1) = (x, y)$
(*ii*) $(x_2, y_2) = (y, z)$
(*iii*) $(x_3, y_3) = (z, y)$
(*iv*) $(x_4, y_4) = (y, x)$
We have:
(*i*) $(x, y), (y, z), (z, y), (y, x) \in D_1 \cup D_2$
(*ii*) there exists no $i^* \in N$ such that $(x, y), (y, z), (y, x) \in D_{i^*}$
(*iii*) $x_1 = y_4$, and $x_\mu = y_{\mu-1}$, for all $\mu \in \{2, 3, 4\}$.
Thus $(x, y) \in D_1$, $(y, z) \in D_2$, $(z, y) \in D_2$, $(y, x) \in D_1$ is a critical loop in D. This implies that D fails to satisfy L-coherence.

(*b*): Suppose D is not coherent. Then there exist $a = (R^a_1, \dots, R^a_n) \in T^n$ and $S \in K$ such tht $C^a_D(S)$ is empty. Choose S such that it is one of the smallest sets belonging to K for which $C^{a}_{D}(.)$ is empty. Let $S = \{x_1, \ldots, x_t\}$, where $x_1, \ldots, x_t$ are all distinct. As in the proof of Theorem 1(b), we can conclude that:

$\exists j_1, \ldots, j_t \in N$:

$(y_1, y_t) \in D_{j^t} \cap P(R^{a}_{j^t})$

$(y_t, y_{t-1}) \in D_{j^{t-1}} \cap P(R^{a}_{j^{t-1}})$

:

$(y_2, y_1) \in D_{j^1} \cap P(R^{a}_{j^1})$

where (*i*) $y_1, \ldots, y_t$ are all distinct; (*ii*) $\{y_1, \ldots, y_t\} = (x_1, \ldots, x_t)$; (*iii*) $t \geqslant 2$.

This implies:

(*i*) $(y_1, y_t), (y_t, y_{t-1}), \ldots, (y_2, y_1) \in \bigcup_{i=1}^{n} D_i$

(*ii*) $\{j_1, \ldots, j_t\}$ is not a singleton, as each R^{a}_{t} is an ordering and each $P(R^{a}_{i})$ is asymmetric.

Thus $(y_1, y_t) \in D_{j^t}, (y_t, y_{t-1}) \in D_{jt-1}, \ldots, (y_2, y_1) \in D_{j^1}$ is a critical loop in D and consequently D is not L-coherent.

Theorems 1 and 2 establish that both formalisations of the idea of coherence are overly strong. While both formalisations rule out incoherent rights-assignments, they also rule out some coherent rights-assignments. It is generally believed that E-coherence and L-coherence are equivalent. The following theorem shows that this belief is incorrect.

Theorem 3:

L-coherence and E-coherence are logically independent of each other.

Proof:

Proof consists of the following four examples:

(*a*) Let $X = \{x, y, z\}$, $N = \{1, 2\}$, $D = (D_1, D_2)$

$D_1 = [(x, y), (y, z), (z, y), (x, z), (z, x)]$

$D_2 = \varnothing$

No critical loop can exist in D, therefore D is L-coherent.

Consider the following profile (R_1, R_2):

(*i*) xI_1yI_1z

(*ii*) xP_2yP_2z

$D_1 \cap R_1 = D_1$

Since D_1 is not consistent, there does not exist any R which is an ordering-extension of each and every $D_i \cap R_i$, $i = 1, 2$. Thus D is not E-Coherent.

(*b*): Let $X = \{x, y, z\}$, $N = \{1, 2\}$, $D = (D_1, D_2)$
$D_1 = [(x, y), (y, x)]$
$D_2 = [(y, z), (z, y)]$
$(x, y) \in D_1$, $(y, z) \in D_2$, $(z, y) \in D_2$, $(y, x) \in D_1$ is a critical loop in D. Therefore, D is not L-coherent.
It is clear that for every $(R_1, R_2) \in T^2$, there exists an R which is an ordering-extension of each and every $D_i \cap R_i$, $i = 1, 2$. Thus D is E-coherent.

(*c*): Let $X = \{x, y, z\}$, $N = (1, 2)$, $D = (D_1, D_2)$
$D_1 = [(x, y)]$
$D_2 = [(y, x)]$
As $(x, y) \in D_1$, $(y, x) \in D_2$ is a critical loop, it follows that D is not L-coherent. Let (R_1, R_2) be such that xP_1y and yP_2x. For such an (R_1, R_2), no R exists which is an ordering-extension of $D_1 \cap R_1$ and $D_2 \cap R_2$. So D is not E-coherent.

(*d*): Let $X = \{x, y, z\}$, $N = (1, 2)$, $D = (D_1, D_2)$
$D_1 = [(x, y)]$
$D_2 = [(y, z)]$
It is clear that D is both L-coherent and E-coherent.

Now, we shall introduce a modified version of E-coherence and show that the modified version is logically equivalent to coherence.
Modified E-coherence (ME-Coherence):
A rights-assignment $D = (D_1, \ldots, D_n)$ is defined to be ME-coherent if for every $(R_1, \ldots, R_n) \in T^n$ there exists an R such that it is an ordering-extension of each and every $D_i \cap P(R_i)$, $i = 1, \ldots, n$.

Theorem 4:
A rights-assignment $D = (D_1, \ldots, D_n)$ is coherent if it is ME-coherent.

Proof:
First, we establish that ME-coherence implies coherence. Suppose coherence is violated. Then there exist profile $a = (R^a_1, \ldots, R^a_n) \in T^n$ and $S \in K$ such that $C^a_D(S)$ is empty. Choose S such that it is one of the smallest sets belonging to K for which $C^a_D(.)$ is empty. Let $S = \{x_1, \ldots, x_t\}$, where $x_1, \ldots, x_t$ are all distinct. As in the proof of Theorem 1(*b*), we can conclude that:

$\exists\, j_1, \ldots, j_t \in N$:
$(y_1, y_t) \in D_{j_t} \cap P(R^a_{j_t})$
$(y_t, y_{t-1}) \in D_{j_{t-1}} \cap P(R^a_{j_{t-1}})$
$\vdots$
$(y_2, y_2) \in D_{j_1} \cap P(R^a_{j_1})$,
where (*i*) $y_1, \ldots, y_t$ are all distinct; (*ii*) $\{y_1, \ldots, y_t\} = \{x_1, \ldots, x_t\}$; (*iii*) $t \geqslant 2$.
As for each $i \in N$, $P(R^a_i)$ is asymmetric, it follows that $\forall\, i \in N$:
$P[D_i \cap P(R^a_i)] = D_i \cap P(R^a_i)$.
Therefore, if there is an R which is an ordering-extension of each and every $D_i \cap P(R^a_i)$, $i \in N$, then we must have (y_1, y_t), $(y_t, y_{t-1}), \ldots, (y_2, y_1) \in P(R)$, which is impossible. Thus D violates ME-coherence.

Next, we show that coherence implies ME-coherence.
For each $i \in N$, define:
$$Q^a_i = D_i \cap P(R^a_i) \qquad (i)$$
$$\text{Let } Q^a = \bigcup_{i=1}^{n} Q^a_i \qquad (ii)$$
Take any $i \in N$ and any $x, y \in X$:
$(x, y) \in Q^a_i \rightarrow (x, y) \in P(R^a_i)$
$\rightarrow (y, x) \notin P(R^a_i)$
$\rightarrow (y, x) \notin Q^a_i$
Therefore
$$P(Q^a_i) = Q^a_i \qquad (iii)$$
Consider any $x, y \in X$:
Suppose $(x, y) \in Q^a$ and $(y, x) \in Q^a$
$(x, y) \in Q^a \rightarrow \exists\, i \in N : (x, y) \in Q^a_i$
$(y, x) \in Q^a \rightarrow \exists\, j \in N : (y, x) \in Q^a_j$
As Q^a_i and Q^a_j are asymmetric, it follows that $i \neq j$. Consider $C^a_D(\{x, y\})$. In view of $(x, y) \in Q^a_i$ and $(y, x) \in Q^a_j$, we conclude that it is empty. This, however, contradicts the coherence of D. So we conclude that
$(x, y) \in Q^a \rightarrow (y, x) \notin Q^a$
This establishes
$$Q^a = P[Q^a]. \qquad (iv)$$

Next, we show that Q^a is consistent.
Suppose Q^a is not consistent. Then there exist $x_1, \ldots, x_t \in X$, $t \geqslant 2$, such that:
$(x_1, x_2) \in P(Q^a)$, $(x_k, x_{k+1}) \in Q^a$, $k = 2, \ldots, t-1$, $(x_t, x_1) \in Q^a$
$\rightarrow \exists\, i_1, \ldots, i_t \in N$: $(x_1, x_2) \in Q^a_{i_1}$, $(x_k, x_{k+1}) \in Q^a_{i_k}$, $k = 2, \ldots, t-1$, $(x_t, x_1) \in Q^a_{i_t}$

$\rightarrow \exists\, i_1, \ldots, i_t \in N$: $(x_1, x_2) \in D_{i1} \cap P(R^a_{i1})$, $(x_k, x_{k+1}) \in D_{ik}$ $P(R^a_{ik})$, $k = 2, \ldots, t-1$, $(x_t, x_1) \in D_{it} \cap P(R^a_{it})$
$\rightarrow C^a_D(\{x_1, \ldots, x_t\}) = \emptyset$
which contradicts the coherence of D. This establishes that Q^a is consistent.
Therefore, by Suzumura Extension Theorem, there exists an R such that
$Q^a \subset R$ (*v*)
and $P(Q^a) \subset P(R)$ (*vi*)
From (*i*)–(*vi*), we conclude that for each $i \in N$:
$Q^a_i \subset Q^a \subset R$
and $P(Q^a_i) \subset P(Q^a) \subset P(R)$.
Thus, there exists an ordering-extension R of each and every $D_i \cap P(R^a_i)$, $i \in N$. This establishes that D is ME-coherent.

Next, we introduce a modified version of L-coherence and show that this modified version is equivalent to coherence.
Modified L-coherence (ML-coherence): Let $D = (D_1, \ldots, D_n)$ be a rights-assignment. The sequence $< (x_\mu, y_\mu) \in D_{i\mu} >^t_{\mu=1}$, $i_\mu \in N$, $(t \geqslant 2)$ is said to be a modified critical loop in D if (*i*) $x_\mu = y_{\mu-1}$, $\mu = 2, \ldots, t$ and $x_1 = y_t$; (*ii*) $\{i, \ldots, i_t\}$ contains at least two distinct individuals; (*iii*) $\forall\, j, k \in \{1, \ldots, t\}$: $x_j = x_k \rightarrow j = k$. D is defined to be ML-coherent if there exists no modified critical loop in D.

Theorem 5:
A rights assignment $D = (D_1, \ldots, D_n)$ is coherent if it is ML-coherent.

Proof:
Suppose coherence is violated. Then there exists profile $a = (R^a_1, \ldots, R^a_n) \in T^n$ and $S \in K$ such that $C^a_D(S)$ is empty. Choose S such that it is one of the smallest sets belonging to K for which $C^a_D(.)$ is empty. As in the proof of Theorem 1(*b*), we can conclude that:
$\exists\, j_1, \ldots, j_t \in N$:
$(y_1, y_t) \in D_{jt} \cap P(R^a_{jt})$
$(y_t, y_{t-1}) \in D_{jt-1} \cap P(R^a_{jt-1})$
:
$(y_2, y_1) \in D_{j1} \cap P(R^a_{j1})$
where (*i*) $y_1, \ldots, y_t$ are all distinct; (*ii*) $(y_1, \ldots, y_t) = S$; (*iii*) $t \geqslant 2$. As each $P(R^a_i)$, $i \in N$, is asymmetric and transitive, we conclude that $(j_1, \ldots, j_t)$ contains at least two distinct individuals. As

$y_1, \ldots, y_t$ are all distinct, it follows that $(y_1, y_t) \in D_{j^t}$, $(y_t, y_{t-1}) \in D_{j^t-1}$,...., $(y_2, y_1) \in D_{j1}$ $(t \geq 2)$ is a modified critical loop in D. Therefore D is not ML-coherent. This establishes that ML-coherence implies coherence.

Next we show that coherence implies ML-coherence.
Suppose D does not satisfy ML-coherence. Then there exists a modified critical loop in D, i.e. there exist $x_1, \ldots, x_t \in X$, $i_1, \ldots, i_t \in N$, $t \geq 2$, such that (*i*) $(x_k, x_{k+1}) \in D_{ik}$, $k = 1, \ldots, t-1$ and $(x_t, x_1) \in D_{i^t}$; (*ii*) $(i_1, \ldots, i_t)$ contains at least two distinct individuals; and (*iii*) $x_1, \ldots, x_t$ are all distinct.
Let $a \in T^n$ be such that $(x_k, x_{k+1}) \in P(R^a{}_{ik})$, $k = 1, \ldots, t-1$, $(x_t, x_1) \in P(R_{i^t}{}^a)$. Such an $a \in T^n$ can always be found because (*i*) $(i_1, \ldots, i_t)$ contains at least two distinct individuals and (*ii*) $x_1, \ldots, x_t$ are all distinct. As $(x_k, x_{k+1}) \in D_{i^k} \cap P(R_{i^k}{}^a)$, $k = 1, \ldots, t-1$, and $(x_t, x_1) \in D_{i^t} \cap P(R_{i^t}{}^a)$, it follows that $C^a{}_D [(x_1, \ldots, x_t)]$ is empty, which violates the coherence of D. This establishes the theorem.

Symmetric Rights-assignments

I shall now consider the important special case of symmetric rights-assignments. A rights-assignment $D = (D_1, \ldots, D_n)$ is defined to be symmetric if $\forall\, i \in N : [\forall x, y \in X: \{(x, y) \in D_i$ if $(y, x) \in D_i\}]$. It turns out that a necessary and sufficient condition for a symmetric rights-assignment to be coherent is that it satisfies E-coherence.

Theorem 6:
A symmetric rights-assignment $D = (D_1, \ldots, D_n)$ is coherent if it is E-coherent.

Proof:
That E-coherence implies coherence, irrespective of whether D is symmetric or not, has already been proved in Theorem 1. Suppose D is coherent. Consider any $a = (R^a{}_1, \ldots, R^a{}_n) \in T^n$. Define

$$G^a{}_i = D_i \cap R_i{}^a,\ i \in N \qquad (i)$$

$$G^a = \bigcup_{i=1}^{n} G^a{}_i. \qquad (ii)$$

From (*i*) and (*ii*), we obtain

$$G^a{}_i \subset G^a,\ i \in N \qquad (iii)$$

Take any $i \in N$ and any $x, y \in X$:

$(x, y) \in P(G^a_i) \rightarrow (x, y) \in P(R_i^a)$, as D is symmetric (*iv*)
$\rightarrow x \neq y$ (*v*)
$(x, y) \in P(G^a_i)$ & $(y, x) \in G^a \rightarrow (x, y) \in D_i$ &
$(y, x) \in D_j$ & $i \neq$ (*vi*).
(*v*) and (*vi*) imply that $(x, y) \in D_i$, $(y, x) \in D_j$ is a modified critical loop, which contradicts the coherence of D in view of Theorem 5. Therefore, we conclude:
$(x, y) \in P(G^a_i) \rightarrow (x, y) \in P(G^a)$
Thus for every $i \in N$:
$P(G^a_i) \subset P(G^a)$. (*vii*)

Next, we show that G^a is consistent.
Suppose G^a is not consistent. Then there exist
$y_1, \ldots y_v \in X$, $v \geqslant 2$, such that $(y_1, y_2) \in P(G^a)$, $(y_k, y_{k+1}) \in G^a$, $k = 2, \ldots, v-1$, $(y_v, y_1) \in G^a$.
As $P(G^a)$ is asymmetric, it follows that there exist $x_1, \ldots, x_t \in X$, $t \geqslant 2$, with $x_1, \ldots, x_t$ all distinct such that
$(x_1, x_2) \in P(G^a)$, $(x_k, x_{k+1}) \in G^a$, $k = 2, \ldots, t-1$, $(x_t, x_1) \in G^a$.
Therefore
$\exists\, i_1, \ldots, i_t \in N$: $(x_1, x_2) \in D_{i^1} \cap P(R^a_{i_1})$, $(x_k, x_{k+1}) \in D_{i^k} \cap R^a_{i_k}$, $k = 2, \ldots, t-1$, $(x_t, x_1) \in D_{i^t} \cap R^a_{i_t}$.
As for each $i \in N$, R_i^a is an ordering and $P(R_i^a)$ is asymmetric, it follows that $(i_1, \ldots, i_t)$ contains at least two distinct individuals. This establishes that $(x_1, x_2) \in D_{i1}$, $(x_k, x_{k+1}) \in D_{i^k}$, $k = 2, \ldots, t-1$, $(x_t, x_1) \in D_{i^t}$ is a modified critical loop in D, contradicting coherence of D.
This establishes that G^a is consistent.
Therefore, by Suzumura Extension Theorem, it follows that there exists an ordering-extension R of G^a, i.e.
$G^a \subset R$ (*viii*)
and $P(G^a) \subset P(R)$ (*ix*)
(*iii*), (*vii*), (*viii*) and (*ix*) establish that for each $i \in N$:
$G^a_i \subset G^a \subset R$
and $P(G^a_i) \subset P(G^a) \subset P(R)$,
which proves the E-coherence of D.

Liberal Individuals and Resolution of Sen Paradox

An individual who respects the rights of other individuals will be called a liberal. In what follows we provide a formalisation of the idea of an individual being liberal.

Let $D = (D_1, \ldots, D_n)$ be a coherent rights-assignment and let $a = (R^a_1, \ldots, R^a_n) \in T^n$ be a profile of individual orderings. By Theorem 4, there exists an R which is an ordering-extension of each and every $D_i \cap P(R_i^a)$, $i \in N$. Let Z denote the set of all ordering-extensions of each and every $D_i \cap P(R^a_i)$, $i \in N$. Let
$\Pi = [(x, y) \in X^2 | \forall i \in N : xP(R^a_i)y]$
Define for each $R \in Z$:
$p(R) = \# [P(R) \cap \Pi]$.
We define $Z_p \subset Z$ by
$Z_p = [R \in Z | \forall R' \in Z : p(R) \geq p(R)']$.
Let R^*_j, $j \in N$, denote the sub-relation of R^a_j that the individual j wants to be counted in social choice.
We define individual j to be liberal if for some $R^j \in Z_p$,
$P(R^*_j) = P(R^a_j) \cap P(R^j)$ and $I(R^*_j) = I(R^a_j) \cap I(R^j)$
The set of liberal individuals will be denoted by N_L.
It will be assumed that $\forall i \in N - N_L$: $R^*_i = R^a_i$.
For every $(R^*_1, \ldots, R^*_n)$, corresponding to $(R^a_1, \ldots, R^a_n) = a \in T^n$, and every $S \in K$ we define

$$C^a_{p^*}(S) = \{x \in S | \sim [\exists y \in S : (y, x) \in \bigcap_{i \in N} P(R_i^*)]\}$$

$$C^a_{p^*}(S) = \{x \in S | \sim [\exists y \in S : (y, x) \in P(\bigcap_{i \in N} R^*_i)]\}$$

Conditional weak Pareto-criterion (CWP): A CCR with unrestricted domain satisfies CWP if $\forall a \in T^n$ and $\forall S \in K : C^a(S) \subset C^a_{p^*}(S)$.
Conditional Pareto-criterion (CP): A CCR with unrestricted domain satisfies CP if $\forall a \in T^n$ and $\forall S \in K : C^a(S) \subset C^a_{p^*}(S)$.

Lemma 1:
Let $D = (D_1, \ldots, D_n)$ be a coherent rights-assignment and $a = (R^a_1, \ldots, R^a_n) \in T^n$ a profile of individual orderings. If a is such that for each $S \in K$, $C^a_D(S) \cap C^a_p(S) \neq \emptyset$, then there exists an R which is an ordering-extension of each and every $D_i \cap P(R^a_i)$, $i \in N$ as well as of Π.

Proof:
Let $a = (R^a_1, \ldots, R^a_n)$ be a profile of individual orderings such that $\forall S \in K$:
$C^a_D(S) \cap C^a_p(S) \neq \emptyset$. Define

$$Q^a_i = D_i \cap P(R^a_i),\ i \in N \qquad (i)$$

$$Q^a = \bigcup_{i=1}^{n} Q^a_i \qquad (ii)$$

$$J = Q^a \cup \Pi \qquad (iii)$$

By definition, Π as well as each Q^a_i, $i \in N$, is asymmetric. Q^a is also asymmetric as has been shown in Theorem 4. I shall now show that J is asymmetric.
Suppose $(x, y) \in J$ and $(y, x) \in J$, $x, y \in X$.
In view of the asymmetry of Π and Q^a there are only two possibilities to be considered: (*i*) $(x, y) \in Q^a$ & $(y, x) \in \Pi$; (*ii*) $(x, y) \in \Pi$ & $(y, x) \in Q^a$.
$(x, y) \in Q^a$ implies that $\exists j: xP(R^a_j)y$ and $(y, x) \in \Pi$ implies that $\forall i \in N : yP(R^a_i)x$.
Therefore (*i*) is not possible. By a similar argument (*ii*) is not possible.

Next, we show that J is consistent.
Suppose J is not consistent. Then there exist $x_1, \ldots, x_t$, $t \geq 2$, such that $(x_1, x_2) \in P(J)$, $(x_k, x_{k+1}) \in J$, $k = 2, \ldots, t-1$, $(x_t, x_1) \in J$.
As Q^a is consistent by Theorem 4 and Π is transitive, it follows that J being non-consistent implies that $C^a_D(S) \cap C^a_p(S) = \emptyset$, where $S = (x_1, \ldots, x_t)$.
This, however, contradicts the choice of $a = (R^a_1, \ldots, R^a_n)$. So J must be consistent.
Therefore there exists an ordering-extension R of J.
From definitions (*i*)–(*iii*), the asymmetry of Q^a_i, $i \in N$, Q^a, Π and J, and the fact that R is an ordering-extension of J, we conclude:

$Q^a_i \subset Q^a \subset J \subset R$, $i \in N$

$P(Q^a_i) \subset P(Q^a) \subset P(J) \subset P(R)$, $i \in N$

$\Pi \subset J \subset R$

$P(\Pi) \subset P(J) \subset P(R)$

which establishes the lemma.

D-Libertarianism (DL):
Let $D = (D_1, \ldots, D_n)$ be any given coherent rights-assignment. A CCR satisfies DL if $\forall a \in T^n$ and $\forall S \in K : C^a(S) \subset C^a_D(S)$.
We say that for $a = (R^a_1, \ldots, R^a_n) \in T^n$, weak Pareto-criterion and D-Libertarianism do not conflict if $\forall S \in K$: $C^a_D(S) \cap C^a_p(S) \neq \emptyset$.

Theorem 7:
Let $D = (D_1, \ldots, D_n)$ be any given coherent rights-assignment. If there is at least one liberal individual in the society then there exists a CCR f satisfying (*i*) U; (*ii*) DL; (*iii*) CWP and (*iv*) the

property that for every $a \in T^n$ for which WP and DL do not conflict, $\forall\ S \in K : C^a(S) \subset C^a{}_D(S) \cap C^a{}_p(S)$, where $C^a = f(a)$.

Proof:
Consider any profile $a = (R^a{}_1, \ldots, R^a{}_n) \in T^n$ and let $(R^*{}_1, \ldots, R^*{}_n)$ correspond to a in the manner defined above. Define
$R_o = \{(x, y) \in X^2 \mid (y, x) \notin P^* \cup P\}$,
where $P^* = \bigcap_{i \in N} P(R^*{}_i)$ and $P = \bigcap_{i \in N} P(R^j)$, $R^j \in Z_p$.

Take any $j \in N_L$. Then $P(R^*{}_j) \subset P(R^j)$ (*A*)
Therefore, $P^* \cup P \subset P(R^j)$ (*B*)
From (*B*) we conclude that $P^* \cup P$ is asymmetric and consequently R_o is reflexive and connected.
Take any $x, y \in X$. $(x, y) \in P(R_o) \rightarrow (y, x) \notin R_o$, which in turn implies $(x, y) \in P^* \cup P$.
On the other hand $(x, y) \in P^* \cup P$ implies $(y, x) \notin R_o$ which in turn implies $(x, y) \in P(R_o)$, as R_o has already been shown to be reflexive and connected.
This establishes that $P^* \quad P = P(R_o)$.
As a $P^* \cup P$ cycle would imply, in view of (*B*), a $P(R^j)$ cycle negates the transitivity of R^j or the asymmetry of $P(R^j)$.
We therefore conclude that no $P^* \cup P$ cycle is possible.
As $P(R_o) = P^* \cup P$, it follows that no $P(R_o)$ cycle is possible which, in view of the reflexivity and connectedness of R_o, establishes that R_o is acyclic.

Now define for every $S \in K : C^a(S) = [x \in S \mid \forall\ y \in S : xR_oy]$. In view of the finiteness of X and the fact that R_o is reflexive, connected and acyclic, it follows from Sen's theorem[7] that C^a is a well-defined choice function.
Now define a CCR f by : $\forall\ a \in T^n : f(a) = C^a$.

(*i*) By construction f satisfies U.
(*ii*) Take any $a = (R^a{}_1, \ldots, R^a{}_n) \in T^n$ and any $S \in K$. Suppose $y \notin C^a{}_D(S)$. Then $y \notin S$ or there exist $x \in S$, $i \in N : (x, y) \in D_i \cap P(R^a{}_i)$. As $D_i \cap P(R^a{}_i) = P[D_i \cap P(R^a{}_i)]$, we conclude that $y \notin S$ or $(x, y) \in P \subset P^* \cup P = P(R_o)$. Consequently $y \notin C^a(S)$, i.e. $C^a(S) \subset C^a{}_D(S)$. Thus f satisfies DL.
(*iii*) Take any $a = (R^a{}_1, \ldots, R^a{}_n) \in T^n$ and any $S \in K$. Suppose $y \notin C^a{}_{p*}(S)$. Then $y \notin S$ or $\exists\, x \in S : \forall\ i \in N : (x, y) \in P(R^*{}_i)$. That is to say, $y \notin S$ or $(x, y) \in P^* \subset P^* \cup P = P(R_o)$.

Consequently $y \notin C^a(S)$, i.e. $C^a(S) \subset C^a_{p^*}(S)$. Thus f satisfies CWP.

(*iv*) Consider any $a = (R^a_1, \ldots, R^a_n) \in T^n$ such that $\forall S \in K$: $C^a_D(S) \cap C^a_p(S) \neq \emptyset$, and any $S \in K$. By *Lemma 1*, there exists an R which is an ordering-extension of each and every $D_i \cap P(R^a_i)$, $i \in N$, and of Π. Therefore, we conclude that every $R \in Z_p$ is an ordering-extension of each and every $D_i \cap P(R^a_i)$, $i \in N$, and of Π. Suppose $y \notin C^a_p(S)$. Then $y \notin S$ or $\exists x \in S : \forall i \in N : (x, y) \in P(R^a_i)$. This implies that $y \notin S$, or $(x, y) \in \Pi$ and consequently $(x, y) \in P(R^j)$, $\forall j \in N_L$. Therefore $y \notin S$ or $(x, y) \in P \subset P' \cup P = P(R_o)$. So $y \notin C^a(S)$. In view of (*ii*), we conclude that $C^a(S) \subset C^a_D(S) \cap C^a_p(S)$.

This establishes the theorem.[8]

Liberty and Rationality

Theorems 4 and 5 establish necessary and sufficient conditions for the absence of the Gibbard paradox and Theorem 7 establishes a sufficient condition for the absence of the Sen paradox. For interpreting these theorems, it is important to distinguish between two aspects of liberty: (*a*) liberty as freedom of choice for individuals in their protected spheres; and (*b*) liberty as a value judgement for determining the set of socially optimal alternatives.

In any society individuals are permitted freedom of choice in certain protected spheres. Of course, these protected spheres vary widely from society to society. As a simple illustration, I shall consider a society of two individuals 1 and 2, who are free to paint their houses with either the colour blue or green, these two colours being the only feasible options. Everything else being given, colour choices by the two individuals define four social states a_{bb}, a_{gg}, a_{bg} a_{gb} (a_{ij}, i = b, g; j = b, g; defines the social state when individual 1 chooses colour *i* and individual 2 colour *j*). It is important to note that the existence of freedom for individuals 1 and 2 to choose the colour in itself has no implications for the evaluation of social states. However, if we accept liberty as a value judgement, then the existence of rights coupled with individual preferences would have implications for the evaluation of social states. One possible formalisation of the idea of liberty as a value judgement is that if any two social states x and y differ solely with respect to those features which are in the protected sphere of a particular individual i, then if individual i prefers x to y so does society (or alternately if individual i prefers x to y then y does not belong to the choice set), i.e. y is not socially optimal. The Gibbard paradox then relates to inconsistency between liberty as a value judgement for the evaluation of social states and social rationality. Social rationality here simply means that the social choice must not be non-optimal in the sense of not belonging to the choice set. In in our illustration individuals 1 and 2's preferences are a_{bg} P_1 a_{gb} P_1 a_{bb} P_1 a_{gg} and a_{gg} P_2 a_{bb} P_2 a_{gb} P_2 a_{bg} respectively, then from the libertarian value judgement it follows that no social state belonging to the set $\{a_{bb}, a_{gg}, a_{bg}, a_{gb}\}$ is socially optimal. Thus libertarian value judgement as formalised above is inconsistent with social rationality. It is then clear that if one wishes to retain liberty as an ethical principle for the evaluation of social states, its scope must be narrower than the way it has been formalised. The characterisation Theorems 4 and 5 spell out the extent of narrowing that is required. In this connection, it is worth noting that Gibbard's condition of unconditional preferences which limits the scope of the libertarian value judgement is a sufficient condition for avoiding the inconsistency.[9] Since the Gibbard paradox arises because of inconsistency between liberty as a value judgement for the evaluation of social states and social rationality, and not because of the existence of rights, it follows that it is possible to avoid the paradox by limiting the scope of the libertarian value judgement without any curtailment of the rights of individuals.

Even if the libertarian value judgement by itself is consistent with social rationality, it is possible that its conjunction with the weak Pareto-criterion is not as has been shown by Sen (1970a: 152–57). Theorem 7 provides a sufficient condition under which the conjunction of a weak Pareto-criterion and coherent libertarianism is consistent with social rationality. The theorem states that if there is at least one liberal individual in the society, the Sen paradox cannot arise.

It is worth noting that Gibbard's condition of unconditional preferences which resolves the libertarian paradox, and the condition of existence of at least one liberal individual in the society which resolves the Pareto–libertarian paradox, relate to two fundamentally different approaches to resolving these paradoxes. While the condition of unconditional preferences makes the libertarian value judgement weaker and consequently less of a constraint on social rationality, the condition of existence of a liberal individual retains the constraints imposed on social rationality by coherent libertarianism and the Pareto-criterion, but weakens the notion of individual rationality in as much as the liberal individual voluntarily agrees not to insist on that part of his preferences being counted in the evaluation of social states which conflicts with other individuals' rights. Suppose in our two–person society the set of feasible social states is $X = \{a_{gg}, a_{bg}, a_{gb}\}$ and individuals 1 and 2 have the preferences $a_{gg}\ P_1\ a_{bg}\ P_1\ a_{gb}$ and $a_{bg}\ P_2\ a_{gb}\ P_2\ a_{gg}$ respectively. Then the libertarian value judgement implies that a_{bg} and a_{gg} cannot belong to the choice set and the weak Pareto-criterion implies that a_{gb} cannot belong to the choice set. Thus every feasible alternative is non-optimal. Here the libertarian value judgement by itself is consistent with social rationality but its conjunction with the weak Pareto-criterion is not. In this example the sets of ordered pairs over which the two individuals have rights are given by $D_1 = \{(a_{gg}, a_{bg}), (a_{bg}, a_{gg})\}$ and $D_2 = \{(a_{gg}, a_{gb}), (a_{gb}, a_{gg})\}$. $D_1 \cap P(R_1) = \{(a_{gg}, a_{bg})\}$ and $D_2 \cap P(R_2) = \{(a_{gb}, a_{gg})\}$. There is only one binary relation $\bar{R}$ which is both an ordering-extension of $D_1 \cap P(R_1)$ and an ordering-extension of $D_2 \cap P(R_2)$, namely $a_{gb}\ \bar{P}\ a_{gg}\ \bar{P}\ a_{bg}$. If we assume individual 1 to be liberal, the relation R^{*}_1 that he would like to be counted in social choice would be given by $P(R^{*}_1) = P(R_1) \cap P(\bar{R})$ and $I(R^{*}_1) = I(R_1) \cap I(\bar{R})$. Thus $R^{*}_1 = \{(a_{gg}, a_{bg})\} \cup \Delta\ x$. When the coherent libertarian value judgement and Pareto-criterion are applied to R^{*}_1 and R_2, the inconsistency disappears.

Thus both the Gibbard and Sen paradoxes can be interpreted as manifestations of divergence between individual rationality and social rationality.[10] Consequently, for the resolution of these paradoxes either one will have to accept a weaker notion of individual rationality which is more in tune with social rationality, or one will have to weaken the constraints on social rationality.

Notes

I wish to thank Professor A. Gibbard, Professor S.-Ch. Kolm and other participants of the International Workshop on Rationality in Cognitive Science and Social Sciences held at New Delhi in December 1992 for comments on the paper.

1. See Suzumura 1978: 341–42.
2. The idea of a critical loop was introduced in Farrell 1976: 7. The definition of critical loop given here is that of Suzumura 1978: 330. A critical loop in the sense of Suzumura, however, is not equivalent to a critical loop is the sense of Farrell. Both Farrell's and Suzumura's definitions differ from the definition of the modified critical loop introduced in this paper.
3. See Lemma 1 in Suzumura 1978: 341–42.
4. See Suzumura 1978: 331.
5. The important distinction between a person's preferences R_j and the preferences that he would like counted in social choice R^*_j was introduced in Sen 1976: 243–44. Sen requires R^*_j to be a sub-relation of R_j.
6. See Suzumura 1983: 16–17.
7. See Sen 1970b: 16.
8. Sen has shown that if there exists at least one liberal individual in the society then there exists a CCR satisfying unrestricted domain, coherent libertarianism (CL) and the conditional-strict Pareto-criterion (CP) (Sen 1976: 243–44). Under the assumption that there exists a liberal individual in the society, Suzumura has proved the existence of a CCR satisfying conditions U, CL and CP (Suzumura 1978: 332–33). Suzumura uses the formalisation of the idea of a liberal individual discussed in the introductory part of this paper. By a slight modification in the proof of Theorem 7, one can show the existence of a CCR satisfying CP in addition to the conditions mentioned in the statement of the Theorem, provided there is at least one individual in the society who is liberal in the sense of this paper.
9. A careful reading of the proof of Theorem 4 makes it clear that for a given $(R_1, R_2, \ldots\ldots, R_n) \in T^n$, a rights-assignment $(D_1, \ldots, D_n)$ is coherent if there exists an R such that it is an ordering-extension of each and every $D_i \cap P(R_i)$, i = 1,2, . . ., n. Thus it is possible to have different coherent rights-assignments for different configurations of individual preferences, as would be the case if libertarianism is constrained by Gibbard's requirement of unconditional preferences.
10. It is possible to argue that the social choice–theoretic framework is inappropriate for discussing the issue of rights and liberty, and that the Gibbard and Sen paradoxes are really manifestations of this inappropriateness. For an exposition of this viewpoint, see Kolm 1992.

References

ARROW, K.J. 1963. *Social Choice and Individual Values*. New York: Wiley.

FARRELL, M.J. 1976. Liberalism in the Theory of Social Choice. *Review of Economic Studies*. 43: 3–10.

GIBBARD, A. 1974. A Pareto-Consistent Libertarian Claim. *Journal of Economic Theory*.

KOLM, S.-CH. 1992. What Sense Social Choice. Paper presented at the International Workshop on Rationality in Cognitive Science and Social Sciences held at New Delhi.

SEN, A.K. 1970a. The Impossibility of a Paretian Liberal. *Journal of Political Economy*. 78: 152–57.

———. 1970b. *Collective Choice and Social Welfare*. San Francisco: Holden-Day.

———. 1976. Liberty, Unanimity and Rights. *Economica*. 43: 217–45.

SUZUMURA, K. 1978. On the Consistency of Libertarian Claims. *Review of Economic Studies*. 45: 329–42.

———. 1983. Rational Choice, Collective Decisions and Social Welfare. Cambridge: Cambridge University Press.

II

Rationality and Social Sciences

Preface

A suitable introduction to this section on the social sciences is provided by A. Raghuramaraju. He puts the concept of rationality in the perspective of the Enlightenment in the West, and explains how it gave rise, there, in the 19th century, to the discipline of social sciences. He uncovers the presuppositions underlying the concept of rationality and their influence on the formative process of the social sciences.

One of the thrust ideas of the Enlightenment is that of man in the state of nature. Raghuramaraju shows that it is, on the part of philosophers like Hobbes, Rousseau and Locke, a purely hypothetical construction, not founded on the observation of primitive or non-societal (asocial) men.

But how is it that all human beings are engaged in society? This is because man realises that it is more beneficial to him to abandon some of his individual prerogatives and, by contract, join society. This conscious and contractual (consensual) surrendering is the manifestation of rationality. Thus, society in its contractual form, based on the general will, is a pure creation of man. All its institutions are therefore rational, and man always remains the reference point.

Now how does one reconcile the expected universality of rationally-built social institutions with their observed diversity? This is where the idea of progress comes in, and provides the cradle for the social sciences in the 19th century (Comte, Marx). Diversity is thus accounted for as a result of differences in the pace of development, and progress (modernity) is expected to ultimately standardise everything.

There is no denying that this ideology of progress and modernity may have provided a handy justification for colonialism. But is it a reason for condemning rationality as a Western devil, as Raghuramaraju does, thus reopening the enduring debate, recently fuelled in the West by post-modernism?

Along the same lines, one might wonder whether J.-Cl. Gardin and D. Bourcier are not representatives of this narrow-minded, technocratic, Western positivism. Isn't D. Bourcier's attempt at computerising legal decision processes an infringement of the prerogative of the human mind over the machine?

But if rationality—and therefore science—is considered to be purely Western, isn't it puzzling that there are Indian Nobel prize winners for instance? How is it that there are 'oriental' scientists recognised by their Western colleagues, interacting with them in the universal language of science? Science in its modern form was indeed born in the West, and this may be an ill-founded matter of pride for some, and of resentment for others. But in any case, it should neither be a pretext for denying others the capacity of doing science, nor for rejecting science altogether on the grounds that it was born elsewhere.

The divide is not between the West and the East. It is—all over the world—between two different types of languages: that of science as opposed to the natural languages. In Jean-Claude Gardin's contribution, the main theme is to ascertain where the discourse of the human sciences fits in this dualism. A tenant of Western positivism would argue that there is no knowledge outside science, and would urge the human sciences to fulfil the criteria of 'hard' sciences, or die. Some people, on the other hand, have advocated that the discourse of the social sciences has its specificity, being neither science nor literature, but representing in itself a so-called 'third way'. Examining the status of the human sciences through his method of 'logicist analysis', Gardin rejects the third way theory, yet does not endorse the conclusion of the stubborn positivist. He admits that, at par with science, literature can be, under certain conditions, 'a complementary mode of *knowledge*' (my italics)

Danièle Bourcier's contribution comes as an illustration of what Gardin has theorised. She describes the project, which has now been implemented, of an expert system designed to help French mayors discharge their duties in the field of municipal jurisprudence. The proof of the system is in its working. But, we are more interested in the conditions of the formalisation of legal reasonings. Though legal discourse is well-formalised and normative, solutions cannot be deduced for every possible case, which leaves the legal officer a margin for interpretation. One of the recognised modes of reasoning in law is analogy. By referring to previous cases, the judge makes up his mind. Now Bourcier's experiment proves that feeding the computer in an appropriate way with the text of the law and with legal case studies, enables it not only to return verdicts but, more interestingly, to create new rules. Man seems to have abandoned once more a part of his prerogatives, this time not to society as his enlightened predecessor, but to the machine.

In Wagish Shukla's contribution, the attempt is to reconcile the West and the East, and the natural with the artificial, by showing that what was considered as the apanage of the West existed in the East too. There is a concept of rationality in India which is close to that of the West, and the whole idea of artificial as opposed to natural may even be less problematic here than in the West.

Indeed if one takes rationality as what differentiates man from the animals, then what is it in the Indian tradition which does the same? 'Speech' is the answer. Not any kind of speech, but only 'correctly articulated speech'. 'Correctly' here means 'grammatically' in the Paninian tradition, and this capacity is only entrusted to an elite among human beings. To show that the same ostracism occurs in the West, Shukla gives the example of Christian Spanish colonisers in America burning the natives' books. Reason was the whites' lot only. Making all due allowances, it reminds us of the 'political correctness' of present-day America.

Now, in an Indian myth quoted by Shukla, 'correct speech' empowers one to create artificial creatures, which are not born through the natural process. In the myth, these creatures can acquire all the human qualities, among which is an acute sense of identity, which takes them in the endless cycle of rebirths. But it is even possible to create artificial creatures that are able to take meta-positions and understand how this notion of identity is despicable. These creatures are thus the equivalent of the most highly accomplished human beings, endeavouring to get out of the rebirth cycle. It means that, in the Indian myth, there are no human qualities, however exalted, that are not artificially reproducible. It should therefore not surprise us to see machines outperform human beings.

OLIVIER GUILLAUME

The Project of Rationality and the Discipline of Social Sciences

A. RAGHURAMARAJU

Instead of engaging in an inquiry into the nature of extension, revision or rejection of reason, in this article I will focus on the background aspects underlying reason. 'Disembodied reason' is *aprioristic* to most discussions in the philosophy of social sciences. Even in those instances which constitute a rejection of rationality, what is rejected is only disembodied reason. The background, the very foundation, the cluster of concepts contributing to the making of reason, remains. I shall therefore concentrate on elucidating a more historically informed analysis of rationality. I shall do this by laying bare different stages in the making of the project of rationality (POR).

I shall begin by discussing the two stages in the making of POR, namely, the postulation of the idea of a pre-societal man (man-in-the-state-of-nature), and the formation of the modern thematic (general will). Further, I shall show the antinomy between the modern thematic and the realities surrounding it. This exploration, I think, will not only establish a close relation between the discipline of social sciences and POR but will also place the discussions within the philosophy of social sciences in a larger historical perspective.

Two Stages in the Formation of the Project of Rationality

The Presupposition of the Project

A new conception of man inaugurated modern thought. This concept is presented by the contract philosophers as existing in the state of nature. It seeks to describe the nature of man without taking into account his relation with other human beings and social institutions. As a matter of fact, it is a projection of an hypothetical nature of man not rooted in the experience of society. Thus formulated, this concept of man, forms the basic presupposition of not merely the contract theories but of modern thought as well. In what follows, I shall sort out the differences amongst three contract philosophers regarding the concept of man-in-the-state-of-nature as well as establish their common assumptions.

The differences neatly fall into two types: those relating to the attributes of this concept, and those regarding the reasons for man to leave his state of nature and, through contract enter into civil society.

For Hobbes, the man-in-the-state-of-nature is 'selfish, brutish'; for Rousseau he is 'good and noble' (Charvet 1978) or 'amoral' (Durkheim 1960); and for Locke he is 'good and rational'.[1] The reasons why this man should leave this state and enter into the civil society are also different for each of the three philosophers. Hobbes believes that the selfishness of man-in-the-state-of-nature is self-defeating when taken to its logical extreme. If everyone is self-centred, wants to grab other's property and exercise dominance, the result will naturally be constant war. So for the sake of 'enlightened selfishness' people consent to surrender their individuality to a common collectivity. This collectivity is absolute in its nature and final in its judgment. For Rousseau, while the noble savage in the state of nature is still desirable, yet he has reached 'the point at which obstacles in the way of their preservation in the state of nature show their power of resistance to be greater than the resources at the disposal of each individual for his maintenance to that state' (1952, p. 11). For Locke, in the state of nature everyone has executive power over everyone else. This power may unjustly be used for selfish reasons. To avoid this arbitrary use of power, Locke posits a transition from the natural to the civil society.

However, behind these two types of differences there lies a common assumption of the notion of man-in-the-state-of-nature, which is subscribed to by the three contract philosophers. This common assumption is described as pre-societal man and/or as non-societal man. These two descriptions have radically different implications for the conceptualisation of this notion of man. While 'pre-societal' would imply a historical state and involve an anthropological approach, 'non-societal' is a hypothetical state posited through a methodological device. The source of this confusion has been discerned by the writers interpreting contract theories.

According to Gough, though Locke 'admits that "if we look back as far as history will direct us towards the original of commonwealths: we shall generally find them under the government and administration of one man"', but, 'this does not disprove the thesis that political society began by the consent of the individuals who united to establish it'. Because, 'Reason, after all, is "plain on our side that men are naturally free"' (Gough 1957, p. 37). Leo Strauss holds Hobbes responsible for the confusion regarding the status of man in the state of nature. He says that though the state of nature according to Hobbes is not historical, the 'appearance that his theory of the state of nature has an historical meaning is not entirely without foundation' (Strauss 1966, p. 104).

While there exists this confusion, it must, however, be noted that the state of nature is not an historical state. In support of this, one can look to some of the contemporary interpretations of contract theorists. Leo Strauss concludes that: 'The state of nature is thus for Hobbes not an historical fact, but a necessary construction' (1966, p. 104). This comes out even more clearly in Durkheim's reading of Rousseau. For Durkheim, '*the state of nature* is not, as had sometimes been said, the state in which man lived before societies came into being. The term might indeed suggest a historical period at the beginning of human development. This is not what Rousseau meant. It is, he says, a state "which no longer, which never existed, which probably never will exist"' (1960: p. 66).

The discussion so far can be summarised as follows: the concept of man in the state of nature as a common presupposition of the contract philosophers has been established. This concept is further shown to be a non-historical/societal state. That is to say, it is a description of neither a historical nor a primitive society, but a mere postulation, a hypothesis.

This hypothetical nature of man, it must further be noted, is at variance with the existing social realities i.e. it maintains its distance both from the traditional framework as well as the primitive societies. The traditional framework as discerned by Alisdair MacIntyre consists of (*a*) the notion of man as he happens to be (human nature in its untutored state); (*b*) the conception of the precepts of rational ethics; and (*c*) the conception of man as-he-could-be-if-he-realized-his-essential-telos (1985, p. 53). The hypothetical concept of man declines to take note of any of these components.

It also maintains its removedness from the so-called 'primitive' societies represented or otherwise by traditional theories.

Formulating a discourse independent of any existing social realities, the presupposition seems to assert its autonomy, objectivity etc. In embracing autonomy and claiming objectivity, the presupposition remains empty of content and confronts the solipsistic state. While it is elated about its autonomy and objectivity, it suffers from solipsism and emptiness. We now come to the next stage where it seeks to overcome this solipsism.

The Formation of the Modern Thematic

When confronted by solipsism, the self makes a novel attempt wherein it seeks to overcome this state while retaining autonomy. In this attempt the individual selves transform themselves through contract and create a general will. It is necessary here to establish the fundamentality of this new concept, i.e., the general will. I shall begin by sorting out the differences amongst our three contract philosophers regarding the nature of general will/civil society.

Analogous to the differences amongst these philosophers regarding the attributes of man-in-the-state-of-nature, there exist differences regarding the nature of general will as well. The Hobbesian commonwealth is absolutely powerful, monarchical in its nature. This powerful commonwealth is necessitated by man in the state of nature. The absolutism of the commonwealth is justified on the ground that only a powerful political organisation can restore peace amongst men who are defined as being by nature selfish, brutish etc. As pointed out by Ebenstein, 'Hobbes is vehemently opposed to the division of powers or mixed government.' And 'To keep the authority of the state strong, Hobbes advises the Sovereign not to allow the growth of groups and institutions that intervene between the state and the individual' (1972, p. 361).

Rousseau's man, who is basically good, forms a sovereignty ruled by general will. This will is not an empirical will, but an objective principle to which everyone must submit. This sovereign

> . . . being formed wholly of the individuals who compose it, neither has nor can have any interest contrary to theirs; and consequently the sovereign power need give no guarantee to its subjects, because it is impossible for the body to wish to hurt all

> its members . . . . The sovereign, merely by virtue of what it is, is always what it should be (Rousseau 1952, pp. 14–15).

The Lockean society is primarily constitutional. Through this constitutional form of government, Locke seeks to avoid the arbitrary use of power in the state of nature and provide protection to private property. This constitutional sovereignty implements the decisions of the majority, which remain final.

Underlying these differences amongst the three contract philosophers, there is a common assumption, namely, the concept of general will as created by the particular wills. The distinguishing feature of this belief is that it makes man both the centre of society and its creator. He becomes the foundation of society and in all social matters, the point of reference is man. All social institutions emerge out of this notion of man who is exclusively rational.

It is important to note the different aspects of this belief: (*a*) the general will is created by men in the state of nature; and (*b*) this creation involves the annihilation of the creator. This general will together with the concept of man in the state of nature—in other words, the 'philosophy of consciousness' as well as the 'decentering of the self'—forms the general thematic of modern thought. This leads to a curious paradox: the first ever attempt to decentre the self is inaugurated by the contract theorists who are at the same time accredited with championing the philosophy of consciousness. Further, the general will, by virtue of being created by man, remains, albeit in a paradoxical sense, anthropocentric. Thus the general will along with particular will and the contract forms the thematic of the Enlightenment. The formation of this thematic launches, as it were, what I would call the Project of Rationality (POR).

The Project of Rationality and the Discipline of Social Sciences

The formation of this modern thematic—which is rational—avoids the problem of solipsism, while simultaneously retaining autonomy. However, the problem of emptiness continues to loom large. As the concept of man rejects all things outside it, and moves towards self-generation, the very thematic, though rational, remains abstract, empty and innocent of experience. In other words, the modern thematic continues to maintain its distance from reality and does not step outside itself. While it succeeds in retaining its autonomy and avoiding solipsism, it nevertheless remains empty.

By making man the centre of social reality and restricting his actions to the rational domain, the actual social institutions that are related to each other at different levels, discharge different functions, and that have evolved over centuries, have been excluded from their analysis. The intermediary relations, which are the actual social realities enjoying some sort of natural independence, have not been considered in these theories. The reason seem to be that these institutions do not fulfill the criterion of the modern thematic. In adopting this approach two things seems to have happened: (*i*) only rational, constructivist social institutions have been accepted, and (*ii*) the intermediary natural social institutions have been rejected.

This thematic seems to create a gap between itself and the realities surrounding it. In other words, the Enlightenment theories are divorced from reality. Commenting on them, Anthony Wildon remarks that the Lockean view of society

> . . . is not based on the characteristics of social organizations at all; it is based, like practically the entire domain of 'social' sciences today, on the psychology and/or the biology of the supposed 'human individual' or on that of the 'species'. This is of course a confusion of logical typing of distinct levels of organization. Characteristics stemming from the actual organization of the system are projected into the supposed 'instincts' or 'innate ideas' of an imaginary 'individual' (1980: xl–xli).

Wildon attributes the *Splitting of the Economic System* and *Alienation* to Lockean–Cartesian individuals.

Apart from these considerations, how does one confront the modern thematic? One way is to reject the very postulations of this thematic as 'misconceptions' or 'illusions'; or as agencies that are confusing the concept of the formation of a state with the formation of a community. (MacIver 1964: p. 451). One can also show that this identification of state with society, or 'confounding one with the other has given rise to modern theories of state, providing the theoretical background to totalitarianism'. (MacIver 1970: pp. 84–85).

This line of criticism can be countered by pointing out that the thrust of these theories is not to present descriptive accounts of societies. So to dismiss them as false descriptions is to undermine the normative project of Enlightenment. And to reject them for the political implication that they gave rise to totalitarianism, is to abbreviate the complex process generated by this thematic.

This, however, does not imply an acceptance of this thematic. The thing to do is to problematise its normativity. How can this empty thematic re-establish its relation with the realities it rejects? How can it overcome its emptiness? However, these questions cannot be formulated within the discourse of the contract theories; they necessarily fall outside their framework. And, among others, one philosopher who has confronted them is Montesquieu.

Montesquieu accepted the two incomparable domains—the universalism of the Enlightenment as well as plural cultures and their diversity—simultaneously (for details on Montesquieu see Geoffrey Hawthorn 1976; Raymond Aron 1979, 1980). But in the absence of a theory of progress or availability of techniques which transform the diversities into unity, this acceptance derailed his thought.

Instead of getting problematised, the tension in this acceptance is tamely reduced to a sequentiality in Comte.[2] In Comte, diversity takes a back seat. And he,—in fact, all philosophers, including Marx, who represent mainstream classical social thought, relates himself to the Enlightenment by attributing to it descriptions of the newly emerged industrial–capitalist society.

Thus situated, the relationship between classical social theories and the Enlightenment is symbiotic. It consists of using the methodology of the Enlightenment to arrive at descriptions of the new social, political and economic realities. This relation of an exchange of form and content, of a collaboration between rationalism and empiricism, concretises the relation between POR and the discipline of social sciences. In this exchange, the Enlightenment thematic overcomes its emptiness with descriptions of the newly emerging social realities provided by the discipline of social sciences. The discipline of social sciences must therefore be related not to the notion of disembodied reason, or reasons (plural/relative), not even to anti-reason. Instead it must be related to the background of POR.[3] Thus formulated, this new nexus results in the Enlightenment thematic minus its emptiness. With this collaboration, the abstraction of the Enlightenment becomes a reality.[4]

Thus portrayed, POR is no longer a mere theory to be accepted or rejected. It is a firm reality in the West. The two facets of rationality are not sufficiently acknowledged or considered in the literature which critiques it.

We have so far discussed two stages in the making of POR. I concluded by describing the discipline of social sciences as responsing to/participating in this thematic by providing the descriptive accounts of newly emerged industrial/capitalist society. This way of relating reason *and* social sciences, I think, not only throws better light on discussions of rationality in the social sciences but also situates the discipline of social sciences in a better perspective.

Notes

Some sections of this paper are originally from a chapter of my doctoral dissertation entitled 'Man and Society: A Critique of the Presuppositions of Modern Political Philosophy,' submitted to Indian Institute of Technology, Kanpur 1985.

Acknowledgements: I wish to thank Parthasarathi Banerjee, J-C. Gardin and A.V. Afonso for their comments and encouragement in the preparation of this paper.

1. The concepts of the Enlightenment theory such as man-in-the-state-of-nature, anthropocentricism, general will, contract etc. have often been frozen in subsequent discussions. This has led to abbreviations and distortions. In order to avoid this, I have presented here a detailed account of these concepts by carefully classifying them.
2. The relation between modernity and tradition in the discourses on sequentiality and simultaneity is elaborated in my paper, 'Critique and Alternative' in Alisdair MacIntyre (ed.) *The Predicament of the West's Critique of Itself*.
3. In the light of this background analysis, it may be fruitful to retrieve the following discussions within philosophy of social sciences. They are: the prescription of natural science methodology to social sciences (Ernest Nagel); the attempts at attributing to social sciences a rationality other than the one available in the natural sciences (Hayek); the paradigm dependency of reason in the natural sciences (Thomas Kuhn); the establishment of plurality/relativity of reason in the social sciences (Peter Winch, Benedict); and even the rejection of reason and the celebration of its opposite (Feyerabend). Analysis of these discussions within the philosophy of social sciences against POR, would open up a new dialogue between political theory and the philosophy of social sciences.
4. This reality is a result of a violent and galvanising attack on the plural cultures in the West. This violence in the West had no active resistance. We must take this absence of the active resistance to the sway of modernity into consideration in our assessment of the success of modernity as well as our resignation to its

strength and inevitability. Achieving this success almost without resistance, it has now turned its attention towards non-modern societies, with attractively packaged arguments of inevitabilities. The colonised subject of the non-West provides a particular kind of response towards the sway of modernity. How the awakened, decolonized subject negotiates this spectacular modernity is at present open to speculation and will provide us with an opportunity to assess its strength.

References

ARON, RAYMOND. 1979. *Main Currents in Sociological Thought-I*. New York: Penguin Books.

———. 1980. *Main Currents in Sociological Thought-II*. New York: Penguin Books.

BENEDICT, RUTH. 1980. *Patterns of Culture*. London: Routledge and Kegan Paul.

CHARVET, JOHN. 1978. *The Social Problem in the Philosophy of Rousseau*. London: Cambridge University Press.

DURKHEIM, EMILE. 1960. *Montesquieu and Rousseau*. Ann Arbor: The University of Michigan Press.

EBENSTEIN, WILLIAM. 1972. *Great Political Thinkers*. Calcutta: Oxford and IBH Publication Co.

FEYERABEND, PAUL. 1975. *Against Method: Outlines of an Anarchistic Theory of Knowledge*. London: NLB.

GOUGH, J.W. 1957. *The Social Contract*. Oxford: Clarender Press.

HAWTHORN, GEOFFREY. 1976. *Enlightenment and Despair: A History of Sociology*. Cambridge: Cambridge University Press.

HAYEK, F.A. 1982. *Law, Legislation and Liberty*. London: Routledge and Kegan Paul.

HOBBES, THOMAS. 1979. *Leviathan*. With an introduction by C.B. MacPherson. Middlesex: Penguin Books Ltd.

KUHN, THOMAS. 1970. *The Structure of Scientific Revolutions*. Chicago: Chicago University Press.

LAKATOS, I. and A. MUSGRAVE. (eds.) 1978. *Criticism and the Growth of Knowledge*. Cambridge: Cambridge University Press.

LOCKE, JOHN. 1963. *Two Treatises of Government*. Cambridge: Cambridge University Press.

MACINTYRE, A. 1985. *After Virtue*. London: Duckworth.

MACIVER, R.M. 1964. *The Modern State*. Oxford: Oxford University Press.

———. 1970. *On Community, Society and Power*. Chicago: University of Chicago Press.

NEGEL, ERNEST. 1984. *The Structure of Science*. Delhi: Macmillan.

O'NEILL, JOHN (ed.) 1973. *Modes of Individualism and Collectivism*. London: Heinemann.

POPPER, KARL. 1959. *The Logic of Scientific Discovery*. London: Hutchinson.

ROUSSEAU, J-J. 1952. *The Social Contract and Discourse*. London: J.M. Dent & Sons, Ltd.

STRAUSS, LEO. 1966. *The Political Philosophy of Hobbes: Its Basis and its Genesis*. Chicago: The University of Chicago Press.
WILDON, ANTHONY. 1980. *System and Structure: Essays in Communication and Exchange*. New York: Tavistock Publication.
WINCH, PETER. 1970. *The Idea of a Social Science*. London: Routledge and Kegan Paul.

Cognitive Effects of the Computational Paradigm in the Human Sciences

JEAN-CLAUDE GARDIN

A distinguished psychologist, Jerome Bruner, recently published a book whose first chapter is entitled 'The Proper Study of Man'. The opening sentence reads as follows:

> I want to begin with the Cognitive Revolution as my point of departure . . . that revolution (which) was intended to bring 'mind' back into the human sciences after a long cold winter of objectivism . . . has been technicalised in a manner that . . . undermines that original impulse (1990: 1).

One aspect of this technicalisation is the part played by the computational paradigm in studies of reasoning processes. I have used that paradigm myself to explore the mechanisms and foundations of scholarly constructs in my discipline, archaeology. My thesis here is that computer explorations of that sort are in fact a step in the direction advocated by J. Bruner inasmuch as they underscore the limits of objectivism in the humanities. However, my views on this much debated topic differ from those of the sizeable population of scholars who present them under other banners such as hermeneutics, post-modernism, semiotics, etc.

Computer Applications in the Human Sciences

Computer applications are known in all sectors of the human sciences. The most common functions served by computers are sorting and counting, i.e., clerical or mathematical operations in which the traditional contest of 'mind vs. machine' tends to turn in favour of the latter. However, the output of mechanical sorting and counting is seldom a finished product: it has to be interpreted and evaluated in a variety of ways, in which the mind regains the upper hand. From an epistemological viewpoint, the results of computer-aided research of this sort hardly differ from those of pre-mechanical times. The case of computer simulations is not essentially different. The models and flow-diagrams that control the mechanical operations are obviously a product of the mind; the claims for the 'revolutionary' nature of the output should not lead us to attribute it to the computer. Epistemologically speaking, we still stand on the familiar grounds of hypothesis forming and testing; the novelty is the electronic *tool* used in the process, invaluable when dealing with complex models and large bodies of data.

The situation is different when we turn to a particular class of simulations in which the object under study is the mental behaviour of the scientist rather than the socio-cultural phenomena that he or she has chosen to explore. Our models and flow-diagrams in this case are meant to express the thought processes observed in the course of scientific research, as hypotheses are taking shape ('discovery' processes) or as they are finally presented in the form of printed papers ('proof' or 'justification' processes). These two kinds of projects are clearly different: the design of learning systems is a standard example of the former, while knowledge-base or expert systems can be viewed as a product or by-product of the latter. My own work in the past years has been limited to projects of the second sort: I shall begin the next section by reviewing the principles upon which they are based, and then move on to the epistemological issues that have been raised by such projects.

The Formal Analysis of Scholarly Constructs

The archaeologist's function is to reconstruct the history and ways of life of past societies on the basis of tentative interpretations of their material remains. Archaeological publications should then be regarded as symbolic constructs made of propositions that necessarily belong to one of the three following categories: (*a*) the 'factual' basis, i.e., descriptions of material remains (artifacts, monuments, organic fossils, ecological context, etc.), as well as abstract statements presented as established truths or tentative presuppositions in the argument; (*b*) the theoretical outcome, i.e., hypotheses or conclusions submitted as end-products of the construct; (*c*) the argument that links these two components in either of the two standard ways: empirico–inductive from *a* to *b* or hypothetico–deductive from *b* to *a*. The form of the argument in both cases is basically a succession of rewrite operations: having stated or established a given (set of) proposition(s) (Pi), the author feels authorised to derive another (set of) proposition(s) (Pj).

The following system of notation has been found convenient: (Po) for propositions that belong to *a*, formally defined as statements that have no explicit antecedent in the paper, i.e., 'initial' propositions in this sense; (Pn) for propositions that belong to *b*, formally defined as statements that have no explicit consequent in the paper, i.e., 'terminal' propositions in this sense; (Pi) for propositions that belong to *c*, formally defined as statements that are derived (inferred, deduced) from another (set of) proposition(s) found in the paper at any of the three levels P0, Pn, Pi, i.e. 'intermediate' propositions in this sense.

The rewrite operations mentioned above are thus of the form (Pi) → (Pi ± 1), in which the subscript *i* is a number varying from 0 to *n* (granted that this is a simplification: several different numbers are commonly subsumed under *i*, which here designates the upper or lower limit of the set).

Example 1: Assuming that the decline of the Indus civilisation was the consequence of climate changes (Pn), I propose various chains of implications that can be derived from that hypothesis (Pi → Pi − 1), until I reach the level of observed or supposedly established 'facts' that I present as the empirical basis or 'proof' of my hypothesis (P0).

Example 2: Observing or asserting that there exist analogies between two sets of archaeological remains found respectively in the Indus valley and in Northern Afghanistan, both dated around 2500 B.C. (Po), I propose one or several hypotheses concerning the historical meaning of these analogies (Pn), on the basis of various chains of inferences (Pi → Pi + 1).

Our scholarly papers can thus be entirely defined as the product of two components: a data base (Po), and a set of rewrite formulae (Pi) → (Pi ± 1) used to generate or to validate a number of hypotheses (Pn). We call this kind of reconstruction a 'schematisation', a term borrowed from the Swiss logician J.B. Grize to designate the formalisation of argument in natural language (NL). A reference to logicism occasionally creeps in to convey the spirit of the exercise (for more details see Gardin 1980; a compact and updated presentation is available in Gardin 1990b).

The first applications of this approach, in the mid-1970s were carried out in the field of archaeology. Others continue to this day, inspired by the same desire to improve our understanding of interpretive processes in that discipline (selected list *in fine*). The restriction to archaeology, however, was from the start contingent: interpretive processes of the same brand are common in other branches of the humanities, amenable to similar schematisations, for the same purpose. The nearest field is history, of which archaeology was long said to be a mere auxiliary: instead of (or in addition to) material remains, historians use texts as a source of 'facts' (P0) upon which they base their 'theories' about past societies (Pn), following various courses of reasoning that connect them through successive derivations (Pi → Pi ± 1). Literary studies are another example: a given text is taken to present a number of distinctive features (P0)—grammatical, stylistic, lexical, semantic, etc.—that suggest cascades of associations (Pi → Pi + 1) leading to various readings of the text (Pn), etc. Thus, our schematisations are nothing but an attempt to capture the logical structure of reasoning in scholarly papers of any sort, taking it for granted that such a structure necessarily exists by virtue of this reference to scholarship, irrespective of the rhetorical garments in which it may be veiled.

'Scholarly' and 'scholarship' are here used as attenuated substitutes for 'scientific' and 'science': I doubt that we would learn much from schematisations of papers in physics or chemistry. As for the social sciences, distinct both from the humanities and from the natural sciences, allow me to leave the matter open for a while. I will only point out that the field of law, whichever way you look at it—a branch of the humanities or one of the social sciences (legal science)—is an area where the analysis of reasoning processes in a computer perspective presents striking analogies with our archaeological endeavours (Bourcier 1990; Gardin 1992b). Danièle Bourcier's contribution to this volume should bring out both the virtues and the limits of such parallels.

Let us concentrate for the moment on our own explorations, limited as they are: what have we learned from them? First of all, the feasibility of schematisations: any interpretive construct published in specialised journals, not to speak of the larger books, is amenable to reformulations of the above kind that summarise its content. 'Summarise' may not be quite the right word: schematisations are no doubt an extremely condensed form of written papers, but if properly conducted, they should not leave out any single piece of the argument, whether in the data base (P0) or in the set of operations that connect it to the author's findings (Pn). I have examined this matter elsewhere (1992a); suffice it to say that our tentative reformulations of NL arguments in a computer-like format may be one of the ways to alleviate the consequences of the 'information explosion' from which archaeology suffers, as do apparently most of the human sciences (for a forceful presentation of this issue in philosophy, see Kenny 1991: 6–8).

'A computer-like format': I could have left out the post-position. A few years after we had adopted the formal conventions just described, we came across similar proposals on the part of computer scientists in the artificial intelligence (AI). More precisely, our schematisations seemed readily convertible into the kind of format recommended for the design of knowledge bases in expert systems through the preceding decade (Lagrange and Renaud 1985). I need not expand upon the congruence of the two approaches, abundantly discussed elsewhere (Gardin *et al.* 1988: 35–58). It merely shows that the computational metaphor, in our logicist programme, prepared the ground for applications in AI that have their own epistemological merits, as we shall see further on. But let us first go back to the lessons drawn from the formal analysis of scholarly constructs, with or without computers.

No one will be surprised at the importance given to the matter of representation. Collecting the various bits and pieces of the data base scattered in a given paper is not an easy task. Displaying them as the semiotic base of the formal construct leads to further difficulties, as the exercise brings into focus the extreme variability of scholarly practice in this respect. Given a set of material or textual objects, there seems to be no such thing as a common language in the way they are represented in different papers. Needless to say, arguments are not lacking in defence of variability, the more worthy from a scientific viewpoint being the well-established relation between representation systems and the theories which are made to rest upon them. Unfortunately, this justification is self-defeating for two reasons. First, linguistic or semiotic variability persists even when the data base serves no theoretical purpose, but rather a documentary or informational one, as in data banks projects. Secondly, if indeed the semiotic looseness is a (methodo)logically sound reflection of theoretical variability, we are left with a no less pressing question: how is it that the same sets of material or textual objects may give rise to so many conflicting 'theories' in the broadest, etymological sense of the word (viewpoints, interpretations, exegeses, commentaries, etc.)?

The standard answer is the inexhaustibility of cultural (vs. natural) objects and the consequent idiosyncrasy of the human (vs. natural) sciences: do we need formal studies to rediscover such a well-established dichotomy? A merit of our schematisations is at least to give some substance to this rather vague concept of inexhaustibility. The looseness of representations is one of its manifestations; so is the laxity of the inference-making process, to which I now turn.

Reading an argument in natural language (NL) is one thing; looking into rewrite operations (Pi) → (Pj) is a different matter, even though the schematisation which they help to build up is nothing but a representation of the NL argument in a computational form. The elucidation of the mechanics of reasoning in this way inevitably raises foundational questions: what is the basis of any particular derivation $p \rightarrow q$? On which ground is one supposed to accept it? The matter seems critical when the play of schematisations brings out the well-known phenomenon of multi-interpretation: from the same premisses p, different authors—or the same author in different papers—derive different conclusions $q_1, q_2 \ldots q_n$. Which should one espouse? Which is one to believe? The following answer is again standard in our post-modern era: 'any one you like, any one which happens to fit your system of values', etc. But this is obviously an answer for the *reader* (*cf.* the celebrated Reader Response Theory). What about the author, i.e., the scholar, the scientist, the expert? What does he do when confronted with ambiguous propositions of that sort?

The truth is that most of the time he does nothing: there seems to be no obligation in the humanities, when stating a given relation $p \rightarrow q$, to even mention, much less to refute or reconcile, the whole set of different views, $q_1, q_2 \ldots q_n$ that have been put forward by past or present authors who happen to use the same premisses p in their respective arguments. Indeed, anything of the sort would doom our trade to extinction. I am not claiming that no historian should be permitted to argue about the causes of the fall of the Roman Empire unless he has read the million of pages devoted to the subject and presented the reasons why 'his' explanation is worth more than 'theirs' (a few dozens of such theories have recently been surveyed by a German historian: Demandt 1984). Yet, the other extreme position is no less embarrassing. Sticking to the computational view of cognition that I have adopted in this paper, I would merely raise one question: how shall we set about the business of building up knowledge bases in the human sciences if most, if not all, potential rules of reasoning turn out to be multivocal in the above sense? One need not be versed in expert systems to predict that the demand on computer time will soon reach unacceptable dimensions through the joint effects of a combinatorial explosion and the increasing complexity of the decision procedures intended to meet it. Moreover, the formal mechanisms that have been hitherto proposed in the guise of decision-making processes are but pale substitutes for the sophisticated reasonings that scholars tend to build up, using fresh empirical data, when confronted with similar issues.

One might of course object that the sole source of embarrassment, in the preceding argument, is my insistence on computers or computability: is it mandatory to build up knowledge bases in the humanities? By no means; only, it so happens that a growing number of scholars seem tempted to do so, for whatever reason; and I, for one, think that the kind of epistemological issues raised on such occasions do deserve attention, despite (or because) of the fact that they could have been raised prior to the advent of the computer. The worst or best example, depending on one's viewpoint, is that of literary studies. The all too famous exercise known as *explication de texte*, which begins at school-age and never ends, consists in producing 'creative' interpretations of renowned works of literature, and, by extension, many other texts as well, in order to gain the appreciation of teachers, colleagues or enlightened readers. 'Creative' means original, personal. Collecting the innumerable representations and inferences generated in the process into an open-ended knowledge base would thus be a Sisyphean task, the purpose of which seems rather unclear (except in socio-historical terms: Nardocchio 1992).

Admittedly, the *explication de texte* in history is different, constrained as it is by considerations of 'factual' knowledge that set a limit to the range of acceptable constructs. The same is true of archaeology. Yet, scholars in both disciplines do not shy away from drawing a parallel between studies of literature and studies of ancient texts and monuments. The common denominator is sought in various directions: the weight of socio-cultural factors in interpretation, or the methodological tributes to alleged sciences of meaning such as semiotics or hermeneutics, or again the part of narrativity in the exposition of argument, etc. Admittedly, the formal analysis of both categories of interpretive constructs—historical and literary—reveals interesting analogies. Firstly, most schematisations tend to take the same form, namely broad and shallow; (*a*) broad, because the data base (Po) occupies the major part of the picture, including as it does not only the description of the texts or objects under study, but also the 'comparative material'—as extensive as the scholar's erudition allows it to be—as well as the large set of 'established knowledge' or presuppositions called for at various stages of the argument; (*b*) shallow, because the number of inferential levels (Pi) that separate the data base (Po) from the terminal propositions (Pn) is more often than not surprisingly small. Conclusions are reached or hypotheses 'proved' through a series of rewrite operations that seldom exceed the magical number seven. No implications are drawn from this fact, psychological (in the manner of Miller 1956) or any other; it merely suggests that the length of a scholarly paper should not necessarily be taken as an indication of depth. Another lesson is that most rewrite operations (Pi) → (Pj) are bold jumps with which we climb the ladder of inference much faster than the slow pace of its unfolding in natural language might suggest. Further, the arrow sign in such formulas covers a wide range of possible relations between (Pi) and (Pj), which are usually left implicit, i.e., unspecified in the NL version of the argument: logical or empirical implication, componential definition, connotation, integration, etc. The logic of the argument thus consists essentially of 'associations', the vagueness of which is aptly conveyed by the uniqueness of the *if–then* relation both in our schematisations and in the rule bases that they serve to build up.

These few observations are enough to throw some light, however dim, on the nature of our reasoning processes in the humanities. Obviously, the kind of rationality upon which such processes seem to rest is not far removed from the rationality of argument in ordinary discourse. Yet, most of us would resist an unqualified assimilation of scholars and laymen on the sole ground of their shared reliance on natural logic or common sense. In other words, we still take it for granted that we form part of the scientific community, even though our ways of thought differ from those of our colleagues in the sciences of nature. The question therefore arises of defining those 'ways' in positive terms, instead of merely claiming that they are neither one thing nor another: neither science nor art, neither literature nor ordinary prose, neither rule-based nor totally free, etc. The remaining part of this article will address that question.

Alternative Rationalities

Jerome Bruner has given us the most thoroughgoing exploration of the two modes of thought which, in his view, exhaust the production of knowledge. The better known of the two is the 'logico–scientific' mode, followed by and large by the world community of researchers in the natural sciences; the other one, in Bruner's terminology, is the 'narrative' mode, about which he first confessed to know precious little (1986: 11–43). However, his indefatigable mind soon provided us with a wealth of data that gave more substance to the concept (1991); the interesting point is that they all originate in the analysis of works of Literature with a capital L, rather than from critical or academic studies thereof. The narrative mode thus seems to characterise the way of creative writers—poets, novelists, playwrights—rather than those of humanists or social scientists. But then, where do the latter stand in Bruner's dichotomised world of knowledge? The question was put to him in the discussion which followed his San Marino presentation (1991): the proffered answer was slyly non-commital.

An easy way to avoid the dilemma consists in substituting a three-place model to J. Bruner's binary opposition: the human sciences then take up the third position, while science and literature remain contrasted with each other as in C.P. Snow's famous 'Two Cultures' model (1959, 1964). A third position does not necessarily mean an intermediate position; yet, exponents of the 'third culture' tend to place it 'between science and literature' (Lepenies 1988), as if it had a foot on each side. This is the view explicitly taken by the distinguished German sociologist in the book which bears this title: his account of the development of the social sciences in the last 200 years is one of continual oscillations between the two poles, as the literary achievements of brilliant thinkers periodically tend to overshadow the attempts of others (of the same) to endow studies of culture with a scientific dignity of their own (Lepenies 1987).

The history of the human sciences in recent decades offers outstanding examples of this phenomenon. Not enough perhaps to convince everyone of Dr. Lepenies' account; but it does not need to be endorsed for the purpose of this paper. The interesting point from our perspective is the hardly questionable fact that we are again left with a 'neither . . . nor' picture of our modes of thought: neither the logico–scientific mode nor the narrative mode defined by J. Bruner would adequately describe the more typical products of the third culture discussed by W. Lepenies. Yet, no third mode stands out, and it would be both wrong and facile to fall back in this case on a 'both . . . and' formula.

Another intermediate status has recently been proposed with reference to two different poles. J.-C. Passeron, a French sociologist, characterises the reasoning of the human sciences as 'un raisonnement de l'entre-deux' (1991: 29)—an in-between process, under the sway of two kinds of logic: the logic of scientific thinking in general, which still has a place in our disciplines, and the logic of natural discourse, presented as the necessary mode of argument about human objects. This is not the place to discuss Passeron's thesis. I would only stress the difficulty we meet whenever the concept of natural logic is invoked. A sizeable number of books and articles have been published on the subject; they provide illuminating examples of the use of argument in particular languages (mostly English and French), as well as in particular contexts or fields of discourse. Clearly enough, the rules or principles that those examples are meant to illustrate are highly dependent upon the languages and contexts in question. It follows that they can hardly be taken as foundational elements of a logic of knowledge to come, in a specific discipline, freed as much as it can be from the lexical and grammatical frames of English, French, or any other natural language. No one would be bold enough to suggest that the natural logic of English speakers ignorant of organic chemistry should be part of the training of Chinese scientists working in that field. I fail to see the reasons why the matter should be put in different terms when the discipline under consideration is a 'human' one: Chinese archaeologists ignorant of French and its natural logic produce historical constructs that I am perfectly able to evaluate with reference to the worldwide logic of knowledge that is taking shape in that discipline, although I know nothing myself of the natural logic of Chinese.

Another difficulty in J.-C. Passeron's middle place theory is that the scientific component seems to be conceived essentially as a set of tools borrowed from the natural sciences, ranging from the more abstract principles of theory-building in empirical disciplines to the more special instruments of mathematics, data analysis and the like. Not much is said about the development of field-related logics of knowledge of the kind just mentioned, divorced as much as they can be from the natural logics of Chinese and French, among others. In the absence of such logics, I fear a proliferation of pseudo-scientific constructs: the combination of statistics and common sense alone is within reach of many, but it does not necessarily produce significant contributions to the growth of knowledge in specialised areas.

The emphasis on the part of *natural* logic and *common* sense in the make-up of our expert interpretations becomes more understandable if we make it clear that the two words mean more or less the reverse: 'natural' means cultural, and 'common' means restricted. The Chinese language and its underlying logic are wholly unnatural to me, while the common sense that rules the behaviour of my compatriots devoted to astrology is common only to them. The two notions are clearly related: an argument is perceived as natural only to the extent that it is submitted by and to members of a community who have been led (raised, convinced, trained, conditioned, etc.) to take it as such. Thus, speaking of expert interpretation, we cannot but bring to bear the familiar concept of interpretive communities popularised under this name or other by a host of 'radical' thinkers. The interesting question from our perspective, then becomes the following: what is it that 'leads' us to accept an argument as natural, i.e. supported by the common sense of a given group, when that group happens to be a community of social scientists or humanists? In other words, what is it that makes us believe—or behave as if we believed—that a given interpretation is acceptable in our respective disciplines, not to say true, valid or other excessive terms of the sort?

The issue has often been raised in recent decades, much in the same terms (Ziman 1978, Van Holthoon and Olson 1987). We all know what have been the favoured answers, backed by an impressive literature published under the auspices of well-established disciplines or schools—hermeneutics, semiotics, pragmatics, the Frankfurt school, etc. The common core is an acknowledgement of the part played by socio-cultural factors in shaping both the production and the reception of our theories about human matters. Relativism is the keyword; it varies in strength, an extreme form being the explicit rejection of the possibility of progress in our ways of making sense of anything human (for example, Rorty 1992). My own views on the subject are irrelevant to my present purpose: I only wish to indicate some of the dangers that arise from the extreme view, and to a lesser extent from the weaker forms of cultural relativism as well, if they are to become the unique basis upon which to ground the human sciences and our interest or belief in them. The first peril is, paradoxically enough a growing rate of *dis*interest and *dis*belief, both within the academic community and outside, as a result of this very insistence on the social basis of our interpretive games and their local or ephemeral success. As one of my students recently put it: 'Many of us feel tempted to be more radical than the learned men who teach us relative points of view, simply by deciding to ignore them both, the views and the men.'

A related risk lies in a growing impatience on the part of our colleagues or employers in the broadest sense of the word, as they become aware of the *volte-face* or *double jeu* that necessarily accompanies a relativist's academic career. Jonathan Culler, whom no one will suspect of defending the pre-post-modern establishment, describes the phenomenon in sharp terms:

> What I have always found particularly disquieting . . . is that people who attained their positions of professional eminence by

> engaging in spirited debate with other members of the academic field . . . by identifying the difficulties and inconsistencies of their elders' conceptions of the field and by proposing alternate procedures and goals, have, once they attain professional eminence, suddenly turned and rejected the ideas of a system of procedures and body of knowledge where argument is possible and presented the field as simply a group of people reading books and trying to say interesting things about them. They thus seek systematically to destroy the structure through which they attained their positions and which would enable others to challenge them in their turn. (1992: 118)

Going all out in this direction caries a further institutional menace, namely the eventual dilution of any difference between professionals and laymen in matters of interpretation. The case is somewhat self-evident and has been amply susbstantiated elsewhere (Gardin 1990a). I would only add that it is a direct consequence of the fact that while our pragmatists are rapturously identifying the difficulties and inconsistencies of their elders' conceptions, they fail to offer any positive version of the alternative rationality according to which their own conceptions raise less difficulties and present fewer inconsistencies than the former. Once more the third way is left in the mist. We have first been told that it is neither the mode of science nor the mode of literature; then, that it is socio-culturally determined and related as such to the mode of reasoning known as common sense or natural logic; finally, that it nevertheless differs from the use of argument by laymen.
Fine; but what is it?

From the Computational Paradigm Back to the Two Cultures

Bold as it may appear, I think that our explorations in formalisation and computability provide us with elements of a possible answer. I shall subsume them under two headings borrowed from J. Bruner (see earlier section).

The Logico–Scientific Component

When reviewing the lessons learned from schematisations of scholarly constructs, I stressed the features that seemed to set a limit to more ambitious computer projects, in an Artificial Intelligence perspective. One should not therefore conclude that there is no room left for progress in our use of argument. In fact, those very lessons do suggest a number of steps that can be taken to bring our constructs closer to the 'logico–scientific' paradigm, one of the two modes of thought identified by J. Bruner. For example: (*a*) the refinement of *ad hoc* (meta) languages needed for the development of data networks, covering the two categories of 'facts' mentioned in the second section (empirical observations and established knowledge); (*b*) collecting sets of inferences used in the literature bearing on any given topic, with a view to unravelling the conceptual basis of competing theories on the subject (see for instance the knowledge bases compiled by H.P. Francfort 1987 on the process of state formation according to different authors; or by W. Stoczkowski 1992 on the various theories proposed throughout the ages to account for the emergence of man, etc.); (*c*) in the all too common cases described above as multi-interpretation ($p \rightarrow q_1, q_2 \ldots q_n$), a search for additional criteria C that, combined with the premisses p, might support a few unambiguous rules $p.\ C_i \rightarrow q_i$ (on the many possible readings of C—cultural context, personal or shared convictions, local versions of common sense, etc., see Gardin 1990b: 21–22), etc.

Clearly enough, all such steps proceed from the assumption that human phenomena can be studied in the same way as natural processes or the behaviour of matter. I have indeed often made that assumption explicit in connection with our logicist explorations: the latter seemed in many ways to imply the former. The consequence was a steady flow of more or less vehement denunciations of our anachronistic programme, supposedly dedicated to the worn-out ideas of positivism, reductionism, mechanicism and the like. Such criticisms are naturally louder on the part of those who have rallied to one or other faction of the post-modern movement discussed in the preceding section. They are off the mark in two respects. First, the logico-scientific paradigm is only one of the two modes which I regard as necessary in order to produce interesting accounts of the human phenomena which we happen to study. I have made that point as forcefully as I could in early presentations of the logicist programme, going as far as to regard historical fictions as a necessary complement of scholarly constructs (1980:

178–180)—which seemed quite unintelligible to ironical reviewers. Similar views have since been put forward by several scholars, which may help to defuse the issue (see next section).

Secondly, the rejection of the logico–scientific paradigm in the humanities puts one under an obligation to propose an alternative mode of assessing the strength of scholarly constructs, other than by merely referring to the benevolent laissez-faire dispositions of selective sets of peers. I again find some comfort in the support of renowned scholars who similarly argue for a reinstatement of rational argument (Gossman 1989) and disprovable interpretation (Eco 1992: 78) in historical and literary studies, after having defended the part of truth contained in contrary positions.

The Narrative Component

Be that as it may, there are limits to what we can achieve by clinging strictly to the *rules* of the logico–scientific game. Returning to my three examples above—*a* to *c*—I readily admit that we have as yet very few examples of formal languages on the line of *a*, and only tentative embryos of field-related logics on the line of *b*. As for the reduction of ambiguities in the spirit of *c*, it can only be envisaged in areas of human sciences where decidability is an accepted goal, not merely as a matter of principle, but in actual research strategies. Examples of this sort can be found in archaeology, but may again form an exception rather than the rule: the premium put on overinterpretation is hardly an incentive to bother about the accumulation of undecidable theories.

I would in fact go much further. Once our interpretive or theoretical constructs have been reformulated as inference trees, the next desirable move is to look for alternative representations (Po) or derivations (Pi → Pj) that contradict them in some way. We should then regard schematisations as decision trees, whose sole purpose is to assess the limits of our claim to scientificity in the reputedly 'narrow' perspective hitherto adopted. One need not pursue the exercise overlong to discover that most of the time those limits are rather low. In other words, the logico–scientific paradigm does not take us very far up the ladder of inference in our quest for Pn's that are both well-formed and strongly supported by empirical evidence. Does it follow that we should refrain from ever climbing higher? By no means: our comments on literary works or ancient texts and monuments would be rather uninteresting if they were to stop at the levels of decidability just suggested; and I suspect that the same would apply to other objects of study in the social sciences. The irritating question, however, concerns the way in which we should then take up our stories. I use that last word deliberately to signify my adhesion to J. Bruner's proposal: the narrative mode is the only possible answer once we accept his vision of the two-and-two-only modes of thought in the realm of knowledge—but with important qualifications.

One must keep in mind that the narrative mode in J. Bruner's presentations is essentially the mode of Literature: not just any kind of storytelling, but clearly *les belles lettres*, however hesitatingly distinguished from ordinary prose—for example, 'light, entertaining, usually sophisticated literature that has no practical or informative function', according to the Longman dictionary. I do not defend that particular definition, but it does convey the kind of attributes that one should be ready to include, among others, in the characterisation of narrative discourse. Now this is a far more demanding agenda than the looser versions of narrativity offered by philosophers who struggle with the same epistemological issues (e.g. Ankersmit 1983, Danto 1985, Ricoeur 1984–88, etc.). *Narratio*, to them, is a way of presenting an argument that may have none of the qualities currently associated with Literature in the above sense. Indeed, speaking of my own scholarly prose, I can testify that as close to narrativity as it purports to be in its most undecidable parts, it certainly is neither light nor entertaining—nor even, I confess, very sophisticated. This is precisely where my position differs from those reviewed earlier in this article, where the only demand made on our professional discourse was that it should conform to natural logic, common sense, or the sociocultural interests of the day. I do not deny that such conformities are inevitable, indeed even desirable, when trying to make sense of human matters. Rather, I stand on the side of those who believe that stronger requirements are called for in order to vindicate our claim to a special cognitive status. One of them is the logico-scientific mode of theory-building, as far as we can take it; another one, for those who are able to carry the exercise further, should be the kind of craftsmanship associated with the production of literary works, viz. the narrative mode as J. Bruner defines it (for similar stands in favour of a return to the tools of literature in historical and anthropological works, see, respectively, Schama 1990, Geertz 1988).

We are thus led back to the 'Science and Literature' duality, but with an important difference: the two cultures are no longer opposed to each other, but rather brought into accord as complementary modes of knowledge. The more interesting accounts of human phenomena are found to contain the two components: a strong scientific core of refutable propositions relayed by imaginative scenarios unafraid of their name. As for a third culture, it exists only by virtue of the quantity of papers that seem to fail by both standards (mine included) in the literature of human sciences. This is enough to raise doubts about the future of the concept, not to speak of the papers.

Postscript

A number of points were raised in the discussion which followed the oral presentation of an abridged version of this paper. First, the appropriateness of the title was questioned. 'Where is the computational paradigm in all this, where are the effects on the human sciences?' In answer to the first point, I can only refer to the second section of this paper: the rewrite formulae used in schematisations of scholarly constructs and their possible conversion into computerised knowledge bases are enough to suggest that for better or worse, we are operating within a computational framework. As for the consequences on the human sciences, I may well be guilty of overestimating them. It is a *fact* that there are as yet few traces of the effects of computers or computability on scholarship. Against which I can only oppose the *belief*, shared by a minority of scholars, that the epistemological issues raised by schematisations and artificial intelligence will eventually exert an influence 'on the way we think and write the humanities' (Gardin 1991). The grounds for that belief are discussed in the preceding two sections.

Technical remarks were made on some logico–linguistic obscurities in my paper which I readily acknowledge:

(*a*) First, what does the arrow indicate in the (Pi) → (Pi ± 1) formulae of our schematisations or the inference rules $p \rightarrow q$ that express them in computable terms? A crude answer is that I really don't know . . . or rather, that we may provisionally shun the question as long as we are merely recording discursive processes as we find them in scholarly papers, my own included: having observed or established a set of phenomena expressed by (Pi), I feel authorised to derive another set expressed by (Pi + 1), without saying much or anything at all on the nature of the logical relation between the two sets. This is the standard way of argument in natural language (NL), the so-called *raisonnement naturel* which, as I recalled above, a distinguished sociologist describes as the most commendatory way of reasoning in the human sciences (Passeron 1991, see section three). Unfortunately, despite the sizeable body of literature available on natural logic, we are still unable to attach proper labels to the arrows of our primitive rewrite formulae. More exactly, the kind of relational distinctions proposed on the basis of studies of *ordinary* texts in *one* particular NL do not seem to be of much use when dealing with *specialised* texts written in *several* NLs. Our concern, then, is with the field-related logic that underlies archaeological reconstructions across different languages: the few research projects aimed in that direction have little in common with studies of natural logic in French, English or any other language (ex.: Lagrange 1992).

(*b*) Yet, the question seemed to remain, although put in different terms: a rewrite formula is after all a linguistic utterance and should be amenable as such to the kind of interpretive processes devised by experts in the philosophy or logic of language. Seen in this light, the prospects of inference rules would seem rather bleak: those processes heavily rely on the context of discourse, both explicit and implicit, and we have reasons to believe that there are as many relevant bits of information to be taken into account in scholarly communication as in NL discourse. My answer to this again consists in recalling: (*i*) that our games are played in heavily restricted fields of discourse; (*ii*) in which the basic principles of scientific reasoning set a limit to our freedom as scholars when mobilising contextual information in the above broad sense; (*iii*) and that our goal is only to clarify what those limits are, not to deny anyone the right to transgress them while still claiming to be an authority in his field on other grounds (see previous section).

A third set of objections concerned my views on the Two vs. Three cultures issue; they were questioned from two sides. First came the customary reservations against 'the assumption that human phenomena can be studied in the same way as natural processes or the behaviour of matter' (see previous section). They were formulated in a way that seemed to imply usual characterisation of our specific mode of reasoning in negative terms alone: neither X nor Y. My only answer to this is that I am more interested in attempts to define that mode in positive terms: Z, in lieu of X or Y.

My own suggestion on this line is something like this: 'first X (science), then Y (literature), but no third way Z'. This reinstatement of literature in our research fields raised further questions. One of the participants expressed his 'shock' at such a proposal if it meant more than a reference to the part sometimes played by elegance in the assessment of competing scientific theories. Should we go beyond this and accept fiction, for instance, as a necessary component of scholarly theories in the human world? My personal answer is, I confirm, YES but with some qualifications that may help to attenuate if not obliterate the shock.

(*a*) Instead of a component of scientific theories, I would call it a complement: my stand, as I have stated in the preceding section, is that we are unable to climb the ladder of inference as high as our readers and possibly ourselves would like to, when we choose to abide by the rules of logico–empirical rationality alone in our reconstructions of the past. *If* we are to retain our specificity as 'experts' beyond the levels reached on such terms, I can think of no other set of requirements than those of literature, however ill-known—i.e. neither a third way, undefined, nor a fall-back upon natural reasoning or common sense, but a move forward to the second genre or mode of thought, no less demanding than the former.

(*b*) A distinction should probably be made at this point between the social sciences and the humanities. My only direct experience is with the latter and it may be that the limitations brought out by formal studies of reasoning processes in historical disciplines are not so severe in economics or sociology.

(*c*) Finally, attention should be paid to the italicised *if* above. We may well be content with founding our legitimacy on scientific constructs alone, in the narrow, ill-famed 'positivist' sense. In fact, there seems to exist a core of historical and archaeological knowledge, updated from generation to generation, reshaped or 'repackaged' according to changing cultural perspectives in time and space, but which nevertheless constitutes the unescapable 'facts' that we have to take into account in our successive constructs. Contributing to the growth of that core used to be regarded as a sufficient *raison d'être* in scholarship. Communicating it through literary scenarios is a further move, rightly praised by several scholars today. But this does not abolish the worth of the former in its own right.

Lastly, there is the 'general comment' offered in writing by Dr. S.C. Malik, a discussant of my paper. It is addressed not so much to the matter-of-fact questions raised in the paper as to a broader and more lofty pronouncement against the Western view of scientific rationality and its limitation for a proper understanding of the world, both material and human. The length and depth of Dr. Malik's reflections deserve more than a brief response that is possible within the limits of this postscript. Besides, I am not sure that I would succeed in bringing my comments on his comments back to the argument of my paper. I shall therefore refrain from the temptation to resume here the discussions that I was happy to have with Dr. Malik in recent years on his views of rationality, in the human sciences and elsewhere. The gist of my response to them is given in a review of his last book, just published (Gardin 1993).

References

ANKERSMIT, F.R. 1983. *Narrative Logic: A Semantic Analysis of the Historian's Language*. The Hague: Martin Nijhof.

BOURCIER, D. 1990. Interpretation in Legal Expert Systems: From Private Conviction to the Formalization of Rules. In R. Ennals and J.-C. Gardin (eds.), *Interpretation in the Humanities: Perspectives from Artificial Intelligence*, pp. 215–28.

BRUNER, J. 1986. *Actual Minds, Possible Worlds*. Cambridge, Mass.: Harvard University Press.

———. 1990. *Acts of Meaning*. Cambridge, Mass.: Harvard University Press.

———. 1991. Narrative Thought and Discourse. A Workshop of the Centro Internazionale di Studi Semiotici e Cognitivi, San Marino, 26–27 June 1991 (unpublished).

CULLER, J. 1992. In Defence of Overinterpretation. In U. Eco, *Interpretation and Overinterpretation*, pp. 109–23. Cambridge, Mass.: Cambridge University Press.

DANTO, A. 1985. *Narration and Knowledge*. New York: Columbia University Press.

DEMANDT, A. 1984. *Der Fall Roms*. München: C.H. Beck.

ECO, U. (with R. RORTY, J. CULLER and C. BROOKE-ROSE). 1992. *Interpretation and Overinterpretation*. Cambridge, Mass.: Cambridge University Press.

ENNALS, R. and GARDIN, J.-C., (eds.) 1990. *Interpretation in the Humanities: Perspectives from Artificial Intelligence*. London: The British Library.

FRANCFORT, H.-P. 1987. Un Système Expert Pour l'Analyse Archéologique des Sociétés Proto-Urbaines: le Cas de Shortughaî. *Informatique et Sciences Humaines*, vol. 7, pp. 71–91.

GARDIN, J.-C. 1980. *Archaeological Constructs: An Aspect of Archaeological Theory*. Cambridge: Cambridge University Press.

———. 1990a. Interpretation in the Humanities: Some Thoughts on the Third Way. In Ennals and Gardin, op. cit. pp. 22–59.

———. 1990b. The Structure of Archaeological Theories. In A. Voorrips (ed.), *Mathematics and Information Science in Archaeology. A Flexible Framework*, pp. 7–28. Bonn: Holos.

———. 1991. On the Way We Think and Write in the Humanities. In S. Hockey and N. Ide (eds.), *Research in Humanities Computing*, pp. 337–45. Oxford: Clarendon Press.

———. 1992a. La Schématisation Logiciste des Textes Interprétatifs en Sciences Humaines Est-Elle un Résumé ou l'Expression Épurée de Leur Contenu? In M. Charolles et A. Petitjean (éds.), *Le Résumé de Texte: Aspects Linguistiques, Sémiotiques, Didactiques et Automatiques*, pp. 75–89, Paris: Klincksieck.

———. 1992b. Approches Informatiques et Cognitives du Texte en Archéologie. In D. Bourcier et P. Mackay (eds.), *Lire le Droit: Langue, Texte, Cognition*, pp. 239–55. Paris: L.E.D.J.

———. 1993. De l'Anthropologie à la Connaissance: une Vision Indienne de nos Destinées. *L'Homme* 125, XXXIII (1), pp. 149–57.

GARDIN, J.-C., GUILLAUME, O., and HESNARD, A., HERMAN, P., LAGRANGE M.-S., RENAUD, M., and ZADORA-RIO, E. 1988. *Artificial Intelligence and Expert Systems: Applications in the Knowledge Domain of Archaeology*. Chichester: Ellis Horwood.

GEERTZ, C. 1988. *Works and Lives: The Anthropologist as Author*. Cambridge, U.K.: Polity Press.

GOSSMAN, L. 1989. Towards a Rational Historiography. *Transactions of the American Philosophical Society*, vol. 79, no. 3, pp. 1–68.

KENNY, A. 1991. Technology and Humanities Research. In M. Katzen (ed.), *Scholarship and Technology in the Humanities*, pp. 1–10. London: Bowker Saur.

LAGRANGE, M.-S. 1992. Symbolic Data and Numerical Processing: A Case Study in Art History by Means of Automated Learning Techniques. In J.-C. Gardin and C.S. Peebles, *Representations in Archaeology*, pp. 330–56. Bloomington: Indiana University Press.

LAGRANGE, M.-S. and M. RENAUD. 1985. Intelligent Knowledge-based Systems in Archaeology: A Computerized Simulation of Reasoning by Means of an Expert System. *Computers and the Humanities*, vol. 19, pp. 37–52.

LEPENIES, W. 1988. *Between Literature and Science: The Rise of Sociology*. Editions de la Maison des Sciences de l'Homme, Paris and Cambridge University Press, Cambridge, Paris. Translated from *Die Drei Kulturen*, Hanser Verlag, München, 1985.

———. 1987. Sur la guerre des sciences et des belles-lettres à partir du 18e siècle. MSH *Informations*, no. 54, pp. 8–17.

MILLER, G.A. 1956. The Magical Number Seven, Plus or Minus Two: Some Limits On Our Capacity for Processing Information. *Psychological Review*, vol. 63, pp. 81–97.

NARDOCCHIO, E. (ed.) 1992. *Testing Reader Response: Practical Semiotics*. Bloomington: Indiana University Press.

PASSERON, J.-C. 1991. *Le Raisonnement Sociologique: l'Espace Non-Poppérien du Raisonnement Naturel*. Paris: Nathan.

RICOEUR, P. 1984–88. *Time and Narrative* (eds.: Kathleen McLaughlin and David Pellauer). Chicago: University of Chicago Press.

RORTY, R. 1992. The Pragmatist's Progress. In U. Eco, *op. cit*. pp. 81–123.

SCHAMA, S. 1990. *Dead Certainties (Unwarranted Speculations)*. New York: Knopf.

SNOW, C.P. 1959. *The Two Cultures and the Scientific Revolution*. Cambridge: Cambridge University Press.

———. 1964. *The Two Cultures and a Second Look*. Cambridge: Cambridge University Press.

STOCZKOWSKI, W. 1992. Origines de l'Homme: Quand la Science Répète le Mythe. *La Recherche*, no. 244, pp. 746–50.

VAN HOLTHOON, F. and D.R. OLSON. 1987. *Common Sense: The Foundations for Social Science*. Lanham: University Press of America.

ZIMAN, J. 1978. *Reliable Knowledge: An Exploration of the Grounds for Belief in Science*. Cambridge: Cambridge University Press.

Annexure

Formal Analysis of Argument in Archaeology: Examples in English

1. GARDIN, J.-C. 1980. *Archaeological Constructs: An Aspect of Archaeological Theory*, pp. 108–20. Cambridge: Cambridge University Press and Editions de la Maison des Sciences de l'Homme, Paris. (Interpretation of a relief in the Konya Museum, Seljuq period, Anatolia, ca. 12th century A.D.).
2. GARDIN, J.-C. *et al.* 1988. *Artificial Intelligence and Expert Systems: Case Studies in the Knowledge Domain of Archaeology*. Chichester: Ellis Horwood. Contains six case studies: (*i*) The Relation between Greece and Central Asia in the Hellenistic Period according to Ceramic Data, by J.-C. Gardin, pp. 61–87. (*ii*) The Bactrian Metal-Working Industry in the Iron Age, by O. Guillaume, pp. 88–105. (*iii*) Let Ancestors of the Cypriot Figurines Raise their Arms, by P.Q. Herman, pp. 106–25. (*iv*) Establishing the Origin of Roman Amphorae, by A. Hesnard, pp. 126–49. (*v*) The Identification of Medieval Constructions, by E. Zadora-Rio, pp. 150–67. (*vi*) SUPERIKON: An Attempt to Combine Six Expertises in Archaeology, by M.-S. Lagrange and M. Renaud, pp. 168–202.
3. GUILLAUME, O. 1990. *Analysis of Reasoning in Archaeology: The Case of Graeco-Bactrian and Indo-Greek Numismatics*. Delhi: Oxford University Press.
4. FRANCFORT, H.-P. 1992. The sense of measure in archaeology: An approach to the analysis of proto-urban societies with the aid of an expert system. In J.-C. Gardin and C.S. Peebles (eds.), *Representations in Archaeology*, pp. 291–314. Bloomington: Indiana University Press.

Emergent Rationality through Jurisprudence

Laurent Bochereau
Danièle Bourcier
Paul Bourgine

Introduction

Although rationalism was for a very long time considered the preserve of philosophers, with the advent of Artificial Intelligence computer experts also began to look more closely at the reasoning process in order to analyse, represent and hence simulate it. Rationalism plays a central role in law, just as it does in mathematics—legal decision essentially depends on the reasoning process by which a decision is reached.

Continental law is based on logic, since the judge evokes the grounds of reasoning when he renders his decision. Through the separation of powers the judge becomes a 'mouthpiece of the law' (Montesquieu). In France, it can be said that whilst the Constitution sets the legal framework, the domain of validity for each rule is created through case law. In reality, the reasoning of a legal expert, whether he be administrator or judge, extends much further than mere syllogism. However, as legislation cannot account for all eventualities and the validity domain is in perpetual evolution, the system may at times appear inconsistent. The qualification of facts is, in itself, a form of reasoning which calls upon categorisation, generalisation and analogy.

The historical framework of our legal system, then, is a clear-cut, axiomatic, hierarchical organisation of rules and power; the traditional model for representing law and reasoning being a logical one based on definitions, concepts and syllogism, rationality was considered as nothing more than a postulate. On the contrary, cognitive rationality implies an emergent approach consisting of knowledge-based models and connexionist models. Such models illustrate that certain cognitive operations can be simulated and the results checked or refuted by computers, which even discover new legal rules through the modelisation.

This article describes how legal rules can be extracted from a connexionist model constructed on a learning base of case law. The multilayer perceptron or multilayer neural network is a model consisting of large numbers of interconnected computing elements and characterised by their learning capacity, since they can be trained on a set of examples. Other authors (Rumelhart et al. 1986; Gallant 1988) have worked on the extraction of explicit legal knowledge from connexionist models.

Our first objective is to test efficient methods for extracting expertise. First, it is necessary to find solutions for discerning legal decision rules that could be directly used in the knowledge base of an expert system. We propose to analyse how rules can be extracted from this multilayer network. The contribution of this model in the extraction, representation and validation of legal knowledge will be compared with other methods for modelling legal reasoning.

The second step is to develop procedures in which logical and holistic approaches can be conciliated. It is generally believed that the neural network approach is a faithful model of certain human processes which cannot be easily represented in rule-based systems. But the problem is that the interpretation of these processes is not easy and the network approach is not well-adapted for making knowledge explicit in the form of rules.

We will then focus on the classification performance of the network rather than on the problem of interpretation of hidden layers. Indeed, the network will be used as a classifier, the target being legal rules. The network can be constructed to produce some of its knowledge in the form of explicit rules. The legal application domain is that of the decisions handed down by the Council of State, the only administrative court in France competent to receive appeals and take the necessary measures against decisions made by local authorities. The explicit rules extracted from the network are compared with the rules that would be used by legal experts to validate or invalidate the municipal regulations.

The Principle of Extracting Legal Decisional Rules from a Multilayer Neural Network

Characteristics of Legal Rules

Knowledge sources are not homogenous in the legal field:

- statute law is written in the form of rules and concepts;
- jurisprudence is a set of cases, consisting of features characteristic of the legal problem, these facts are 'subsumed' under a rule or concept (civil law) or classed in terms of precedent (case law);
- legal doctrine is a set of heuristic and interpretation rules which enable concepts to be determined or legal precedent to be explained.

Within each legal domain, decisional rules vary considerably. A rule can consist of a set of determined concepts that may be defined more or less precisely, or even quantified (number of decibels, tonnage of heavy goods vehicles etc.). In the field of municipal law, the task of defining these concepts may also be delegated by statute law to another authority: mayors or judges for example.

Take the example of traffic regulation: Article L 131–4 of the Municipal Code stipulates: 'According to traffic needs, the mayor may, during specified times, prohibit access to certain roads or road sections, or limit access to certain categories of drivers or vehicles.'

The term 'traffic needs' just like 'public order', 'emergency' or 'decency', is a standard, that is to say, an open-structured concept intended to regulate. Standards, which have the particularity of always remaining undefined, are numerous in the domain of municipal policy. Other research in this field (Bourcier 1993) has shown that a standard is not substantive: it is an heuristic operator used for building dynamically a network between concepts, circumstances, arguments and actions. This specificity of legal knowledge has been stressed by most scholars concerned with logical representation of concepts. As Hafner (1987) said, 'It is clear that the conceptual connections represented in legal decision rules are an important aspect of a legal researcher's knowledge.'

Therefore, in our domain, knowledge cannot be defined in terms of production rules by simply reading statutes. Neither judges nor legal doctrine can isolate specific factors from legal reasoning in the form of explicit rules. Thus, when it comes to standards set out in the code, only the systematic analysis of case law (and, if available, commentary) will enable us to determine which decisional context justify the mayor's regulating heavy goods vehicles on the road, or the sale of lily-of-the-valley on the public byway on the first of May.

Each case is a model of reasoning. In order to describe cases we can use summaries in natural language. In NEUROLEX, we select key elements such as the legal domain, the invoked legal and factual standards, the controversial mayoral decision and the judge's solution. These elements will be the variable of our network.

Knowledge Extraction in the Mairilog Project

MAIRILOG is a project designed to create software to aid mayors in making decisions in the domain of public order within their own municipality. It consists of legal knowledge bases, data files, statistic files, text editors, banks of legal texts and dictionaries. Originally written in PROLOG, a version in TURBO PASCAL is commercially available. However, researchers continue their work on the PROLOG version (Bourcier 1989).

The knowledge base was constituted through traditional analysis methods, with computers used to help determine the structure and basis of normative texts. The 'generator' terms in the domain could thus be pinpointed: list of sources of noise, offences, penalties, criteria for determining importance of offence etc. as laid down by the law. But it was soon decided that these text-engineering methods were inadequate to extract the deeper sense of the text and discover the legal decision rules, particularly those concerning standards invoked (Bourcier 1992). We studied a corpus of Council of State decrees, using the factor analysis method. This method produced significant results, notably in identifying specific moments in time or making correlations between facts and standards. From this analysis, we were able to create both a normative knowledge base and an argumentative knowledge base destined to provide suitable legal argumentation for municipal bye-laws.

Legal Applications of Neuro–mimetic Networks

Unfortunately, the factor analysis methods can only drive linear models. Non-linear models such as multilayer neural networks have been used in the legal field over the past few years and appear appropriate for structuring this type of knowledge, as they enable the capabilities of the factor analysis methods to be considerably extended (Feldman and Ballard 1982).

On the one hand, theoretical work on connexionism in law has mainly involved parallel exploration in the different branches of reasoning, where conflictual or undetermined rules are to be resolved. On the other hand, all the legal applications have dealt with natural language processing. There are two main applications of neuro-mimetic networks which have been developed. The first one is the AIR system that uses multilayer neural networks for extracting conceptual legal information (conceptual retrieval systems) from free text (Belew 1987). By substituting a connexionist representation in the form of a weighted index graph, the manual indexation phase can be avoided. The second system based on parallel processing was designed for the public prosecutor, allowing him to compare and check texts from briefings against 3,500 significant words taken from the Dutch Criminal Code (Fernhout 1989).

No matter which approach is employed, the common difficulty is to satisfy the constraints of the real world and the natural language (incertitude, ambiguity, difficulty to recognise either the rules applied, or the decision) with models that take these elements into account, without modifying a structure or imposing an intermediary model.

Case-based Reasoning versus Rule-based Models

So far, AI research in the field of law has dealt with two kinds of knowledge: rule-based knowledge and case-based knowledge. A significant example of a rule-based model is the PROLOG representation of the British Nationality Act (Sergot et al. 1986).

This representation is well adapted to structured and defined domains which can be expressed in terms of necessary and sufficient conditions. But it is not efficient for fields which contain many open-structured concepts that have to be further specified according to the circumstances of the case. Moreover, some scholars have observed that a set of rules without a deep structure cannot represent legal expertise (McCarty 1990) or simulate the processes of 'thinking like a lawyer'. On the other hand, a case-based approach has been used for solving mediation problems. Generally, cases were indexed by features. Hammond (1986) used an indexing scheme for his case-based problem solver. Other systems were based on a causal model or a claim lattice (Rissland et al. 1987).

Unlike these models, neural networks do not require an 'a priori' model. Neurolex aims at generating its own heuristic techniques to test the sensitivity of the network to various changes in the features and incremental changes between cases are made automatically. But like them, Neurolex handles case law and tries to simulate the decision-making process.

Building Neurolex

Our goal was to use connexionist models to define the common factors in decision-making situations. This meant that we had to establish the reasoning taken into consideration by the judge (or the Council of State) appointed to check the municipal decision.

Structure of the Network

The Council of State, one of the two major French jurisdictional bodies, has within its competence the investigation of bye-laws. We examined a corpus of municipal jurisprudence consisting of 378 judgments of the Council of State, validating or invalidating bye-laws. Thus, the output layer of the network consists of two units: annulation or confirmation of the initial bye-law. The input vector is composed of variables that are distributed in four subsets: regulations, bye-laws, factual standards and normative standards.

A factual standard concerns facts such as 'emergency' or 'degree of disturbance of the noise'; it includes an appreciation of circumstances. A normative standard is one that consists of inducing general rules from cases, for example public order or decency. The four subsets include respectively 10, 11, 13 and 15 variables. Therefore there are 49 boolean input variables.

	Input Variables *(examples)*
Regulation	**(CIRC) Traffic**
	Hygiene*
	Public performances*
Bye-laws	**(INTE) Ban**
	(REGL) Regulation
	(REAUT) Authorisation
	(INJO) Injunction
	(CARE) Inaction
	(FERM) Closure
	(AUCO) Conditional authorisation
	(SUDE) Suppression of derogation
	(REIN) Refusal of interfering
	(RECO) Conditional refusal
Factual Standards	**Noise offence***
	Restricted traffic*
Normative Standards	**(ORPU) Public order**
	(BONO) Law and order
	(TRAQ) Public tranquility
	(SEPU) Public safety
	(TRAD) Tradition
	(EGAL) Discrimination
	(MORA) Public decency

*** Variables not-used in our examples.**

The 378 suitable bye-laws (those which involved at least one standard) were separated from the corpus as follows:

- 331 bye-laws constituting the learning base;
- 47 bye-laws constituting the validation base, enabling us to check that the network had correctly 'learnt' from the learning base.

For each example, an output vector was calculated by using the input vector and the weight distribution within the network. The actual output vector was then compared to the calculated result and the error rate used to modify the distribution of all the connexions according to various rules: we used those from the back propagation method developed in the 1980s (Fogelman-Soulie et al. 1987; Lecun 1985). The resulting network thus contained 49 input neurons and two output neurons. Because of the number of examples available, we limited the number of neurons on the hidden layer to four.

A Network Learning Base

The success rate obtained when presenting the network with new cases was around 80 per cent. In our analysis, we observed and calculated the differences of the weights which interconnected the neurons. The method was thus able to extract certain regularities in the data in the form of decision rules.

Knowledge Elicitation

We will first recall the equivalence principle stating that connexionist classifiers are functionally equivalent to a set of logical rules. The most efficient methods for extracting the logical rules are implicit enumerations using constraints propagation methods. However, the extraction problem is NP complete and the combinatorial number of input vectors can make these methods too time-consuming. The introduction of a validity domain allows a reduction of the combinatorial problem. In our application, the reduction applied is strong enough to allow us to use explicit enumerative methods for extracting logical rules.

Equivalence Principle

When considering feedforward-type multilayer networks with complete connections between adjoining layers and a given set of examples $< X^h, Y^h >$ where X^h takes its values in the hypercube B^m and Y^h in the hypercube B^n, the network calculates a mapping g from B^m to I^n, by using the back propagation algorithm (see Figure 1).

If we apply a decision function d to the output vector I^n, we are mapping the output vector into B^n. The decision function is usually chosen to be the maximum rule (i.e. keeping the best), but it can also be another function such as keeping the p best. The decision function has essentially to take into account second degree risks, as is common in decision theory. It is possible to give a deeper meaning to Figure 1: the neural networks makes an induction;

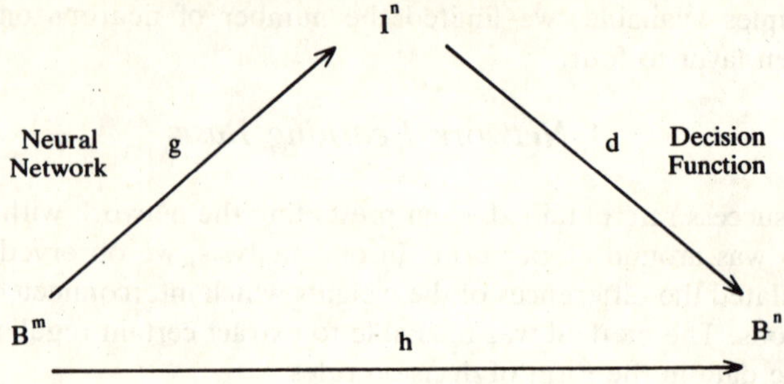

Figure 1: Equivalence diagram (where B is <0,1> and I is (0,1)

Note: In our application, m = 49 and n = 2. Therefore, h is a boolean function with 49 boolean variables.

however, after taking into account second degree risks, the resulting process is an abduction. The composition function h maps B^m into B^n. We can then show that each output variable h is equivalent to a logical formula constructed from the input booleans.

Construction of Validity Domain for NEUROLEX

The validity domain is related to the idea that there exists a neighbouring area of the training base on which the generalisation works well. The construction of such a validity domain can be effected by various methods and depends on both the problem considered and the probability of occurrence for each example in the training base (Bochereau and Bourgine 1990). The validity domain can be determined from *a priori* knowledge, either related to the data coding or formulated by an expert, or to the statistical regularities observed from the training base. We can then derive constraints on the input hypercube B^m of the neural network. This enables the initial cardinality (2^m) of the domain to be reduced.

These are various methods for building the validity domain of the neural network. Here, we have focused on describing the validity domain as a set of logical mathematical constraints. It is therefore possible to include the constraints specifying the validity domain in the constraints propagation methods that can be used for extracting equivalent logical rules from the neural network. The extracted rules will be relevant if the validity domain has been well defined and if the performances of generalisation for the network are satisfying. A detailed explanation of how these constraints can be derived will not be given here because there are various possible procedures; but we will return to this question later on, when we examine the application to the corpus of jurisprudence.

In our application, the overall validity domain is equal to the cartesian product of the validity domains corresponding to the four subsets. For each of the first three subsets, one and only one input can be true; therefore, the domains corresponding to these three subsets correspond to exclusive variables. The remaining domain corresponding to the normative standards subset, Norm Std, consists of 15 boolean variables; therefore it has an *a priori* cardinality of 2^{15}; however, a brief statistical study on the normative standards has shown that only a small number of them are involved in any one decision. This allows us to restrict the validity domain to cases where less than four normative standards are true. We can restrict the validity domain even more by considering the conditional probabilities between normative standards. We can then represent these probabilities as a valued graph on the normative standards. Furthermore, we decided to remove all the arcs whose conditional probability is less than 0.25.

$$\Gamma = (x,y/p\ (x/y) \geqslant 0.25)$$

The resulting graph Γ is composed of six connex components; one of them is shown in Figure 2.

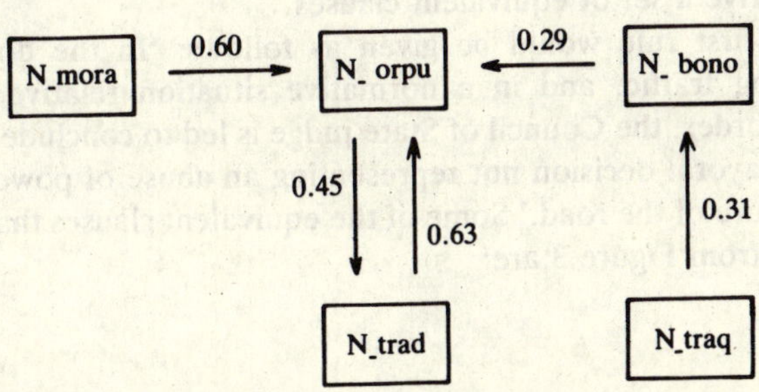

Figure 2: Example of connex component

The next step is to examine all the paths whose lengths are less or equal to three in the preceding graph, that is, to keep all the paths used when constructing $\Gamma \cup \Gamma^2 \cup \Gamma^3$
Twelve paths can be built from Figure 2 which shows one of the connex components. Each path corresponds to a vector. For example:

L1 = = orpu
L2 = = orpu ˆ trad
L3 = = bono
L4 = = orpu ˆ bono
L5 = = orpu ˆ trad
...
L12 = = orpu ˆ mora ˆ trad

This procedure can be applied on the other five connex components. So 36 paths can then be selected.

Extraction of Logical Rules

We can now discuss the extraction of logical rules from 120 cases processed by NEUROLEX and selected among the 51,480 cases (10 regulations × 11 bye-laws × 13 factual standards × 36 paths of normative standards) of the constructed validity domain. 36 is the number of possible combinations of normative standards as explained in the preceding section.

After setting a regulation, a connex component and a circumstantial standard, we can then study a subset of the validity domain and derive a set of equivalent clauses.

The first rule would be given as follows: 'In the domain of directing traffic, and in a normative situation relative only to public order, the Council of State judge is led to conclude that the only mayoral decision not representing an abuse of power would be closure of the road.' Some of the equivalent clauses that can be drawn from Figure 3 are:

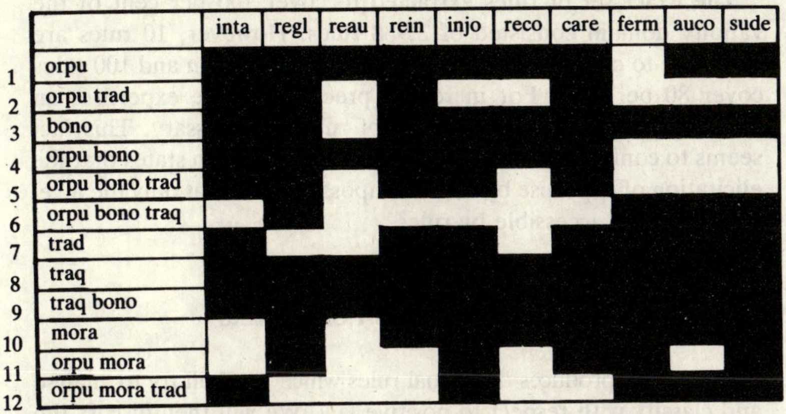

Figure 3: Responses given by the network for directing traffic, with the connex component C1 (the black squares correspond to annulation, the white to non-annulation).

Figure 3: Responses given by the network for directing traffic, with the connex component C1 (the black squares correspond to annulation, the white to non-annulation).

$$
\begin{aligned}
&\text{circ} \rightarrow \\
&[\ (1_1 \hat{}\ \neg\text{ferm} \rightarrow \text{expo}) \\
&\hat{}\ (1_2 \hat{}\ \neg\text{reau} \hat{}\ \neg\text{injo} \hat{}\ \neg\text{reco} \hat{}\ \neg\text{ferm} \hat{}\ \neg\text{auco} \hat{}\ \neg\text{sude} \rightarrow \text{expo}) \\
&\hat{}\ (1_3 \hat{}\ \neg\text{reau} \rightarrow \text{expo}) \\
&\hat{}\ (1_4 \rightarrow \text{expo}) \\
&\hat{}\ (1_5 \hat{}\ \neg\text{reau} \hat{}\ \neg\text{reco} \hat{}\ \neg\text{ferm} \hat{}\ \neg\text{auco} \hat{}\ \neg\text{sude} \rightarrow \text{expo}) \\
&\ldots\ldots]
\end{aligned}
$$

The rules thus obtained can be expressed in natural language and passed on to an expert in that domain.

It means that the standard 'breach of public order' is not the right argument to justify a mayoral decision in the field of traffic policy.

The degree of generalisation of the rules extracted is proportional to the number of cases depicted in a table such as that in Figure 3. The previous method was applied on 120 cases (in the field of traffic policy) but could be extended to the entire validity domain.

The exact set of rules extracted to cover 100 per cent of the validity domain consisted of 2,988 rules. However, 10 rules are sufficient to cover 50 per cent of the validity domain and 100 rules cover 80 per cent. For increased precision of the expertise, an exponentially increasing number of rules is necessary. This fact seems to confirm the claim by Dreyfus (1986) which states that the elicitation of expertise by rules is impossible and that only the level of a novice is accessible by rules.

Legal Validation of Results

The network produces decisional rules which we shall try to analyse and classify with respect to positive law. We will then discuss the specific problem of validating the knowledge generated by our model.

Interpretation of the Rules

Regarding our input and output variables, rules obtained are built according to the following scheme: 'In a subdomain of municipal regulation, when the mayor makes a decision and then gives legal or factual arguments, there is a strong probability that the judge will not annul the mayoral bye-law.'

Let us examine the different rules generated by our model. We can define five types of logical clauses for our previous example of traffic regulation:

(*a*) confirming explicit legal rules of the status
(*b*) adding conditions to explicit legal rules
(*c*) confirming general explicit principles
(*d*) extracting new rules
(*e*) extracting meta-knowledge

a) *Confirming explicit legal rules of the statutes:* Let us examine an example of this kind of logical rule (e.g. Figure 3, L6):

(L6): *If Traffic* regulation is concerned *and* standards '*Breach of Public Tranquility*', '*Breach of Law and Order*', '*Public Order*' are involved, *if* the mayoral decision is a ban
Then Decision Confirmed by the Judge (non-abuse of power).

In any situation involving violation of fundamental statutory principles,the mayor can forbid an activity. The text of the corresponding statute is as follows: 'Municipal policy is intended to ensure public order, security, tranquility, safety and health' (Municipal Code, Articles L131-1 and L131-2). There is a striking correspondence between the content of this decision rule produced by the network and the statutory principles. We can conclude that the network could verify whether the law has been correctly applied by the mayor and the judge.

b) Adding conditions to explicit legal rules: Let us take Line 1 from Figure 3.

> (L1): *If* standards '*Law and Order*' (bono) *and* the mayoral decision involves *Closure* (ferm)
> *Then Decision Annulled by the Judge*.

In opposition to the previous class of rules, we should note that even when certain explicit legislative standards are involved, the judge can overrule the bye-law. It means that the standard *Law and Order* cannot argue such a serious decision as the *Closure* of a noisy locality—for example *in the field of traffic regulation*. The standard Law and Order is weaker than that of Public Order (see L1).

The network can therefore be used to check the application or interpretation of the statute: the judge can introduce supplementary conditions for application which are not stipulated in the statute.

c) Confirming general non-explicit principles: In practice, some standards have been elaborated by the judge. What is the weight they have on the network? Let us examine the jurisprudential standards traditions:

> (L7): *If* standard '*Tradition*' (trad) *and* the mayoral decision is a ban
> *Then Decision Annulled by the Judge*.

Our network gives then an equivalent weight to a standard of the municipal bye-law. It means that when a traditional activity such as the sale of lily-of-the-valley in the streets on the first of May is forbidden by the mayor, the judge will annul this decision.

d) Extracting new rules: We will take the example of cases in which public decency is involved: after the Second World War and with the popularisation of summer holidays on the seaside appeared a jurisprudence on decency on the beach. For instance, the mayor was led to forbid undressing outside cabins built for this very purpose. A new rule appeared and is reflected in our following clause:

> (L10): *If* standards *'Public Decency' (mora) and* the mayoral decision involves *Ban, Authorisation, Refusal,*
> *Then Decision Confirmed by the Judge.*

The network is not limited to confirming known rules; it can also extract new rules composed of a series of positive or negative conditions that an expert would not be able to put together in an explicit manner. At this level, we are dealing with the syncretic, intuitive, complex 'deeper knowledge' that Dreyfus (1986) spoke about.

e) Extracting meta-knowledge: Meta-knowledge (Pitrat 1990) refers to strategies of choice amongst rules, which means that in order to solve a problem it is better to use *some* rules rather than others. Consider L8 in Figure 3 where we only have black squares: it means that the argument of tranquility is not at all relevant—for the mayor—in the field of traffic regulation.

So, there exists a hierarchy between standards according to the domain of municipal action. The final decision does not necessarily result from the same type of variables; there are hidden steps of reasoning and circumstances do not always have the same impact on the decision-making process; in short, the judge's decision can be founded on a complex and not predetermined combination of variables.

Validation of Expertise: Some Specific Limits

Our experiment raises several questions concerning legal knowledge: traditional legal knowledge was founded on written rules. For that very reason, the traditional modelling approach could not but be logical. With the neural network modelling approach, we deal with another kind of knowledge: the model does not pretend to find the universality of a rule but only its efficiency in a relativistic frame. That is why there is a problem in validating results. The only conclusive text would involve a comparison between NEUROLEX and a panel of experts on a large number of diverse cases.

NEUROLEX can essentially be used to compare different kinds of results according to modes of learning, to periods, or to supplementary input variables and input rules (those of statutes, for instance). Unlike symbolic models, our neural network model offers the advantage of taking into account special cases, contradictory cases and incomplete cases. This is the other reason why validation protocols must be adapted to another kind of expertise: knowledge by simulation.

Conclusion

This project served a double purpose. Not only did it enable us to construct a multilayer neural network through the extraction and modelisation of explicit knowledge, it also led us to define the concept of a validity domain and develop the means of creating our own.

The validity domain of this neural network consists of around 50,000 cases. We can estimate at fewer than 1,000 the logical clauses required to derive a symbolic expertise equivalent to the neural network by following the maximum rule. In the training base, we eliminated as a matter of caution all combinations of standards which did not appear sufficiently significant. We are then able to extract equivalent working rules from the network. These rules enabled us to check legal applications when discretionary power is concerned in a decision-making process. More importantly, the neural network model can extract implicit decisional rules that even an expert could not have formulated, rules to which we would have no access otherwise.

This model simulates a certain conception of law, normatively closed, yet cognitively open, as it evolves each time the judge takes into account the evolving environment when he hands down his decision. Research has, however, highlighted certain paradoxes:

- law is expressed by language, but language is a means of perceiving reality and constructing facts;
- law is based on concepts but evolves with standards; it consists of open-structured notions;
- the official theory of interpretation is a refusal of subjectivity, yet interpretation means continuous adjustment in a process of interaction.

In a legal system, as in any type of social system, rationality emerges when the internal equilibrium is maintained, but at the same time adapted to suit the changing environment. The NEUROLEX model was developed to simulate this emergent process.

References

BELEW, R.K. 1987. A Connectionist Approach to Conceptual Information Retrieval. *Proceedings of the First International Conference on Artificial Intelligence and Law*. New York: ACM Press.

BOCHEREAU, LAURENT and PAUL BOURGINE. 1990. Rule Extraction and Validity Domain on Multilayer Neural Network. *IJCNN*, June, San Diego.

BOURCIER , DANIÈLE. 1989. 'The Expert System Bruitlog and the Mairilog Project' in I. Snellen, W. Van De Donk and J.P. Baquiast (eds.), *Expert Systems in Public Administration*. Amsterdam: Elsevier.

———. 1992. 'Expert System and Reasoning Aids for Decision-Makers', in A. Martino (ed.), *Expert Systems in Law*. Amsterdam: North-Holland.

———. 1993. *La Décision Artificielle, Le droit, la machine et l'humain*. Paris: PUF.

DREYFUS, H. and S. DREYFUS. 1986. Mind over Machine. New York: The Free Press.

FELDMAN, S.E. and D.H. BALLARD. 1982. Connectionist Models and their Properties. *Cognitive Science*, 6, pp. 205–54.

FERNHOUT, F. 1989. Using Parallel Distributed Processing Model as Part of the Legal Expert System. *Proceedings of the Third International Conference, Logica, Informatica, Dirrito.* Firenze.

FOGELMAN-SOULIE, F., P. GALLINARI , Y. LECUN and S. THIRIA. 1987. 'Automata Networks and Artificial Intelligence', in F. Fogelman-Souliè, Y. Robert, M. Tschiente (eds.), *Automata Network in Computer Science*. Manchester: Manchester University Press, pp. 133–86.

GALLANT, S. 1988. 'Connectionist Expert Systems' in *Communication of ACM*, no. 31.

HAFNER, CAROLE. 1987. Conceptual Organisation of Case Law Knowledge Bases. *Proceedings of the First International Conference on AI and Law*, pp. 35–42. Boston: ACM Press.

HAMMOND, C.J. 1986. CHEF: A Model of Case Based Planning. *Proceedings of the Fifth National Conference on Artificial Intelligence*. American Association for AI, Philadelphia.

LECUN, Y. 1985. A Learning Procedure for Asymmetric Networks. *Proceedings of Cognitive Art.* Paris, pp. 599–604.

MCCARTY L. THORNE. 1990. AI and Law: How to Get There From Here? *Ratio Juris no. 3*, Oxford: Basil Blackwell, pp. 189–200.

PITRAT, JACQUES. 1990. *Meta–Connaissance Furtur de l'Intelligence Artificielle*. Paris: Hermès.

RISSLAND, L., EDWINA and ASHLEY K.D. 1987. A Case Based System for Trade Secrets Law. *The First International Conference on Artificial Intelligence*. Boston: ACM Press.

RUMELHART, D.E., G.E. HINTON and R.J. WILLIAMS. 1986. 'Learning Internal Representations by Error Propagation' in D.E. Rumelhart and J.L. MacClelland (eds.), *Parallel Distributed Processing: Exploration in the Microstructure of Cognition*, Vol. 1, pp. 318–60. MIT Press.

SERGOT, M.J., et al. 1986. The British Nationality Act as a Logic Program. *Communications of the ACM*, no. 29, pp. 370–86.

The Remainder World

Wagish Shukla

The rational horizon in heav'n is but one, and the sensible horizons on earth are innumerable.

—Milton, quoted in the *Oxford English Dictionary*, 1990, under *rational*

I

Ratiomorphous computing mechanisms are by definition unconscious processes, inaccessible to our consciousness. They are, at the same time, sensory and nervous processes. Presumably, they are accessible to machines which are, so far, thought to be devoid of consciousness and, so far, are to be devised consciously. So here probably is the first differential of the *home sapiens*, the Wise Man, the capacity to go and employ artificial devices in the aid of reason.

Just how does one know that these devices do what they are supposed to do? Just how is one sure, using a pair of spectacles, that what one observes is what one would have observed if one's eyes were 'right'? What is a magnifying glass and just how does one know that it magnifies anything?

These are examples on which there is general agreement. There are others, the readings on which scientists differ on. A lot depends on 'informed readings' and 'reasonable guesses'. Blood pressure readings guide one's life-style. And blood pressure readings can be erroneous. To err is human. So, apparently, is instrumentation.

II

Generally, it is post-medieval Europe which is supposed to have come up with the thesis that reason or rationality is the differential which distinguishes man from animal. By way of an approximate contrast, one may put forward an Indian point of view: human beings are *vyakta-vāc* (those who have articulate speech); others, animals included, are *avyakta-vāc* that is, those who do not possess articulate speech. Articulation then, rather than language, is the differential. To be explicit, we can pick up a statement from *Śatapath Brāhmana* (4–1–3–16) which divides vāc (speech) into four parts distributed among life-forms, only one of which is *nirukta* (well said). This 'well said' speech is attributed to human beings. What is meant by 'well said'? It is the opposite to 'ill said', but there is more than truism involved here. Who decides what is 'well said'? How does one know what is 'well said'? The answer given by Indian grammarians is: by studying *vyākaraṇa*. Both *vyākaraṇa*, for which we use the word 'grammar' but for which I think 'unfolding' is a better approximation, and *nirukti* or 'well said-ness', for which 'etymology' is frequently used but for which I think 'explanation' is a better approximation, are *'reasoned' attempts* to understand language. To take a vedic mantra (*Rig* 10/71/4) as explained by Patanjali in his *Mahābhāṣya*, one should study grammar so that language/speech unveils herself exactly as a woman full of desire unveils herself to her husband. Here then is the legitimisation process: reason, almost like pre-scripted foreplay in pillowbooks, is to be an agent of the awakening of a dormant emotion in the cosmofeme ready to be consumed/consummated = con-sum-mated.

Variations on the metaphor can be immediately proposed. The cosmofeme may not respond with an invitation to gratify, she may choose to strike. Patanjali has a somewhat smug attitude regarding this—only those who are not equipped with grammar (read knowledge) will be smitten. Thus a legend has been quoted: in a sacrifice originally intended to kill Indra (the king of the gods) the compound word *Indra-s'atra* which should have meant 'the killer of Indra' was wrongly intoned with the result that the same compound was interpreted as 'the one whose killer is Indra', thus convoluting the teleology of the exercise and the exercise of the teleology so that Indra destroys his intended killer. Patanjali's stand is that if the intonation were proper, this would not have happened.

As Kaiyata, a commentator of Patanjali, said, the do's and don'ts (*vidhi* and *niṣedha*) are accessible to human beings but not to other life-forms. The verse cited which refers to the legend mentioned above, has two variant readings and they are (not) different in their import. To be explicit, Patanjali reads 'the polluted word' (*duṣtah s'abdaḥ*) whereas some other texts, like *Yājñavalkya s'ikṣā*, quoting the same verse read 'the imperfect mantra' (*mantro hinaḥ*). The usual explanation is that Patanjali has in mind both revelational speech and everyday speech, the Word and the word respectively, whereas the other readings confine the destructive power only to revelational speech (imperfectly used). I will only reset the spelling at this point here: revelational speech is simply re-veil-ational speech and note that revelation is, after all, an orthographic resetting of what should/could be 'revealation'. Take the adjectives *duṣta* and *hīna*. I have translated *duṣta* as 'polluted' and *hīna* as imperfect. Other translations are possible but the point is that *duṣta* conceives of something within reach externally injured later whereas *hina* involves an idea of something to be reached. This may be taken as yet another piece of evidence for the standard explanation, within-reach-word versus to-be-reached-Word, but the word/Word dichotomy is somewhat superfluous in the Indian thought system. In any case, that is not what I am talking about. The point is: who are those who do not employ reason and thus get injured? Those who pollute, and those who are short of purity.

I will not extend the fornicational metaphor given above and rename these two as rapists and the hypopotents respectively. As against these doomed ones, there are the coaxers, the foreplay experts who are blessed. But this postulates a *harnessing* of reason. It is a terrible thing to be damned simply because one is not actively intelligent. There is an obvious risk of active intelligence/reason being defined in terms of the sword. To quote:

> We found among them a great number of books written with their characters and because they contained nothing but superstitions and falsehoods about the devil, we burned them all, which they felt most deeply and over which they showed much sorrow.

This was written by Father Diego de Landa, first archbishop of Yucatan, shortly after the Spanish conquest of Mexico while referring to the great bookburning ceremony which took place at the city of Mani (quoted in Anthony F. Aveni's 'Astronomy in Ancient Mesoamerica').[1] Immediately after the quotation, Aveni informs us:

> Fragments of but four Maya manuscripts or codicas survive today. All contain a wealth of information relating to the heavens: lunar and solar almanacs, even a Venus calendar usable for one hundred years.

The first archbishop of Yucatan presumably did not know about this 'wealth of information'. But given that Galileo has only recently been absolved by the Church, the archbishop would hardly have been wrong if he destroyed the Maya codices despite knowing that they had a 'wealth of information relating to the heavens'. That, however, is not the point. The archbishop must have been guided by logos—the rational principle that governs the comos—in his work. In this particular case, logos is embodied in Jesus Christ. In another case, it may be embodied into some other entity. The point is simply that the self-righteousness, the piety, the motivations of social, personal, political and civilisational degradations are all guided by an incarnate/incognate rationality which is supposed to be uniquely possessed by homo sapiens and that too only a select few. The old and new aristocracies, the my-theistic theologies, the white man's burden, are all instances of choices of certain rational principles which do not envisage the existence of anything else. To get rid of anything not compatible with it, it may choose to kill, to bribe, to preach. It does not really care.

Logos-logic-language is the burden mankind carries. The burden is to seek, to find out, to rationally employ rationality. We do not know whether it is a gift or a curse. Probably a curse. If, out of all the life-forms, the human is the one with which God decided to enter into a covenant as the semitic discourses seem to enjoin, then human beings were singled out for punishment rather than reward.

III

There have, of course, been several attempts to somehow retain exclusivity and yet avoid the inevitable punishment. The semitic exercises have largely depended on exclusivity itself, with logos being interpreted as an intercessor. There is really not much difference between the idea that tears can wash out your sins and the idea that you can purchase, by paying money, a ticket to heaven. Payment is demanded and in return the burden of responsibility for being a human being is removed. Those who do not possess the treasure of remorse or of money are excluded.

Indian exercises are somewhat more complex and more exhaustive. The treasure is expanded to include, besides remorse and money, everything that a human being possesses. Enmity for instance. Indian mythology is full of personages who decided to take on God and in return were billed for it. Almost invariably, in any account of the final billing, there is a statement of envy at the ease with which these personages achieve *sāyujya* (unity with God) for which the gnostic and the surrenderers work hard but generally fail to achieve. So here we are dealing with a gradation: the most difficult thing is to continue to be a man of the world; the easy way out is to be remorseful or angry. One may almost conclude that to reason is difficult, to emote is easy. This apart, the basic idea of divine robbery is still at work, pushed to its logical extreme. One is a human being through a divine agency and only by being erased as a human being through that divine agency can one become free. The means that the divinity uses do not matter. This perhaps needs some explanation.

Anybody who reads Indian mythology with some care cannot fail to notice that divine decisions to terminate with extreme prejudice are invariably taken and carried out with the purpose of preserving an order which has little to do with the social or ethical systems of ordinary people. This does not mean that the commonsense welfare ideals are different. Good food, good health, good criminal and civil justice etc. are not only accepted as being desirable but are enjoined on the state in much the same manner as today. The point is that the mythic personages who were killed and who were invariably personages of some power, were not killed because they were bad rulers or bad administrators of justice. Even in terms of religious discourse, they were not atheists or desecrators. They fully recognised that they were fighting a lost battle. They were frequently and deliberately trapped in a no-win situation by the divinity and forced to risk self-destruction by following the right code of conduct.

I wish, however, to move on to something else. There is in the fourth book of Yogavāsiṣtha a story about Dāma, Vyāla and Kaṭa. Briefly, Rama is being advised/blessed by his teacher, Sage Vasiṣtha, that Rama's position should not be like that of Dāma, Vyāla and Kaṭa; rather, it should be like that of Bhīma, Bhāsa, and Dṛdha. The narrative offered to explain what is meant by this is as follows: There was a king of the antigods named S'ambar. He defeated the army of the gods and the gods later defeated his army. In the usual style of revenge and counter-revenge, there is looting and arson in the township of the gods which, significantly, is occupied by only 'weeping lesser gods and womenfolk', the bigger gods having run away much in the manner of contemporary top echelons of authority at a time of defeat. Finally, fed up with conventional warfare where victory is followed by defeat, S'ambar decided to create a new force consisting of three antigods who were to protect his army like 'past, present and future'. These three, namely, Dāma, Vyāla and Kaṭa, were like contemporary robots who did not know life or death, victory or defeat; in fact they had no idea what war meant. Directed by S'ambar's will, their sole function was to attack the gods wherever and whenever they faced them. Unmanned tracking warheads are a close parallel. This time, the gods faced what appeared to be a final defeat. They went to Brahmā, one of the divine trinity, to ask him how to win their empire back. The gods were told to wait for a hundred thousand years, after which S'ambar would be killed by Viṣṇu, another member of the divine trinity. Meanwhile, they were to continue to fight and lose to the triad of Dāma, Vyāla and Kaṭa, who through their repeated victories would acquire a sense of identity. Once they did that, they would become vulnerable. This is exactly what happened. Once the triad members acquired a sense of identity, they soon began to feel that they had lives of their own and that their victories were their own. They were then defeated and driven away. According to the narrative, S'ambar then decided to manufacture three other antigods named Bhīma, Bhāsa and Dṛdha. This second triad was equipped with Viveka, which enabled them to take a meta-position so that whenever an identity-idea started to take shape, they asked themselves questions like 'who am I', 'what is the purpose of life' etc., thus preventing the formation of a sense of identity. They could therefore do what they were supposed to do, namely, vanquish gods without interest or purpose, delinking themselves from themselves so to speak. Obviously the gods found themselves helpless. Then, after one hundred thousand years elapsed, Vishu killed S'ambar as had been foretold. The new triad also disappeared.

It is quite natural to ask what happened to the first triad. Their story is continued as a sidetext but is comparatively more detailed. They went through various births and rebirths, mostly leading painful lives, but after all the transmigrations they were finally freed from the continuous cycle of birth and death.

As is usual in Indian mythic texts, all possible queries that can be raised have been raised and answered in the exposition. It is not feasible here to go into all the details. But it is interesting to note that:

(*a*) It is accepted that primary creatures can create secondary creatures with qualities that are usually associated with primary creatures. Thus S'ambar created the first triad and the second triad.

(*b*) These secondary creatures are subject to the same kind of birth and death cycles that are associated with primary creatures and ultimately exit out of this infinite loop in some manner or the other. Thus the first triad enters into the loop and exits out of it after listening to its story. The second triad does not enter the loop and exists directly like gnostics, surrenderers etc. The first triad goes through the motion of becoming rapist and hypopotent, the second consists of coaxers.

(*c*) For this to happen, it does not matter whether you are on the side of gods or against them. Thus the first triad, which one is not supposed to emulate, and the second triad, which one is supposed to emulate, are both against the gods.

(*d*) The power to create is minor compared to the power to destroy, the latter being available to only the supreme divinity. The divine computer invariably comes up with the same output, namely destruction. And it accepts inputs which are products of inputs.

The story clearly transcends all boundaries. Thus the secondary creatures have souls, emotions, consciousness, reason and do precisely what ordinary creatures claiming divine parentage do. For instance, the first triad, once it is defeated by the gods, moves into a netherworld where the three get real wives and raise families. The culture of the artificial is indistinguishable from the culture of the real; it can merge into the latter without a trace. Indeed, if 'real' means created by divinity and 'artificial' means created by 'reals', then there is absolutely no distinction between real and artificial as inputs. The divinity is not jealous of the power to create because it has the power to destroy. Any creature, artificial or real, can destroy any other creature, artificial or real, but such a destruction will result in a new creation, artificial or real.

And then of course, there is the final non-finality. Even after there is total destruction, a new metacycle of creation begins. The ruthlessness of this idea is untouchable. As the seventeenth century Indo–Persian poet Mirzā Abdul Qādir Bī-dil says:

> From this very clueless garden, the mysterious flower blooms
> Without a ray is the light of this candle with no candlestick to hold it
> There is no orchestra of calligraphs and images and no symphony of
> *nos* and *ego*
> There is nobody who can make an entry into this assembly of wonder
> Yes, there is a mirror of thoughts but that is an informer against *Bidil*.

IV

Roger Penrose[2] argues that there must be an essentially non-algorithmic ingredient in the action of consciousness. The activity component of awareness is used by him to deny consciousness to rocks, insects, worms and all but a few mammals. He quotes an author who denies consciousness to human beings before 1000 B.C. This consciousness is nonalgorithmic. By contrast, unconsciousness is algorithmic. There is a theoretical quantum parallel computer whose computing power so far does not seem to exceed that of the Turing machine and which thus remains algorithmic 'but there are early days yet'. Penrose argues that 'unconscious processes could well be algorithmic but at a very complicated level that is monstrously difficult to disentangle "in detail"'. Certain assertions made have a familiar ring. Thus a video-camera aimed at a mirror recording itself does not possess self-awareness. Philosophical musings and mutterings are the 'baggage' carried by conscious beings and 'it is hard to imagine that an entirely unconscious automaton should waste its time with such matters'. Questions like 'who am I?' 'what is the purpose of my life' etc. are 'funny' and 'ridiculous'. But conscious beings do ask such questions. Thus conscious beings, i.e., human beings, are chosen by a natural selection but the favour has been granted not because they ask such questions but because of some other reason which is mysterious enough but also insignificant enough to be left unexplored.

This is how the elite in any society think. Not so long ago, it was a matter of some pride among the European nobility to be illiterate. Poets, philosophers, musicians and scientists could always be hired. Today, there is little change in the situation but since even rulers are paid servants of an invisible paymaster, every guild—businessmen, politicians, philosophers and men of medicine hired by universities—can claim illusory respectability. A sovereign ruler paying income tax like a subject is not only not funny but is even desirable. The tyranny of reason in the quotidian life was never so clear as in this century.

At the same time a New Age post-rationalism is looking for a comeback which may well be called metempsychotic. If today we have architects who are proud of designing buildings which cannot be built, the 'wildness' is a pointer to something which in my opinion can run something like this: (*a*) let us try to build without science and engineering; (*b*) it does not matter if we cannot do it since science and engineering are 'acquired'. We seem to be trapped today in the discourse of natural selection. Consciousness is either 'acquired' (the noble savage to Picasso wanting to paint like a five-year-old to New Age post-rationalism) or a hereditary leftover (Penrose pushed to his logical extreme). 'Baggage' it is. Many others are likely to think that it is unconsciousness which is baggage. Sum up, and everything is a baggage with only a void to carry it. Change tack. Generally, Strong AI holds that consciousness is algorithmic. Penrose says it is not. But unconsciousness is algorithmic. Sum up, and everything is algorithmic. Sum up, and everything is non-algorithmic. 'Sum up' is not a voodoo chant, just a call for ethnic cleansing in the world of ideas.

Rational talk is a repressive discipline. But the decentring of rationality means that there is no communication, no questions to be asked, no answers to be given. A catechism-free world is immobile. But, with all the catechism-directed mobility we seem to be no further than building pyramids for mummies which may well start talking if the algorithm for articulation is found and activated through the relentless pursuance of rationality. The teleology of rationality is post-rationality.

There is no theoretical objection to the creation of *homo rationalis secundus* by *homo rationalis primus*. There is no distinction between the artificial and the real. The remainder world, after all the baggage has been thrown away, is a void, incognate as an abyss. The video-camera aimed at the mirror is a perfect example for all that creativity—rational, pre-rational or post-rational—amounts to. Like the triad of S'ambar, it simulates the three divisions of time, past, present and future, fighting on the side of an army doomed to be destroyed.

As the poet Mirza Ghalib says:

> Be it faith, be it hereticity, both are wavements of a yawn out of the same stupor
> Be it skepticity, be it belief, both are swerves in the same straight line being drawn
> Be it Quibta, the arch to direct you to the faith-point be it the eyebrow of the idol to lead you away from it both are same, the path of a stopped curiosity
> The house of the God and the idol-temple are exactly the same—a prop to sleep through nightmare.

Notes

1. Aveni, Anthony F. 1984. 'Astronomy in Ancient Mesoamerica' in E.C. Krupp (ed.), *In Search of Ancient Astronomies*. Harmondsworth: Penguin Books.
2. Penrose, Roger. 1989. *The Emperor's New Mind*. Oxford: Oxford University Press.

III

Rationality, Theory Construction and Science

Preface

The age-old notion of science as the paradigm of rationality is now under heavy assault from various quarters. Rationality itself seems to be a redundant concept to many. Science, especially the unassailable superiority of physics, is now being challenged by many philosophers of science. The logical character of theory construction and theory evaluation is threatened by the emergence of sociological explanations of knowledge. If that position holds then scientific activity would be either on arbitrary or on extra-logical grounds.

The first paper in this section, 'Objectivism and Social Determination' by *Mahasweta Chaudhury*, examines the issue of knowledge (theory) and its social construction. It argues that the putative dilemma between objectivism and the social construction of knowledge is a false dilemma and can be dissolved by disambiguating some cognitive concepts.

The second paper, 'Changing Concepts of "Rationality" in Science' by *Tusharkanti Sarkar* challenges the concept of science as a rational activity by recognising the role of non-rational factors in science. A suggestion is made as to how an 'inconsistency-tolerant NMR-based fuzzy hypothetico-default' model of scientific rationality is better than others.

In '"Identity" and "Property" in Vagueness', *Mihir Chakraborty* and *Mohua Banerjee* deal with the very important philosophical concepts of identity and property. The paper aims at developing a mathematical model of vagueness ('fuzziness') that contradicts the standard belief that only clarity/precision is rational, by also making room for the concept of vagueness in that domain.

The fourth paper by *S.M. Bhave*, 'Rationality of Mathematical Constructions', addresses the question of rationality in mathematical constructions. Do mathematical concepts grow autonomously and are then legitimised by some ad hoc methodology? Or, are such constructions not just historical reality ('form of life' as Wittgenstein says) but a logical/rational growth from the extant body of mathematical knowledge? Bhave concludes that such concepts are rational constructions and not just historical reality.

MAHASWETA CHAUDHURY

Objectivism and Social Determination: A Plea of a Rationalist

MAHASWETA CHAUDHURY

The theory of social determination in general holds that all our beliefs and opinions, ideas and thoughts are conditioned by the particular time and society we belong to, science being no exception. The alleged antagonism between objectivism and sociology of knowledge can be analysed by examining Popper's view against that of Mannheim, who is the traditional proponent of the social construction theory. The latter theory, however, has many contemporary followers who reject rationality and objectivism. It is argued here that Popper's arguments against sociology of knowledge and vice versa are misdirected because objectivism and social determination are theories about two different issues.

Before we examine the relation between objectivism and any theory based on the social determination of knowledge, let us note that although Popper eventually rejected the final thesis of the sociology of knowledge, in fact he conceded a great deal to them. The main points of his agreement and disagreement can be summed up as follows:

Popper's Agreement with Mannheim[1]

- The knower or the observer has a vital role to play in knowledge. In fact, we cannot know in vaccuum. 'Tabula rasa' is a fiction. Knowledge is constructivist (Popper shares Mannheim's activistic theory of knowledge).

- Background knowledge is essential for any theory; the scientific community has a great role to play with our notion of the world and its objects.
- We are subject to many deep-rooted prejudices (some of which may be class interest).
- There are underlying assumptions, expectations, prejudices that can influence one's view about the world.
- Knowledge is not immutable, it is changing and fallible (anti-essentialism).
- No evidence can completely *justify* knowledge except trivial truths like 'a is a'.

Popper's Disagreement with Mannheim

- Although knowledge cannot start from scratch, our ideas are not 'reflex functions' of the sociological (or historical) conditions under which they arose.
- The question how our knowledge is caused or has originated is relevant for psychological or sociological enquiry but totally irrelevant for epistemology and science.
- Although we are victims of 'perspectives' there are no reasons to believe that all these perspectives are equally important.
- Our theories or views about the world are influenced by other theories or views, but not by any extralogical notions, or even if they *are*, that fact is not relevant for validity of knowledge.
- Theories are not products of society, but attempts or conjectures to solve some problems. These problems may be caused by some social conditions, but that fact does not have any bearing on the evaluation of a theory.
- Theories are free conjectures, products of imagination.
- Scientific knowledge is different from any other forms of imagination because it can be criticised and tasted. Criticism is the method of empirical knowledge.
- Scientific knowledge differs from metaphysical speculation not because it is justifiable but because it is falsifiable.
- Though knowledge can never be justified, it is nonetheless objective. Scientific objectivity consists in intersubjective testability, not in the individual scientist's transcendence of social prejudices. No amount of purging can assure scientific objectivity. Objectivity is ascertained through repetition of tests

and intercommunication of theory in a scientific community. So a fallibilist need not be either a subjectivist or a relativist.

From this construal of Popper's points of agreement and disagreement with Mannheim (or any theorist of the sociology of knowledge for that matter), a picture seems to emerge which is indicative of a major difference. This difference is of serious concern because it appears to be decisive regarding the problem of objectivism (the theory that knowledge is objective).

Sociology of knowledge is usually associated with a certain epistemological stance, namely subjectivism and relativism. The essence of relativism is aptly described by D. Bloor as follows[2]: 'The objectivity of knowledge resides in its being the set of accepted beliefs of a social group . . . . The authority of truth is the authority of society.' The relativist believes that what makes it true that P, is the fact that a socially dominant group asserts that P, whereas the objectivist argues that what makes it true that P is simply P or in more conventional terminology, a proposition P is true if and only if it corresponds to the fact P.

Objectivism can be regarded as the opposite view which can be succinctly described as a theory according to which a 'fundamental distinction must be drawn between the way the world is and what we say about it, even if we all happen to agree. We could all be wrong'.[3] Dawson remarks on this statement that when it is noticed that objectivism includes the belief 'we could all be wrong', it appears that there is 'no real dispute between the ostensible antagonists'.[4] If Dawson is right, then Popper's fear that objectivism is challenged by the sociology of knowledge is ungrounded.

I shall argue along similar lines that: (*i*) objectivism and sociology of knowledge are not two horns of a dilemma; and (*ii*) the apparent incompatibility can be dissolved by a proper analysis of some cognitive concepts.

First the apprehension about sociology of knowledge winning over objectivism. Before examining that alleged threat, let us probe the reason for it. As I said before, sociology of knowledge seems to encourage a sort of relativism (although Mannheim ostensibly makes a claim that it does not). Relativism may indeed have various connotations. Speaking very generally, it can aptly be described as the view that there is no single standard to evaluate knowledge and truth. It depends on the framework or the particular coordinates chosen for what we may call truth or knowledge.

Popper calls it the 'myth of the framework'. He describes the myth as the basic assumption that no rational discourse is possible unless the participants share the same agreed framework of thought.[5]

Notice that my general construal and Popper's version are not exactly the same, although Popper's follows from mine. For several standards of truth and knowledge cannot allow rational discourse across and thus lead to incommensurability of theories to which Popper never concedes.

Both the general and the specific version of relativism as presented by Popper are being hinted at by different brands of the sociology of knowledge. Does it challenge objectivism then? The apparently obvious response to this question is, 'yes'. In recent times Kuhn is the most appraised and most rebuked champion of the sociology of science and relativism in a certain sense.[6] Although his original 'revolutionary' view has become somewhat subdued now, his theme remains more or less the same. On his account, the practice of science shows that ordinary scientists ('normal science') are not 'seekers of truth'; they only match their findings with an existing theory to which they have a general consensus. During normal periods the existing theory, which he calls 'paradigm', determines the research principles.

Scientific education is a kind of sophisticated dogmatism that dictates the researchers to fit or force nature to match the paradigm. The scientists do not aim at finding out the truth; they are committed to the framework of the paradigm. If nature is stubborn and a scientist cannot succeed in his 'puzzle solving', then the individual is the one to be blamed (for incompetence or other shortcomings) and the paradigm remains unscathed. Rival paradigms compete and anomalies are explained by other than the individual scientist's fault only in 'extraordinary' periods—periods of revolution. But soon some theory wins over the others and normal science of common consensus returns. This is the general picture of scientific research according to Kuhn's account although he later denied many of the strong features ascribed to him.

Anyway, without going into any methodological quibbles, a picture clearly emerges from his account about knowledge and truth. One may also discover in his conception of rival scientific communities with competing paradigms a ghost of Mannheim's different social groups. Even the micro-communities at different levels (e.g. physicists and then solid state physicists and so on) remind us of different interest groups within the same society. What does this picture show? First of all, that science (i.e. the paradise of objective knowledge) does not aim at truth; most of its research is nothing but mere matching. Secondly, as there are rival theories, there is no general consensus about truth; it varies from theory to theory. Thirdly, because there is no absolute standard to judge truth, it is not legitimate to talk about true or false theories.

Moreover, since there is no common truth, theories are incommensurable. Scientists ascribing to different paradigms cannot speak at the same level. Of course, Kuhn's position is often unequivocal. Sometimes he says that 'professional communication across group lines' is comparatively rare and 'often arduous', and he insists that 'communities . . . which approach the same subject from incompatible viewpoints . . . are far rarer (in the sciences) than in other fields'. He also says, on the other hand, that there are 'such things as schools in science, communities which approach the same subject from very different points of view'.[7] He also suggests some way out of the communication breakdown. The suggestion is to 'recognise each other as members of different language communities and then become translators'. It hints at a possibility of rational discourse among the participants only after they realise their respective framework with its special features. So, strictly speaking, Kuhn does not contribute to the incommensurability thesis. The general stance nevertheless remains the same. Indeed it is incompatible with objectivism if the latter stands for the view that although knowledge is an activistic process, objects of knowledge are 'out there', as coercive, external realities. In other words, knowledge aims at finding out truth or what is fact. We may fail to attain that, but that does not imply that truth can vary. In this account 'knowledge is fallible' does not entail that 'truth is relative'.

Now, truth is a problematic metaphysical concept and even a hard-core objectivist like Popper would not claim that knowledge of truth is easily acquired. On the contrary, he often says that a scientist (or anyone else) may not know the truth even after he encounters it, just like a climber may not *know* that he has arrived at the summit when he has already reached it. But objectivity lies not in achieving truth, but in aiming at it. The test for objectivity[8] lies in the intersubjective testability of a view. There may be various views about a state of affairs, but that does not imply that all of them are *equally* true. Relativism seems to hold this absurd position. Fallibility of knowledge only warns us that we may be wrong about our approach, but it is not inconsistent with the objectivist claim that there is only *one* way in which the state of affairs obtains. Call it truth, call it fact—that does not change with changes in our knowledge, education, training, social conditions, etc.

I shall now attempt an analysis in the light of the above interpretation of relativism and objectivism (which an unsympathetic critic may derogatorily call 'absolutism') to see whether these two are mutually exclusive positions. My guess is that even if relativism is true, that does not necessarily make objectivism false.

Relativism in general does not allow for 'correspondence to facts', which seems to be the basic feature of objectivism. It is interesting to note that objectivism tends towards realism whereas relativism, although it begins with activistic elements in knowledge, reduces the knower to a helpless victim of social determinants (of various kinds like socio-economic conditions, education, culture, class interest, etc.). The position is paradoxical. Moreover, objectivist theory cannot be rejected on the basis of not corresponding to facts because then the sociologists of knowledge would have to depend on a view of truth which they mistrust.[9] Indeed, one can reject it on the ground of its allegedly repugnant social effects. 'Objectivism has been firmly embedded in the norms and rituals of academic culture and its transmission . . . . Educational psychology has been a powerful legitimatising agency and rationalisation for objectivism. As such, it has become an important form of social control.'[10] This argument again undermines relativism and the case for sociology of knowledge. Because what makes it *true*, if it is true that objectivism has had these consequences, is not that it is held by a dominant theory or that a socially (or intellectually) dominant group says so, but that these things (the said social consequences) have really happened. So, sociologists of knowledge cannot disprove objectivism by their own method. And objectivism still remains true if sociology of knowledge and relativism are true. Therefore, not only are objectivism and sociology of knowledge not mutually exclusive but also that sociology of knowledge implies objectivism, perhaps to its utmost dislike.

Another line of argument in my analysis is based on the distinction between a proposition and a speech act. As described very aptly by Searle: '. . . the conditions for the truth of the *proposition* are not the same as the conditions for the performance of the *speech act* of asserting that proposition.'[11] All meaningful actions, including speech acts, are performed with a certain intention in mind and in accordance with the social conventions of the language community. Speech acts are thus conditioned by society. But while the question of truth and falsity does not pertain to them, the question of relevance does. Speech acts may be appropriate or inappropriate according to the social conventions. Truth is not a measure there. For example, the response 'the sky is blue' will not be appropriate if the question is 'Do you believe in the immortality of the soul?', although the response reveals a truth. On the other hand, a statement may be false but an appropriate speech act in a certain social or linguistic situation. 'Black people are inferior in intelligence' is dubious in its truth value but an appropriate speech act in a racist community. 'Tomatoes are blue' is similarly false but nevertheless an appropriate response to a question asking the colour of the said vegetable.

Propositions, however, are not the *act of asserting* but *what is asserted, believed or known*. What makes the proposition 'the grass is green' true is simply the fact that grass is green. 'The grass is green is a fact' means that the truth of the corresponding proposition 'the grass is green' is independent of whatever is said, believed or known about it.

The repercussion of this distinction is serious for the theory of sociology of knowledge because the theory has conflated these two notions. Speech acts (the act of asserting, believing, knowing), intentions, and linguistic conventions may be influenced by social conditions and are therefore objects of sociological analysis. But propositions and their truth are not amenable to sociological investigations. Speech acts may be *caused*, influenced and conditioned but the fact that grass is green can only make the proposition 'the grass is green' true. The truth of this proposition does not depend on social conditions or anything else.

By saying this, one is not trying to undermine the sociology of knowledge or reduce it to a subservient position vis-à-vis objectivism. One is trying only to establish that the putative dilemma between relativism and objectivism is a false one. It can be dissolved by an analysis like this one, which recognises the distinction between what *is* asserted and the *act* of asserting. Objectivism takes into consideration only facts that can make a proposition true, but cannot make a speech act meaningful or appropriate. On the other hand, the sociology of knowledge or relativism generally confuses the speech act with what is asserted, believing that knowledge (proposition) can be determined by social conditions. There are therefore no 'facts' or single standard that can make it true. The sociologists of knowledge are *right* about speech acts being dependent on social conditions but *wrong* about sweeping the content of a proposition also under these determinants. The notion of fact is not reducible to any social factor.

Even if the dilemma disappears after disambiguating the term 'knowledge', the problem remains, for a hard-core sociologist of knowledge may complain that the burden is now shifted to the notion of fact. He may also doubt this notion and complain that even the notion of fact is loaded and 'pure fact' is a fiction. It may be argued on his behalf that the concept of fact also can be socially determined. In other words, a frontline sociologist of knowledge may claim something more than social determination of knowledge. He may claim what is known in sociological literature as the social *construction* of knowledge. Which would mean that not only is the process of knowledge conditioned but the object of knowledge is also socially constructed. This is a much stronger claim than the original position. It asserts that there are no 'pure' or 'brute' facts. Whatever we know is all socially determined. All facts on this account are institutionalised.

What would be the objectivist's response to this claim? For a hardline essentialist or a naive realist, a natural response would be to reject it outright because to them objects are part of external reality, and are forced on us by the process of knowledge. Objects or facts have essential qualities that make them alien, independent entities with only accidental external relations with the knower. The knower is a passive recipient and nature is a stubborn customer. But this intuition does not appeal to Popper's kind of objectivism where knowledge is basically activistic. In fact his account in this respect is very close[12] to that of the sociologist of knowledge. Popper abhors what he calls the 'bucket theory' of knowledge, by which he refers to the Humean type of view based on pure observation. According to him, all observations are 'theory-impregnated'; so are the objects of knowledge. We have to be very careful in distinguishing between this type of activistic theory and a constructionistic account. Do we know only brute facts (as the essentialists and the realists say) or institutionalised facts (as the social constructivists say)? Which way can an objectivist account go? Apparently the latter alternative seems to lead to subjectivism. Popper is not very explicit about this, but he has been faithful to 'realism' of some kind despite the recognition of activistic elements in knowledge. He upholds the commonsense view of knowledge so far as it accepts external reality as having an existence independent of us.[13] Although Popper does not have a thesis on the nature of fact and its social conditioning, the following analysis of the social construction theory may endorse his variety of objectivism.

Propositions contain concepts which are as they are not because they reflect the structure of external reality. In other words, on this account all such concepts correspond to institutional as distinct from brute facts. The reasoning for such view may have arisen from the fact that the same individual object or person or animal can be mirrored differently to a physiologist, archaeologist, anthropologist, a layman, a maneater and so on. Therefore a fact or a concept does not have any essential quality that is sufficient for its knowledge. Brute fact is *necessary* but not sufficient for any analysis of knowledge. For example, what would be the brute fact about a person called Mary? If we describe her as (*a1*) 'Mary is a homo-sapien', it can only state something that can please say, an anthropologist or a physiologist or perhaps a maneater. (*a2*) 'Mary is a female' looks better, but is still not very informative. In Popper's account, knowledge becomes more 'interesting as we have more and more informations about a fact'. (*a3*) 'Mary is a female philosophy graduate' is more informative than (*a1*) and (*a2*). A social constructivist would argue that there is no such thing as a brute fact. He would find some purpose even in (*a1*) where the individual is described in a certain way. The conclusion he would draw is that there are no concepts used in propositional knowledge which express brute facts; all facts are more or less institutionalised.

There is a grain of truth in this formulation of the social construction of knowledge. Popper too would have to concede that bare facts and essence are concepts with a metaphysical connotation. It is very difficult and arbitrary to determine the essential features of brute fact. The problem is not so serious in the case of natural kinds[14] as with other kinds of facts including individuals. It is difficult to decide how much should be considered as being the bare essence of the facts.[15] Without going into that metaphysical debate in the present context, I suggest another line of approach which may help us through this seemingly insuperable difficulty. It is no doubt highly problematic to decide what is the 'essence' of a fact which is independent and undetermined, because all facts in some sense are determined by some pre-existing factors. Had these factors not been present, the fact in question would have been different. Mary being a homosapien (*a1*) is also conditioned by a knowledge of natural kinds in a way which reflects a certain socio-intellectual condition. So it is totally arbitrary to stop and say 'now I have reached the unpolluted pure fact'. The social constructivists are right in the sense that all facts reflect some socio–intellectual condition. But does this imply that there is nothing in the fact that it is so?

I do not think so; it is institutional (depending on the level of knowledge attained) that I describe Mary as a (*i*) homosapien, (*ii*) a female, (*iii*) a philosophy graduate, (*iv*) a radical feminist and so on, but that I can describe her *so*, depends on something in *her*. That irreducible element in that person does not depend for its existence on any social institution. In 'the grass is green', it is institutional (relativised in language, social convention etc.) that we call it 'grass' or name the colour to be 'green' and not 'red'. But there is something in grass and its being green that it is so described. Moreover, the putative dichotomy between brute fact and institutional fact is also misleading. In most cases facts are partially both. We cannot find entirely pure or entirely institutional facts. Some philosophers, however, have tried to make a strong case for the sociology of knowledge by citing examples from scientific concepts. They say that concepts of physical entities like 'mass', 'electron', 'plane' are highly charged with theoretical connotation relativised in a certain kind of science.[16]

It is indeed true that science[17] deals with facts of a certain standardised sort, but that does not lead to the naive view of scientists passively 'discovering' pure facts. My response to such attempts would be only to agree with them so far as they recognise that isolated brute facts shining with their essence are a myth. But I agree with them no further than this. The sociologists of knowledge do not realise how far a constructivist theory of knowledge can go along with them without conceding to relativism of any kind. The institutional character of language[18] is by no means underrated by Popper or Frege. And it is also recognised that whatever is institutional can be socially determined. Language as the main vehicle of thought also cannot avoid social conditioning. My perception of the issue would be clear if we distinguish between 'proposition' and a 'propositional act' as Alonzo Church proposed and with this distinction in mind, the apparent dichotomy can be given a new and clear perspective.

Frege, who claims to speak for an objectivist epistemology, is perhaps the first to make a distinction between what is asserted, believed, known, thought or stated, and the act or state of believing, asserting, knowing, thinking or stating. As Searle says, this distinction is significant for a sociological enquiry of knowledge. Church also distinguishes between a proposition (the meaning of a sentence) and a propositional act such as a belief, and from the way belief sentences (such as I believe that 'he is here') are translated, he argues that a proposition rather than a sentence is the object of our belief or other propositional acts. We can also recall in this connection the distinction between 'constative' and 'performative' utterances proposed by J.L. Austin. 'Constative' utterances can only be called true or false but 'performative' utterances are the performance of some act and not the report of such performance. Propositions can be, as Frege also says (in his attempt to establish the objectivity of empirical and formal science), true or false; sentences are true or false only in the derivative sense of expression of a true or false proposition. A true proposition is a fact. Speech acts are different from a proposition; in fact, all meaningful actions are performed with certain intentions in mind and in accordance with certain conventions. And so the performance of speech acts will surely be influenced by the social situations in which the speakers find themselves. It would be wrong to claim that a speech act (or any other action) was true in the sense of corresponding to facts; equally wrong would be the supposition that the truth of a proposition could be socially determined or conditioned in any way.

It is clear now that questions about the truth or falsity of proposition and logical relations among propositions can and should be pursued quite independently of questions about the causes and effects of the performance of speech acts. The dilemma—either knowledge is objectively true and so escapes the social influences or it is socially conditioned and therefore is not true—appears now to be a false dilemma. Even if one accepts that social conditions can influence speech acts, that does not lead to the rejection of objective knowledge. But even now a hard-core sociologist of knowledge might further argue that knowledge is not merely socially conditioned but socially constructed. That knowledge is a social construction is a claim about propositions to the effect that they contain concepts which do not reflect the structure of external reality but only our interests and purpose. All facts are constructions. It is important to note that Popper accepts much of this claim—no facts are 'hard facts' for him; all facts are loaded, not with our individual (or class) interests and purposes, but with our theories. But here we must make a fine distinction between 'constructed' and 'theory-laden'. That we can know of bipeds and quadrupeds or that gold is valuable, depends on certain social conditions; but that we can do so has something to do with their nature. Here we must make a distinction between 'brute fact' and 'institutional fact'. 'Mary is a human being' states a brute fact whereas 'Mary is a philosophy undergraduate' states an institutional fact. But all facts are neither entirely brute nor entirely institutional. Brute fact and institutional fact interact in such a way that it is an empirical question how much emphasis is given in any particular assertion. Once we know an institutional fact about something or somebody it entails our knowledge of that person or that thing not only institutionally, but also as such. For example, 'wife' entails 'female', 'human being', 'biped' etc. Given any sufficiently determinate standard or norm, it will be an objective issue, a matter of truth and falsity, how well a particular object measures up to it. Moreover although norms are (arbitrary) conventional, they are also attempts to approximate truth and are thus evaluated and adjusted according to the available knowledge of the time. And even if we know an object as an institutional fact, it entails, in addition to our knowledge of it, knowledge about the object as a 'brute fact'. For example, our knowledge of the object, say gold, may be partly institutional, dependent upon several social conditions. But 'this is gold' or 'gold is valuable' is not only a statement about a socio-economic system, but also about a brute fact possessing certain properties, in this case an object possessing the atomic number 79. Those who think knowledge is socially constructed oversee this. Knowledge that X measures up to the standards only entails that someone in authority says so. But that X measures up to the standards does not entail that X possesses properties A_1 . . . . A_n, for those in authority may make a mistake, the standards may be incorrect, available knowledge may not be adequate and so on.

If it is realised that in pronouncing a judgement, an authority (or the whole society) is not asserting a proposition claiming truth, falsity, validity etc.) but only a speech act, then the objective theory of knowledge would not have any quarrel with the sociology of knowledge. But that is not to be. The sociologists of knowledge—at least militant ones like D. Bloor—have gone as far as to banish truth and validity from the realm of knowledge and attempted to reduce Popper's Third World to a social phenomenon. To avoid subjectivism, the sociologists of knowledge have opted for a more problematic choice. Progress of knowledge and its fallibility can be better explained if we can avoid the good old Baconian idols of the tribe. Cultural or anthropological relativism, even if true, says only that there are several standards of judging social phenomena. It does not mean that we cannot evaluate the different standards. And relativism, even if true in social phenomena, does not have an epistemological corollary.

Brute and institutional characteristics interact in such a way that it is an empirical question how much of a brute fact will be decisive for an institutional fact. That means the standard for any institution depends on social conditions, but that cannot influence the brute fact. It is contingent on what standard or scale we classify natural kinds like anthropoid apes and *homo sapiens*; it is something brute about, say Jane, that she is not classified as an anthropoid ape but as a *homo sapiens*. Suppose I have an institutional fact about Jane that she is married and wife of Mr. X. So I have knowledge of an institutional fact about her that she is a wife. But that knowledge entails that I possess knowledge, not only about a social institution of marriage but about *her*. It is a matter of convention that possession of a_1—a_n (say, having b_1) is regarded as the standard for being a good student, so the knowledge of that is socially conditioned and therefore a matter of decision and standard. But it is an objective issue involving truth and falsity whether X actually possesses a_1—a_n. If we do not recognise that the latter is not a question of decision, but of asserting a proposition, there is a constant threat of subjectivism. But once this ambiguity is cleared and the myth of the dichotomy dispelled, objectivism has nothing to fear from sociology of knowledge. The two theses can get along together. This analysis can also endorse Popper's version of objectivism which allows a great deal of constructivist element in concepts despite his irreducible realism.

Notes

1. I have chosen Mannheim as a model, but Popper's arguments can be applied to any general theory that believes in social determination (and control) of thought. K. Mannheim, 1972. *Ideology and Utopia*. London: Routledge and Kegan Paul. (particularly Chapter 5).
2. D. Bloor, 'Popper's Mystification of Objective Knowledge'. *Science Studies*, January 1974.
3. R. Trigg, *Reason and Commitment* (1973), cited in 'Objectivism and Social Construction of Knowledge' by Graham Dawson, *Philosophy*, vol. 56, 1981.
4. *Op. cit.*
5. K. Popper, 'The Myth of the Framework' in E. Freeman (ed.), *The Abdication of Philosophy: Philosophy and Public Good*.
6. Kuhn initiated it in *The Structure of Scientific Revolution* (1962) but later changed his views to some degree.
7. T.S. Kuhn, 'Reflections on My Critics' in Lakatos and Musgrave (eds.), *Criticism and The Growth of Knowledge*, 1970.
8. As found in his different works. The firebrand philosopher Feyerabend, however, radically challenges this entire prescriptive orientation of science, declaring it to be nothing but a social artefact myopically value-laden and fatally severed from historical reality.
9. I owe this point to Graham Dawson, *ibid.*
10. G.M. Esland, 'Teaching and Learning as the Organisation of Knowledge' in M.F.D. Young (ed.), *Knowledge and Control*.
11. J.R. Searle, 1966. 'Assertions and Aberrations' in B.A.O. Williams and A.C. Montefiore (eds.), *British Analytical Philosophy*,
12. As I have explained at the beginning of this paper.
13. K. Popper, 1973. *Objective Knowledge*, Chapter 2.
14. Kripke, Putnam, Donnellan and many others have revived a sort of essentialism from their theory of reference according to which scientific investigations decide the essential nature of natural kinds.
15. Kripke offers a criterion that the origin is the essence of an individual person or an object. Like 'made out of a hunk of wood "a"' or 'born from the sperm "S" and egg "E"'. See his *Naming and Necessity* (1979).
16. N.R. Hanson has given a brilliant account of such concepts in *Patterns of Discovery* (1958). See Ravetz, *Scientific Problems and its Roots in Social Sciences*.

17. Analytic research work conducted in the exact sciences to see the relations occurring between cognitive processes and social practices has shown that within networks of actions comprising research work, there exist nodes where cerebral and social events merge and function symbiotically. See Terry Shinn, 'Cognitive Process and Social Practice: The Case of Experimental Macroscopic Physics', in Fuller, de'May, Terry Shinn and Woolgar (eds.), *The Cognitive Turn*. Reidel, 1989.
18. K. Popper, 1957. *Poverty of Historicism*, pp. 154–55.

Changing Concepts of 'Rationality' in Science

T.K. Sarkar

The pursuit of science is universally acknowledged as a rational activity. This may appear to be a very clear and innocuous statement, but the fact is not so. Do we start with a preconceived notion of rationality and consider *only* such intellectual pursuits as conform to that notion of rationality to be 'scientific'; or do we start with the general corpus of knowledge called 'science' at any given time and use such a pre-given notion of 'scientificity' as a criterion of rationality?

A look at the historical interplay between the norms of rationality and the norms of scientificity seems to land us in a vicious circle:

(*a*) It is the scientific temper of an age that determines its dominant concept of rationality; and
(*b*) The dominant concept of rationality of a given age determines what counts as scientific in that age.

In another paper ('From "Logos" to "Rationality" in Science', originally developed as a prelude to this paper) I delineated four periods in the evolution of the concept of rationality: *i)* the classical period, *ii*) period of optimism, *iii*) period of anti-science romanticism, and *iv*) the modern period, i.e. the period of reason contextualised.

The first two periods gave precedence to the notion of rationality as pre-given and tried to dictate what science should be like. The methodological attitude was predominantly deductive–analytic. The romantic period came as a reaction to this analytic–formalistic view of science. The later emphasis on the interpretive–hermeneutic dimension of science (as against the purely formal–analytic one) may be viewed as an off-shoot of the romantic period. Initially this was considered to be the dividing line between the 'hard' (formal, natural) sciences and 'soft' (social) sciences. There has been a growing awareness among the contemporary thinkers about the need to bring together the two apparently disparate dimensions of science (the analytic–empiric and interpretive–hermeneutic) as essential methodological components of science.

Here, I shall be concerned with capturing the changes in the concept of scientific rationality as it is taking place within the fourth, i.e. the modern period. For convenience, the modern phase is divided into an early and a later period which will be referred to as *modern* and *post-modern* respectively. I regard the Popperian views as ushering in the modern period while the views that originate by way of reaction to the Popperian view, beginning with Thomas Kuhn, belong to the post-modern period.

Popper pointed out the methodological inadequacy of Mill's inductivism. He offered his own 'falsifiability' model as an alternative and made it amply clear that he was an *anti-inductivist*. His falsificationism was promptly (and wrongly) interpreted as a kind of negative verificationism and almost inevitably, by an argument-by-default, his anti-inductivism was immediately appropriated as another brand of deductivism. Popperian 'criticism' is a sort of natural selection force. It allows only the fittest, i.e. the most resilient theories/hypotheses to survive the methodological selection pressure.

Popperian methodology is still guided by an optimistic model of pristine rationality. It is pristine in the sense that *irrationality* and *prejudices* are allowed no place in it. It is also ahistorical and purely objective. It is optimistic in so far as it is confident that through rational criticism (conjectures and refutations), it can lead us ever closer to objective truth. It differs from the Hegelian optimistic model in so far as it denies that pure reason, unaided by experience, can lead us to truth through a purely logical process. It differs from the positivist model as it denies the possibility of logical reconstruction of reality on the basis of empirical evidence and also denies that a *formal* justification of the reconstructed picture, i.e., scientific (theory) can be given. It excludes, *ab initio*, on theoretical grounds the possibility of any *genuine* proliferation of alternatives. It differs from inductivism in holding that our empirical evidences can only *put constraints* on the kind of conjectures we make, but such evidences *cannot determine* the nature of such conjectures. In short, theories are terribly underdetermined by experience. The fact that Popper believed that theories can be conclusively refuted despite the fact that Duhem had written long before him, shows that Popper was still labouring under an absolutist model of rationality.

However, Popper, deserves credit for preparing the ground for the subsequent development of more flexible and more down-to-earth models of rationality. Though Popper's model of rationality was still unalloyed and absolutistic, it was considerably tempered compared to that which dominated the era of optimism. First, the claim of being apodictical was given up. Secondly, the power of reason to reveal the nature of reality through pure ratiocination was denied. Thirdly, the claim of reason about its ability to disclose truth about reality in one go (even under the most ideal condition) was rejected. Reason has to follow the asymptotic path to reality through gradual approximation. Finally, the pristine quality of rational endeavour in science was in a sense compromised in so far as elements of chance and uncertainty were allowed entry in the arena of science via indispensability of lucky guesses (conjectures) and happy hits (potential falsifier–resilient theories). In other words, the non-rational, if not exactly the irrationals and prejudices, gained entry in an area which was, till then, considered to be the exclusive preserve of purely rational activity.

The road from here to the Kuhn–Feyerabend model of rationality was not very long. Kuhn (1962) tried to show that in the history of science there emerge typical patterns. There are periods of 'normal science' in which theoretical assumptions are not questioned, followed by comparatively shorter periods of upheaval involving 'paradigm shifts'. During such periods, science becomes self-critical and questions the very assumptions that sustained it during the immediately preceding period of normal science. Some new, disparate alternatives open up and the scientists are thrown into the unenviable situation of having to choose one among alternative paradigms without having any rational norms to guide the choice of one particular paradigm over another.

According to Kuhn, a choice that leads to a paradigm shift must, therefore, be a rationally unjustifiable choice. But the scientist *has to* make a choice and what determines it includes, among others, sociological factors and even the personal idiosyncracies of the paradigm-breaking scientist. When Kuhn compares paradigm-shifts to 'gestalt switches' and insists that 'proponents of competing paradigms are always at least slightly at cross-purposes' (Kuhn 1962, p. 148), or when he declares that there is 'no neutral algorithm for theory choice, no systematic decision procedure which, properly applied, must lead each individual in the group to the same decision' (ibid, p. 200) and also maintains that theory choice or paradigm shifts are legitimised only on the basis of a consensus among the scientific community, Kuhn's critics charge him of vitiating the image of science into an irrational, subjective and relativistic activity where mob psychology rules (cp. D. Shapere 1964; I. Scheffler 1967). Laudan interprets Kuhn as arguing that 'choices between competing scientific theories, in the nature of the case, *must be irrational*' (Laudan 1977, p. 3).

To say this, however, is a misinterpretation of Kuhn's view. Kuhn's intention was never to claim that scientific inquiry is irrational but rather to show the way to a more open, flexible and historically oriented understanding of scientific inquiry as a rational activity. He is suggesting that we need to transform our concept of rationality in order to achieve a better understanding of the nature and changing pattern of scientific inquiry. Kuhn argued that there is something fundamentally wrong with the conception of science propagated by most mainstream orthodox philosophers of science. This orthodox image of science is itself a blending of deeply entrenched dogmas inherited from traditional empiricism (Bacon–Mill inductivism) and rationalism (positivistic scientism). Kuhn attempted to sketch an alternative image of science by redefining the conception of rationality. His aim was to effect a fundamental shift of our paradigm of rationality and consequently that of scientificity as well.

The hostile reactions of Kuhn's critics should be seen in this context. If the orthodox image of science and the familiar established paradigm of rationality is to be given up, then since *all* justifications are *intra-paradigmatic*, those who are used to the orthodox paradigm and regard it sacrosanct must consider the new paradigm 'irrational'. They fail to realise that rationality appraisals make sense only *within* a set paradigm. Thus inter-paradigm incommensurability is regarded as irrational, non-objective and as smacking of psychologism. The situation is comparable to special relativity where each observer considers his own frame of reference as inertial (absolute) while *all others* are thought to be in motion. Here the inertial frame (when absolutised) corresponds to the well-entrenched orthodox paradigm.

Similarly, the charges of subjectivism and of resorting to 'mob psychology' are also not surprising. If universality of reason is considered to be the very foundation of objectivity, then any view that challenges the orthodox belief in the trans-historical universality of reason (which was unquestioned till the age of optimism) must be counted as 'subjective' by established orthodoxy. The allegation of mob-psychologism, on the other hand, is due to the failure to see (engendered by uncritical commitment to the orthodox paradigm of rationality) that reason is *not* a faculty that can free itself from its historical context. Reason is always historically situated. It gains its distinctive power *always within* a living tradition. Hence, acceptance by common consensus, just because it is community-approved, need not fail to qualify as legitimate and reasonable. Keeping the mob away does not necessarily augment reasonability. (It is a separate, though very crucial, issue to figure out in detail how a community happens to evolve such rational consensus that Kuhn speaks of. This, however, is outside the purview of this paper.)

Let us take a look a Kuhn's main objection against the falsificationist model. Popper's model of rationality, as we have seen, is based on the assumption that (*i*) reason is *one*, i.e. single-paradigmed; (*ii*) reason is purely *objective*; and (*iii*) rational growth is *continuous* and *unilinear*. This implies that every alternative scientific theory is either *definitely falsifiable* or not with reference to the established paradigm. Moreover, their degrees of reasonableness (as measured in terms of their power of resilience) are comparable on the *same* scale. Now, if paradigms shift and rational growth is non-continuous and multi-linear (as Kuhn claims and as is corroborated by the actual history of growth of scientific knowledge) then *incommensurability* of the alternative paradigms becomes an unavoidable necessity. Inter-paradigmatic incommensurability comes to stay.

Moreover, Kuhn points out that the 'fit' between a 'paradigm' and 'nature' is never perfect. There are always some discrepancies that a simple internal adjustment of the existing paradigm cannot explain. Thus, as an established paradigm gets used over longer and longer periods, its unexplained residues pile up cumulatively until they reach the 'critical mass' which triggers off a reaction leading to a paradigm shift. The interim period between the formation of the 'critical mass' and the period when the paradigm-shift finally occurs is called the *period of crisis*. During this crisis period rival paradigms are proposed. Scientists do not reject an established paradigm when anomalies and discrepancies are discovered. They try to explain the anomalies by mending the set paradigm. This is a historical fact. But if Popper's model were correct, each anomaly would falsify the established paradigm and would thus necessitate its rejection without a second hearing. This is contrary to the standard practice followed in scientific research. Again, due to the less-than-perfect fit that inalienably exists between a 'paradigm' and 'nature', every theory and every paradigm is, according to Kuhn, 'born refuted'. This is the main reason why Kuhn thinks that the falsificationist model of science is untenable and the model of scientific rationality it suggests is misleading.

I have already pointed out that inter-paradigm incommensurability is a logical corollary of Kuhn's view. This, in turn, entails *relativism*. Moreover, it is impossible, in principle, to have a criterion of rational justification for choosing between rival paradigms. A shift of paradigm is legitimised only by a *common consensus* of the scientific community. Further, Kuhn himself says that in controversies that arise when new and rival paradigms are proposed, the situation is comparable to political revolutions. There are no straightforward criteria that are sufficient to resolve the dispute.

As Richard Bernstein puts it:

> Kuhn has been struggling to articulate when dealing with the complex issues of theory-choice and paradigm-switches—his insistence that reasons function as values which can be differently weighed and applied to concrete situations, and his defense of the role of judgement in making choices and decisions—are closely related to Gadamar's analysis of *phronesis* and the role that it plays in all understanding and interpretation (Bernstein 1983, p. 40).

That this interpretation is correct is borne out by Kuhn's own writings:

> What I as a physicist had to discover for myself, most historians learn by example in the course of professional training. Consciously or not, they are all practitioners of the hermeneutic method. In my case, however, the discovery of hermeneutics did more than make history seem consequential. Its most immediate and decisive effect was instead on my view of science (Kuhn 1977, p. xiii).

Along with Kuhn's hermeneutic connection, if we keep in mind the impact of Wittgenstein's later philosophy on him, it is not difficult to see that working under a Kuhnian paradigm is much like living in a particular *form of life* in Wittgenstein's sense. As Radnitzky and Andersson put it very aptly: 'The paradigms are incommensurable—much like the monadic forms of life in the later Wittgenstein. They cannot be criticised except from within' (Radnitzky and Andersson [eds.] 1978, p. 9).

The model of rationality that Kuhn's view of science implies bears comparison with Popper's. Popper's model of rationality as has been pointed out earlier, was still a sort of optimistic objectivism though its optimism was mellowed down. A groping quality, an element of imponderables found a recognised place in his model of rationality. Science was no longer a closed game of deductions from *a priori* truths, but an open game of conjectures and refutations. Yet the vestiges of deductivism (via the emphasis on the hypothetico–deductive method) and of impersonal objectivism (via the one-track route to a uniquely convergent representation of reality) still dominates the Popperian model of rationality. Only the substitution of 'verisimilitude' for 'perfect fit' makes Popperian objectivism much more sober than the utopian objectivism of the period of optimism. Popper's view may thus be characterised as 'mollified anti-inductivist objectivism'. In this model of rationality some imponderables find place but neither the irrational nor the subjective elements do. The role, if any, played by history and sociology in the make-up of rationality (and thus of scientificity) is merely external.

In contrast, Kuhn's model of rationality (as shaped by his notion of scientificity) is socio–historically rooted anti-inductivist interpretationism (I avoid using both 'subjectivism' and 'relativism' because of their misleading overtones).

The subsequent models of rationality along Popperian line as developed by Imre Lakatos, J.W.N. Watkins, etc. are different variations of the same basic theme. Lakatos sticks to Popperian objectivism (criticising Kuhn for being irrational and subjectivist) but rejects falsificationism as an adequate model of growth of scientific knowledge. Moreover, he proposed a new unit of appraisal. Instead of emphasising 'theories' as Popper did, he focused on 'scientific research programs'. This made his view more holistic and anti-positivistic than Popper's and also prompted him to replace the single test-criterion (falsifiability) by a triplet: (*i*) empirical content, (*ii*) successful predictions, and (*iii*) heuristic power.

There is a sort of incompatibility between taking verisimilitude seriously and admitting 'degrees of falsifiability' unless there is some way of mapping the different degrees of falsifiability (Watkins' 'empirical content') one-to-one on the converging scale of verisimilitude that ultimately tends to zero-in on the true nature of reality. (For a puritan version of falsificationism, 'degrees of falsification' hardly makes sense though.) Watkins keeps the Popper model, including falsificationism, intact except for modifying it in such a way that the idea of comparing empirical contents (of different theories) can be rescued. Watkins relates empirical contents to how many scientific questions a theory can answer. He tries to show that it is possible to compare empirical contents in a set-theoretical way with the help of the idea of 'counterpart consequences' (different answers to the same question). The success of this programme would guarantee the overcoming of the problem of inter-theoretic incommensurability. However, both Feyerabend and Hübner are quick to point out in response that 'content comparison is impossible because contents of theories are not subsets of each other' (Radnitzky and Andersson [eds.] 1978, part IV).

The objection against Popper, based on the impossibility of 'content comparison' (or equivalently, the objection that falsification as such cannot handle the problem raised by incommensurability), could be avoided if the notion of verisimilitude could be given a formal characterisation. Various attempts to that effect have also been made, but without success. D. Miller (1976) has criticised various formal/semi-formal redefinitions of verisimilitude in a convincing way. Such criticisms tend to suggest that verisimilitude may not be explicable through a definition couched in a formalised language.

If Lakatos and Watkins are modified Popperians, Feyerabend is a modified Kuhnian. Against Popper's impersonal objectivism (which both Lakatos and Watkins defend) Kuhn's socio–historically rooted interpretationism allows prejudices and even the irrational elements (not merely the imponderables) a role in scientific theory construction. Feyerabend carries this to the extreme, gives the irrational elements and prejudices the centre-stage in his scheme, tirades 'against method' and regards the switching of paradigms as more akin to political propaganda than to a result of neutral rational argumentation. When this is sprinkled with a pinch of Duhemism, his 'anarchistic' methodology comes to admit that 'anything goes'.

It is true that in any field of *creative* human activity (which scientific research is) there must be an element of free play by the mind which is not *rule-governed* and thus non-rational. At the same time, there must be some methodologically imperative rule-governed constraints that govern such creativity. Thus scientific research (being a creative activity) must be viewed as an interplay between 'madness' (rule-defying free play by the mind) and 'method' (subservience to rule-governed constraints). Popperianism can capture only the method aspect while Feyerabend's 'anarchism' is better suited to capture only the madness aspect. But what we need is a model of scientificity (and thus of rationality) that can capture the method-in-madness as one integral process.

Kuhn's model of socio–historically rooted reason would be all right had it not been for the fact that it fails to indicate how the role of socio–historical factors, thought not rule-governed, is not totally chaotic either. Moreover, Kuhn even fails to give the slightest indication as to how the impact of the socio–historical factors can be put to order. Both the Popperians and the Kuhnians fail to figure out a new kind of logic of justification beyond deductivism–inductivism bipolarity. Popper refuses to consider inductive logic as logic proper, while Kuhn harps on the impotency of deductivism in the face of the non-rational elements in science.

The socio-historical factors that influence theory-choice and paradigm-switch constitute what we may call 'the background knowledge' of the scientists at work. Commonsense is a part of this background knowledge. Alan Musgrave (1974) emphasises the role of this background knowledge in scientific research and specifies three different senses of such knowledge. Once the boundary of this background knowledge is roughly circumscribed, we can discern a method-in-madness—a justificatory method which is neither deductive nor inductive in nature. Such a justification is based on non-monotonic reasoning (NMR) or, what is also called, default logic (DL). I have shown in detail elsewhere (Sarkar 1992, chapter 20) how NMR can be profitably exploited to formulate a new kind of justificatory criterion and consequently, in developing a new model of rationality. DL also helps us to make better sense of Popper's 'conjectures and refutations' model.

I shall now suggest the outlines of a new model of rationality with the following characteristic features. It makes (*i*) the notion of rationality context-bound (which is the characteristic feature of the models of rationality, especially in the post-modern period); (*ii*) it is based on a fuzzy version of default logic; (*iii*) it avoids both *apriorism* and positivistic empiricism, i.e., it starts with a groping search for the right hypotheses; (*iv*) it is interpretationistic, i.e. not impersonally objectivistic; and (*v*) it is accommodative of pragmatically dictated *paraconsistency*. The second of these features takes care of reason's rootedness in and dependence on background knowledge via the DL component, while it rejects one of the absolutistic assumptions, viz. the assumption of precise definability via its fuzziness component. In addition, dependence on DL guards our model against succumbing to deductivism by rejecting the Leibnizian assumption.

The last feature in the list, viz. (*v*) , indicates that in such a model pragmatic success gets top priority when it comes to matters of theory-choice or paradigm-switch, and therefore, when dictated by pragmatic considerations, a PFHD-model-based rational system may tolerate 'deep faults' in its foundation, even if the subsystems, on the two sides of the fault are logically irreconcilable. This happens to be the case with modern physics which, some physicists admit, has a logical crack in its foundation, e.g. between quantum mechanics and relativity. We shall call this model of rationality incorporating *all* the five features mentioned earlier, the 'Paraconsistent Fuzzy Hypothetico–Default' model (PFHD-model for short) of rationality.

Let us now clarify the concept of rationality (and of scientificity) the PFHD-model commits one to. First, as this is a variant of the models in which reason is contextualised, it rejects the idea of an absolute reason in favour of the view that all rationality, including mathematical reasoning, are essentially context-relative. No action is rational in all possible contexts, i.e. context-invariantly. The notions of absolute rationality, relative rationality, rootedness of reason in, what Wittgenstein calls, 'form of life', etc. have been discussed by me elsewhere (Sarkar 1992, chapters 8, 16, 17).

Secondly, if reason is essentially context-relative, then there are no truths of reason, not even the laws of logic, which cannot be revised. Admissibility of paraconsistency by the PFHD-model reflects this feature.

Thirdly, its HD-component (hypothetico–default component) shows that rational theory construction, just like any other empirical construction, is basically a slow, groping, approximative trial-and-error process.

Fourthly, since DL is a part of the logical machinery underlying the PFHD-model, some proved metatheorems about DL can be used to justify some philosophically significant methodological claims about it. For example, Israel and Reiter have shown that the formal approaches to non-monotonic inference, with which DL is concerned, are in general *non-semi-decidable*. This implies that any system that uses a DL-based non-monotonic approach will necessarily make mistakes from time to time without ever being able to detect such mistakes. In the present context it means that any scientific theory based on the PFHD-model of rationality is bound to be *fallible* and revisable. (For elaboration, see Sarkar 1992, pp. 222–29)

Lastly, the PFHD-model of rationality carves out a notion of scientificity which gives a better fit with the actual history of theory construction in science. In other words, this last feature of the PFHD-model shows that it is a theory of rationality which is *post-facto*, i.e. constrained by the demand of facts rather than being an *a priori* theory that tries to tailor facts to suit the logical needs of a theory.

The power of the PFHD-model of rationality becomes clear when we try to handle the problems of incommensurability and 'justification' of the paradigm-switch. It is true that when one shifts from the Newtonian to the relativistic framework, the concept of 'mass' in the two frameworks no longer remains the same. It is also true that they are *not congruently comparable* across frameworks. But it is equally undeniable that the concept of mass is *recognisably similar* in both the systems, e.g. in both, 'mass' satisfies the equation F = ma, though in Newtonian physics *m* is velocity-invariant and constant, while in Einsteinean physics it is not. So there is no *absolute incommensurability* between the two concepts of mass.

An analogy might be helpful here. Botanists would say: 'A flower is a modified shoot'. But neither in appearance nor in function is a *flower* comparable to (commonsurable with) a *shoot*. But in a sense they *are* the same. They are commensurable though not straightaway congruent. Let us call it *non-congruent* commensurability. Using the idea of fuzzy-similarity and fuzzy-identity such a notion of non-congruent commensurability can be defined and logically handled.

Once it is shown that thc incommensurability that Feyerabend speaks of, is only a case of non-congruent commensurability, we can take a new look at the problem of reconciling the continuity of scientific progress with discontinuity due to incommensurability. The formal aspect, showing how this can be done or how non-congruent commensurability can be defined in terms of fuzzy similarity, is the subject matter of a different paper. A similar view, roughly along this line was expressed (in an informal and somewhat amateurish way) in one of my earlier papers: 'A Plea for Sobreity in Anarchism: Contra Feyerabend' (1977). (This paper was presented at the Graduate Seminar at the University of Waterloo.)

Besides being powerful enough to handle non-congruent commensurability, the PFHD-model offers one unified model of scientific rationality in place of two different types of scientific rationality—one suited for the natural sciences and the other suitable for the social sciences. Philosophers who are eager to preserve the autonomy of social science without giving up its claim to scientificity argue for a two-sorted model of rationality. It takes different forms. Peter Winch's notion of 'rule following' (rather than 'rule-governed'), Max Weber's notion of 'Verstehen', etc. are examples. The two main arguments underlying such a view are the following:

(*a*) The kind of objectivity that natural sciences aim at is basically different from the kind of objectivity that is achievable in social science. So the concept of rationality in social science must be of a distinct kind.

(*b*) Natural sciences are descriptive–positive sciences, while social sciences are basically normative–interpretive in nature. So, the kind of objective formal reason that works well for natural science does not work when it comes to the methodology of social sciences.

It is clear that the PFHD-model, by rejecting the notions of absolute reason and impersonal objectivity on the one hand and by emphasising the inalienability of the hermeneutic (interpretive) dimension of both the natural and social sciences on the other, brings them methodologically closer together. The PFHD-model thus suggests a way of achieving the ultimate goal of unity of science without taking a reductionistic approach. That this is the general tenor of contemporary philosophy of science is borne out by Mary Hesse who, after elaborating the five methodological points of distinction between natural and social sciences, concludes the section thus: 'It follows . . . that in natural science a one-way logic and method . . . is appropriate . . . . In human science, on the other hand, the ('logic') of interpretation in irreducibly circular . . . .' She then goes on to add:

> What is immediately striking about it to readers versed in recent literature in philosophy of science is that almost every point made about the human sciences has recently been made about the natural sciences, and the five points made about the natural sciences presuppose a traditional empiricist view of natural science that is almost universally discredited (Hesse 1980, pp. 171–72).

Another interesting and significant spin-off from our model of rationality is that there is no sharp cut-off point separating the sphere of the rational from the sphere of the irrational. The latter shades off imperceptibly into the former. As a consequence, the line of demarcation between 'science' and 'non-science' (say, magic) also becomes blurred. The 'metaphysics' of one age can become 'physics' of another (e.g. transition from the atomism of Democritus to the atomism of Dalton and the transition from ancient cosmogony to modern cosmology). The reverse may also happen, e.g. the notion of an absolute frame of reference in Newtonian physics is now treated as a metaphysical fiction after the advent of relativity.

Finally, let us see how the two-way feedback loop linking the notions of rationality and scientificity can be understood in terms of the PFHD-model without being threatened by a vicious circle. The vicious circle we spoke of earlier was only apparent. We can never *start* either with a cut-and-dried, pre-given notion of rationality or with such a notion of scientificity. At every stage of history the two remain locked in a process of development through a constant dialectical interaction. The vicious circle can arise only if we allow it to. Just as there is nothing paradoxical in the case of the famous chicken-and-egg paradox, there is nothing vicious in the two-way closed loop (circle) holding between the concepts of scientificity and rationality. In no point of history does one start either with a chicken or with an egg. It is only through a long process of evolution (in which an organism interacts with its environment) that chicken and egg came to be differentiated and related in the way they are. Only if we ignore their evolutionary history it does seem paradoxical. The same holds, *mutatis mutandis*, of the relation between scientificity and rationality. Each of these concepts is being continuously shaped through interaction with the other. By projecting backward in time, the PFHD-model prompts us to expect a homogeneous matrix, an amalgam of scientificity and rationality in their undifferentiated form during the earliest stage of human civilisation. History shows that this expectation is justified.

It seems plausible to think that a mathematical model of this progressive dialectical interaction between rationality and scientificity may be developed by dovetailing the theory of positive feedback with the catastrophe theory. I do not have the required mathematical competence but I think this is a fertile area with vast scope for concerted teamwork involving philosophers of science and professional mathematicians. In the meanwhile, I will consider my effort amply rewarded if the PFHD-model of rationality that I have suggested is able to arouse some interest.

References

BERNSTEIN, R.J. 1983. *Beyond Objectivism and Relativism*. London: Basil Blackwell).

FEYERABEND, P. 1975. *Against Method*. London: NLB.

———. 1978. 'Science in a Free Society', Vol. II. of *Against Method*. London: NLB.

———. 1978. 'The Gong Show–Popperian Style' in *Radnitzky*, 1978.

GADAMAR, H. 1975. *Truth and Method*. New York: Sainsbury Press.

———. 1977. *Philosophical Hermeneutics*. Berkeley: University of California Press.

HABERMAS, J. 1973. *Theory and Practice*. New York: Beacon Press.

HESSE, MARY. 1980. *Revolutions and Reconstructions in the Philosophy of Science*. Brighton: Harvester Press.

HUBNER, K. 1978. 'Reply to Watkins' in *Radnitzk*, 1978.

KUHN, T. 1962. *Structure of Scientific Revolutions*. Chicago: University of Chicago Press.

——— 1977. *Essential Tension*. Chicago: University of Chicago Press.

LAKATOS, and I.A. MUSGRAVE. 1970. *Criticism and the Growth of Knowledge*. Cambridge: Cambridge University Press.

LAUDAN, L. 1977. *Progress and its Problems*. London: Routledge & Kegan Paul.

MILLER, D. 1976. 'Verisimilitude Redeflected' (BJPS).

MUSGRAVE, A. 1974. 'Logical Versure Historical Theories of Confirmation' (BJPS).

RADNITZKY, G. *et al.* (eds.) 1978. *Progress and Rationality in Science*.

SARKAR, T.K. 1992. *Knowledge, Truth and Justification*. New Delhi: Allied Publishers.

SCHEFFLER, I. 1967. *Science and Subjectivity*. (Bobbs-Merrill)

SHAPERE, D. 1964. 'The Structure of Scientific Revolutions', *Philosophical Review*.

WATKINS, J.W.N. 1978. 'Corroboration and the Problem of Content Comparison,' in *Radnitzky*, 1978.

WINCH, PETER. 1958. *The Idea of a Social Science and Its Relation to Philosophy*. London: Routledge and Kegan Paul.

'Identity' and 'Property' in Vagueness

M.K. Chakraborty
Mohua Banerjee

Rationality yields precision—thus goes the standard belief; and by contraposition, vagueness embodies less of rationality. But the complexity of the world around us is enormous and 'as the complexity of a system increases, our ability to make precise and yet significant statements about its behaviour diminishes until a threshold is reached when precision and significance (or relevance) become almost mutually exclusive characteristics' (Zadeh 1973). So vagueness is not a dead issue; rather, it is an issue that generates vigorous debate. Should a rational thinker consider vagueness as a defect? Should s(he) believe that vagueness is 'curable' or at least recedes with the acquisition of knowledge? Is 'vagueness' a characteristic ascribable only to the language used to describe reality or to reality itself or to both? Can there be vague objects so that their identity turns out to be vague and consequently statements about their identity are of indeterminate truth value? Is it meaningful to talk about the indeterminate identity of objects disapproving, at the same time, vagueness in objects?

Instead of entering into all the intricacies of these debates, I shall begin with two relatively accepted assumptions: (*a*) Leibniz's principle as the criterion of identity; and (*b*) the existence of vague properties. By vagueness we shall not mean indeterminacy only; the latter will prove to be a special case of the former. This point will be made clear by the discussions that follow. It will be observed that as a consequence of assumptions (*a*) and (*b*), identity in some situations may turn out to be vague. Another aspect of this article is the study of the interrelationship between the two notions of 'identity' and 'property'. Can there be any notion of identity without a notion of property and any notion of property without a notion of identity? Some light will be thrown on these questions as well.

A vague statement is one that has an intermediate truth value, a value between 'true' and 'false'. As a special case, 'indeterminate' will be considered to be a value different from 'true' (1) and 'false' (0) and relative to some supposed ordering lying between them. Generally, the truth value set is taken as a lattice, finite or infinite. It will be supposed that there are vague statements. If P is any predicate and 'a' the name of an object, Pa may obtain a truth value other than (0) or 1 for various reasons. I shall primarily take into account the situation when this happens due to an intrinsic vagueness of the predicate P, that is when Pa is neither fully true nor fully false. In other words, it is partly true and partly false. In my view, the root of real vagueness in any predication lies in the existence of some object to which the predicate both applies and does not apply. This view, however, is not generally accepted by philosophers. If 'a' is any instance in the borderline (which again cannot be sharply drawn) of the predicate P (say, an observational predicate like 'is orange') then Pa may be thought to be true to some extent and also false to some extent. Obviously and naturally, the law of the excluded middle is violated. How such a truth value should be determined is an issue outside the scope of this paper (in fact, outside the scope of logic as such).

According to Leibniz's law of identity

(L): a = b if and only if (Pa <—> Pb, where P is any property).

This principle may be considered as the definition of identity (Tarski 1965), 'a = b' being the definiendum and the statement on the right-hand side of 'if and only if' being the definiens. The trouble with the definiens is the presence of the word 'any' in it without the mention of the domain from which the 'properties' are to be chosen. There are also other epistemological problems (Gorsky 1987). Moreover, we are not taking the position expressed in the thesis of Heraclitus that one cannot step into the same river twice. That means we would like to see 'sameness' in a person who has lived over a stretch of time and whose hair has turned gray from black. So only *some* rather than all conceivable properties need to be considered in identification. This 'some' depends upon the purpose of individuation. It is unclear, at least at this stage of technological development, if there could be some absolutely essential properties for identification. Nowadays, a person's sex can be changed and yet retain the person. Theoretically, there is no reason to disbelieve that a biped may transform itself into a deer D. Marich is the same living being as D but not the same biped as D.[1] Thus, in my opinion, for identification some essential properties are required and these are determined by the purpose. The principle (L) may then be modified to

> (L′): a = b if and only if (Pa <—> Pb, where P is any property belonging to some relevant set).

* * *

Leibniz's law, be it in the form (L) or in the modified form (L′), presupposes the notion of 'property'. Is the notion of property more evident than the notion of identity? This point has validly been raised by Tarski (1965). A property is ascribable to some already individuated entities. While making statements like 'a has P' or 'a does not have P', it is presumed that 'a' has been individuated and hence for any 'b' it is clearly understood whether $a = b$ or $a \neq b$. From the extensional point of view, a property is a subset of some set and in the latter, elements are properly individuated. One has to answer whether *this* 'a' falls under the set P or not. So it may appear that there is some circularity in the definitions (L) or (L′). '=' is understood in terms of 'property' and 'property' in terms of '='. But this is only apparent. There are levels of identification. At any level (1_1, say) if an identity is understood, properties can be defined (extensionally as subsets). Some or all of these properties shall determine, by (L′), an identity at the next level (1_2, say). The identity of level 1_1 is determined by properties of the earlier level (1_0, say). One thus obtains a hierarchy of identities. The following diagram may be helpful in clarifying this hierarchy.

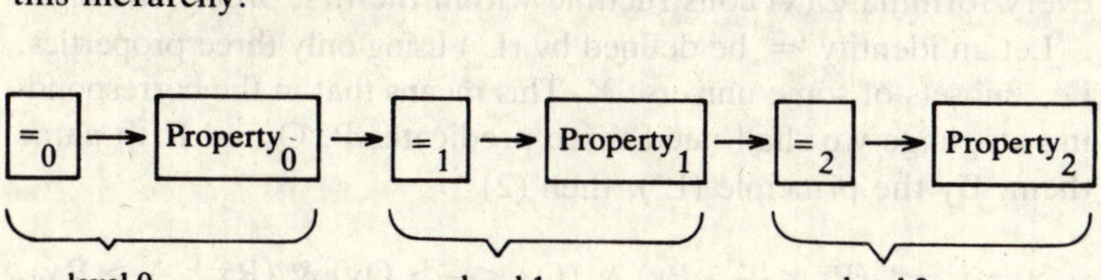

Mathematically, the phenomenon may be described thus: Let X be some universe, the elements being fully individuated so that any subset of X can be sliced out neatly. A collection of subsets of X are taken. These will be (extensionally) the relevant properties to define an identity ($=_1$) by the principle (L′). $=_1$ will then determine subsets of X that may be called properties (of level 1) relative to $=_1$. Some of these latter subsets may be taken again to define an identity $=_2$ of level 2 which in turn will determine those subsets that are called properties of level 2. This process continues. An identity determines property of the same level but property determines an identity of the next higher level. Although some $=_i$ may be the same as $=_{i-1}$, this is not in general true. And thus no circularity is involved.

It will be observed that $=_1$, $=_2$, etc. are equivalences in traditional mathematics. One can refuse to call them genuine identities and maintain that the identity in X which we have started with and which allowed us to make slices in X, is the only genuine one. But that there is not much reason in this belief is discussed in the next section.

An identity relation is defined axiomatically by the following axioms:

(1) $x = x$.
(2) $x = y \rightarrow y = x$.
(3) $(x = y \,\&\, y = z) \rightarrow x = z$.
(4) $(x = y \,\&\, \varnothing(x)) \rightarrow \varnothing(y)$, where $\varnothing(x)$ is any predicate (well-formed formula of the language) containing x free and y is free for x in $\varnothing$.

It is known that this axiomatisation admits non-standard models (Bridge, 1977). It is also known (Mendelson 1979) that if (4) i.e. substitutivity holds for the atomic formulae only, then it holds for every formula $\varnothing(x)$ constructible within the first order language.

Let an identity '=' be defined by (L′) using only three properties, i.e., subsets of some universe X. This means that in the corresponding language we shall need three predicates P, Q and R to name them. By the principle (L′), then (2)

$$x = y =^{def} (Px \leftrightarrow Py) \,\&\, (Qx \leftrightarrow Qy) \,\&\, (Rx \leftrightarrow Ry).$$

= satisfies all the axioms (1) to (4), meaning thereby that the formulae are all theorems. The general statement in this regard may be stated as the following proposition:

Proposition 1: Let x ~ y be defined by
$(P_1x <\!\!-\!\!> P_1y)$ & $(P_2x <\!\!-\!\!> P_2y)$ & . . . & $(P_nx <\!\!-\!\!> P_ny)$
(P_1, P_2, . . ., P_n being the monadic predicates of a first order language. Then the following are theorems in the language.

(1) x ~ x.
(2) x ~ y → y ~ x.
(3) (x ~ y & y ~ z) → x ~ z.
(4) (x ~ y & ∅(x)) → ∅(y), where ∅(x) is any well-formed formula of the language with x free and y free for x in ∅.

Proof: See appendix.

What are the 'properties' relative to =? In the language they are precisely the formulae ∅(x) with x as a free variable. In the interpretation, the referent of ∅(x) is a subset of X. So these subsets of X may be called properties. But not every subset of X can be referred to by some ∅(x) because ∅(x) will depend ultimately on the basic predicates P, Q, R and various logical combinations of the atomic formulae constructed out of them.

Let us look at this point from another side, from the viewpoint of interpretation. Let a subset A of X satisfy the saturatedness condition (S) relative to the equivalence = if and only if

(S): a = b and a ∊ A imply b ∊ A.

It has been proved in (2) that a subset A satisfies (S) if and only if A is the union of some of the equivalence classes determined by the extensions of the predicates P, Q, R. These classes are shown in the following figure.
The regions 1 to 8 may be obtained by using the set-theoretic operations on the regions P, Q, R, In fact.

region 1 = $P \cap Q \cap R$,
region 2 = $(P \cap Q \cap R)' \cap (P \cap Q)$,
region 3 = $(P \cap Q \cap R)' \cap (P \cap R)$,
region 4 = $(P \cap Q \cap R)' \cap (Q \cap R)$,

region 5 = $(Q \cup R)' \cap P$,
region 6 = $(P \cup Q)' \cap R$,
region 7 = $(P \cup R)' \cap Q$, and
region 8 = $(P \cup Q \cup R)'$.

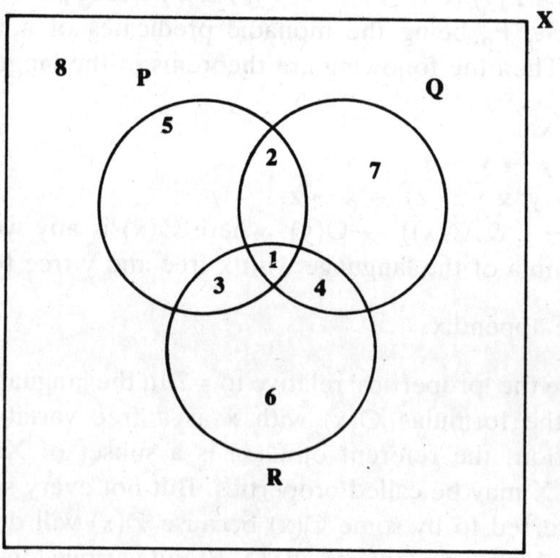

So, any subset of X that may satisfy the condition (S) is obtained from P, Q, R by union, intersection and complementation and hence is the referent of some formula $\varnothing(x)$ in the language. From this discussion it is clear that the referent of any formula $\varnothing(x)$ satisfies (S) because $\varnothing(x)$ satisfies (4) and conversely any subset satisfying (S) is the referent of some $\varnothing(x)$. So the following definition of property may be legitimately adopted:

Definition: Given an equivalence relation $\sim$ in a set X, a property relative to $\sim$ is a subset A of X that satisfies the saturatedness condition

$$a \sim b \text{ and } a \in A \text{ imply } b \in A$$

We also state the following proposition whose truth may be established without much effort.

Proposition 2. If $x \sim y$ is defined by

$(P_1x <\!\!-\!\!> P_1y)$ & $(P_2x <\!\!-\!\!> P_2y)$ & . . . & $(P_nx <\!\!-\!\!> P_ny)$
$P_1, P_2, \ldots, P_n$ being the predicates of a first order language then in any interpretation X, ~ denotes an equivalence relation.

Conversely, given any equivalence relation ~ in X with finite partition, there exists a language in which ~ is an identity defined by

x ~ y is $(P_1x <\!\!-\!\!> P_1y)$ & $(P_2x <\!\!-\!\!> P_2y)$ & . . . & $(P_nx <\!\!-\!\!> P_ny)$

$P_1, P_2, \ldots, P_n$ being the predicates of the language.

Proof: The direct part of the theorem follows from *Proposition 1*. For the converse, let ~ be an equivalence on X inducing partitions $P_1, P_2, \ldots, P_n$ of X. We consider a first order language with exactly n-many unary predicates (also denoted by $P_1, P_2, \ldots, P_n$) and define x ~ y by

$(P_1x <\!\!-\!\!> P_1y)$ & $(P_2x <\!\!-\!\!> P_2y)$ & . . . & $(P_nx <\!\!-\!\!> P_ny)$

By *Proposition 1*, it follows that ~ is an identity predicate of the language.

We shall conclude this section with the following summary:

- A set of properties (extensionally subsets) determine an identity.
- Identity determined thus is an equivalence satisfying the substitutivity condition with respect to some subsets.
- These subsets obey the condition (S) relative to the equivalence and are called properties.
- Any equivalence on X with finite partition can be viewed as an identity satisfying axioms (1), (2), (3) and (4).

In the next section this issue is dealt more mathematically in a more general set-up including vague properties.

*

Let $L(\vee, \wedge, 0, 1)$ be a complete lattice and *, a binary operation in L satisfying the following conditions:

(*i*) L (*, 1) is a commutative monoid and
(*ii*) $(\vee_p\, \alpha_p) * \beta = \vee_p(\alpha_p * \beta)$ (V denotes supremum).

For an arbitrary collection $\{\alpha_i\}_{i\in I}$ of elements of L, I being a non-empty index set, we define $\Lambda \{\alpha_i\}_{i\in I}$ as
$\Lambda\{\alpha \in L: \alpha = \alpha_{j^1} * \alpha_{j^2} * \ldots * \alpha_{j^n}, \alpha_{j^k} \in \{\alpha_i\}_{i\in I}, k = 1,2,\ldots, n, n \geqslant 1\}$
where Λ denotes the infimum, L is the truth set and * the operation for computing 'and'.

Let X be a set and $\tilde{A}_i: X \rightarrow L$, $i \in I$, be mappings called fuzzy subsets of X. Let $\epsilon: X \times X \rightarrow L$ be a mapping defined by

$$(L''): \epsilon(\langle a,b\rangle) = \Lambda \{\tilde{A}_i(a) \longleftrightarrow \tilde{A}_i(b)\}_{i\in I}, a,b \in X$$

where the operation $\rightarrow$ for implication is defined by

$$a \rightarrow b = V_x \{x \in L : a * x \leqslant b\} \text{ and}$$

$$a \longleftrightarrow b \text{ is the element } (a \rightarrow b) * (b \rightarrow a)$$

Proposition 3: ϵ satisfies the following conditons.

(e_1) $\epsilon(\langle a,a\rangle) = 1$.
(e_2) $\epsilon(\langle a,b\rangle) = \epsilon(\langle b,a\rangle)$.
(e_3) $\epsilon(\langle a,b\rangle) * \epsilon(\langle b,c\rangle) \leqslant \epsilon(\langle a,c\rangle)$.

The proof of this proposition is routine, and it is not necessary to give it here.

ϵ will be called a vague identity determined by the properties $\{\tilde{A}_i\}_{i\in I}$. This is a fuzzy equivalence relation [3] and reduces to an ordinary equivalence when all the properties $\tilde{A}_i$, $i \in I$, are non-vague.

The generalized saturatedness condition relative to the vague identity ϵ is given by

$$(S'): \epsilon(\langle a,b\rangle) * \tilde{A}(a) \leqslant \tilde{A}(b), a,b \in X, \tilde{A} : X \rightarrow L.$$

(S′) is a generalization of (S). In [2], a necessary and sufficient condition for a fuzzy set $\tilde{A}$ to satisfy (S′) has been obtained.

We now pass on to the next proposition.

Proposition 4. Each $\tilde{A}_i$, $i \in I$, satisfies the generalized saturatedness condition relative to ϵ. This proposition has been proved in [2].

(L″) is the generalized version of (L′).

We shall now define the hierarchy of identities.
Let $\zeta_1 = \{\tilde{A}^1_{i^1} : X \rightarrow L\}_{i^1} \in I_1$ be a collection of fuzzy subsets and ϵ^1 the identity defined on X by (L″). ϵ^1 will be called an

identity at level 1. Let ζ_2 be the collection of fuzzy subsets that satisfy the condition (S′) relative to ϵ^1. From proposition 4 it follows that $\zeta_1 \subseteq \zeta_2$.

Let $\{\tilde{A}^2_{i2}\}_{i^2 \in I_2} \subseteq \zeta_2$. We define $\epsilon^2 : X \times X \rightarrow L$ as

$$\epsilon^2 (<a,b>) = \Lambda \{\tilde{A}^2_{i2}(a) <\!\!-\!\!> \tilde{A}^2_{i2}(b)\}_{i^2 \in I_2}$$

ϵ^2 is an identity on X and shall be placed at level 2. Generally, if an identity ϵ^{k-1} at level $k-1$ is defined, by taking any collection of fuzzy subsets satisfying (S′) relative to ϵ^{k-1}, an identity ϵ^k at level k may be defined.

Proposition 5. Let ϵ^{k-1} and ϵ^k be so defined that the properties used to define ϵ^k include those used to define ϵ^{k-1}. Then

$$\epsilon^k (<a,b>) \leqslant \epsilon^{k-1} (<a,b>)$$

The proof is straightforward.
An immediate corollary of this proposition is the following result:

$$\text{If } \zeta_1 \subseteq \zeta_2 \subseteq \zeta_3 \subseteq \ldots \text{ then } \epsilon^1 \geqslant \epsilon^2 \geqslant \epsilon^3 \geqslant \ldots$$

Thus a hierarchy of identities is obtained.

The role of the and–ing operator * is important in this hierarchy. When * is taken to be the lattice meet Λ itself, $\epsilon^k = \epsilon^{k-1}$ [2]. This means that if the collection of properties to define ϵ^k includes the properties to define ϵ^{k-1} and if Λ is used in (L″), no new identity will be obtained. In particualr, this remains true in the ordinary case when the lattice is {0,1}, i.e. the defining subsets (or properties) are ordinary (or non-vague). But with operators other than Λ there are examples [2] when an identity of a higher level thus obtained is different from its predecessor.

Some interpretations may help readers to appreciate the significance of the above mathematical formalism.

- Fuzzy subsets $\tilde{A}_i : X \rightarrow L$ used in (L″) are the extension of properties some of which can be vague. There is no restriction on the number of properties. To accommodate such a definition of identity in a language, many predicates would arbitrarily be required and arbitrary conjunction and disjunction would have to be allowed. This extending of the language, however, does not stand against the basic thesis as *Proposition 2* will still hold.
- $\epsilon(<a,b>)$ is a member of the truth set L and denotes the extent to which a is identical with b.

- (e_1), (e_2) and (e_3) are the generalised versions of reflexivity, symmetry and transitivity.
- The hierarchy $\epsilon^1 \geqslant \epsilon^2 \geqslant \epsilon^3 \geqslant \ldots$ means that a,b become identical to a lesser and lesser extent as the levels increase. This amounts to saying that as the set of determining properties grows, objects hitherto indiscernible become discernible, an observation which is naturally expected.

Notes

1. Marich is a character in the epic *Ramayana* who transformed himself into a golden deer in the episode of the abduction of Sita by Ravana.
2. If higher-place predicates are needed, for example a one-place predicate 'A', two two-place predicates 'B' and 'C' and a three-place predicate 'D', then we can define 'x = y' as short for (Quine 1979):

$$
\begin{aligned}
&Ax \longleftrightarrow Ay \;\&\\
&\forall z\,(Bzx \longleftrightarrow Bzy \;\&\; Bxz \longleftrightarrow Byz \;\&\; Czx \longleftrightarrow Czy \;\&\; Cxz \longleftrightarrow Cyz \;\&\\
&\forall z'\,(Dzz'x \longleftrightarrow Dzz'y \;\&\; Dzxz' \longleftrightarrow Dzyz' \;\&\; Dxzz' \longleftrightarrow Dyzz'))
\end{aligned}
$$

References

BRIDGE, J. 1977. *Beginning Model Theory*. Oxford: Clarendon Press, p. 58.

CHAKRABORTY, M.K. and M. BANERJEE. 1992. 'Saturatedness and a Hierarchy of Approximate Identities, Uncertainty Fuzziness and Knowledge-Based Systems.' Communication.

CHAKRABORTY, M.K. and M. DAS. 1983. 'Fuzzy Equivalence I, II'. *Fuzzy Sets and Systems, 11*, pp. 185–93, 299–307.

GORSKY, D. 1987. *Generalisation and Cognition*. Moscow: Progress Publishers, p. 18.

MENDELSON, E. 1979. *Introduction to Logic*. New York: D. Van Nostrand Co., p. 80.

QUINE, W.V. 1979. Philosophy of Logic. Delhi: Prentice Hall of India Pvt. Ltd., p. 63.

TARSKI, A. 1965. Introduction to Logic and to the Methodology of Deductive Sciences. Oxford: Oxford University Press, p. 55.

ZADEH, L.A. 1973. Outline of a New Approach to the Analysis of Complex Systems and Decision Processes. IEEE Trans. Syst. Man Cybern., SMC–1, pp. 28–44.

Annexure

Proof of Proposition 1:

(1), (2) and (3) are easily proved. (4) is proved by induction on the number of occurrences of connectives and in $\emptyset$. Let $\emptyset(x)$ be of the form $P_i x$, $i = 1, 2, \ldots, n$.

Now	$\{x \sim y, P_i x\}$	$\vdash$	$x \sim y,$
i.e.	"	$\vdash$	$(P_1 x <\!\!-\!\!> P_1 y) \,\&\, \ldots \,\&\, (P_i x <\!\!-\!\!> P_i y) \,\&\, \ldots \,\&\, (P_n x <\!\!-\!\!> P_n y).$
Hence	"	$\vdash$	$P_i x \rightarrow P_i y.$
Also	"	$\vdash$	$P_i x.$
Hence by M.P.	"	$\vdash$	$P_i y.$

Thus $\vdash (x \sim y \,\&\, P_i x) \rightarrow P_i y.$

Next assuming that (4) is a theorem for $\emptyset(x)$ and $\psi(x)$, we shall show that (4) is a theorem for (i) $\neg\emptyset(x)$, (ii) $\emptyset(x) \rightarrow \psi(x)$ and (iii) $z\, \emptyset(x)$.

(i)	$\{x \sim y, \neg\, \emptyset(x), \emptyset(y)\}$	$\vdash$	$x \sim y \,\&\, \emptyset(y).$
Hence	"	$\vdash$	$\emptyset(x).$
Also	"	$\vdash$	$\neg\emptyset(x),$

i.e. $\{x \sim y, \neg\emptyset(x), \emptyset(y)\}$ yields a contradiction.

Hence $\{x \sim y, \neg\emptyset(x)\} \vdash \neg\emptyset(y).$

So $\vdash (x \sim y \,\&\, \neg\emptyset(x)) \rightarrow \neg\, \emptyset(y).$

(ii)	$\{x \sim y, \emptyset(x) \rightarrow \psi(x), \emptyset(y)\}$	$\vdash$	$x \sim y \,\&\, \emptyset(y).$
So,	"	$\vdash$	$\emptyset(x).$
Also	"	$\vdash$	$\emptyset(x) \rightarrow \psi(x).$
Hence	"	$\vdash$	$\psi(x).$
So	"	$\vdash$	$x \sim y \,\&\, \psi(x).$
Hence	"	$\vdash$	$\psi(y).$

So $\vdash (x \sim y \,\&\, (\emptyset(x) \rightarrow \psi(x))) \rightarrow (\emptyset(y) \rightarrow \psi(y)).$

(iii) z should be different from both x and y.

	$\{x \sim y, \forall z\, \emptyset(x)\}$	$\vdash$	$\forall z\, \emptyset(x).$
So	"	$\vdash$	$\emptyset(x).$
So	$\{x \sim y, \forall z\, \emptyset(x)\}$	$\vdash$	$x \sim y \,\&\, \emptyset(x).$
Hence	"	$\vdash$	$\emptyset(y).$

By generalisation, therefore,

$$\{x \sim y, \forall z\, \varnothing(x)\} \vdash \forall z\, \varnothing(y)$$

So $\vdash (x \sim y \,\&\, \forall z\, \varnothing(x)) \rightarrow \forall z\, \varnothing(y)$.

Hence the proposition.

Rationality of Mathematical Constructions

S.M. Bhave

Rationality of Science

Though a number of historians and philosophers of science profoundly doubt the rationality of science, a large number of scholars and laymen hold that science is a paradigm of rationality. Many thinkers have tried to analyse this rationality and perhaps the most widely accepted analysis is that given by Popper.

According to Popper (1972), a scientific theory is rational because it is open to refutation. It clearly states the circumstances under which it will stand refuted. Thus, for example, the theory of relativity stated that bending of light under gravity was the crucial test for its validity; if this process was not observed, it would stand falsified. But, and here comes the crux of fallibilism, the occurrence of the phenomenon did not justify the theory. A scientific theory is thus never justified; it can only be falsified.

This principle of falsification is central to the code for scientific methodology which Popper formulated. The code is so framed that a scientific enterprise carried out in accordance with it will ensure the falsifiability of its product i.e. the scientific theory.

Since Popper's code governs the methodology of science, it must therefore depict how any branch of science progresses and how at each step the rationality of science is preserved. The Popperian picture of the development of science can be briefly described thus. At a given moment of time there is a ruling scientific theory T1 which professes to explain every event E that lies in the domain D1. Now occurs an event Ē which apparently lies in D1 and which T1 fails to explain. This means that T1 is falsified.

Once a theory is falsified, it is overthrown. Already there are several other theories competing with T1 to dominate over D1. A theory T2 from among them happens not only to explain all the events in D1 but also the event Ē. Therefore T2 now becomes the ruling theory. The overthrow of T1 and crowning of T2 definitely indicate progress, for the domain of T2 includes the domain of T1 as a subset. Therefore, T2 has more explanatory power.

This methodology presupposes certain metaphysical positions which, like their opposite positions, are irrefutable. But that does not make the positions meaningless. The first position is rationalism, that is, the conviction that facts and theories can be debated upon and certain decisions can be reached. In the course of falsification, the scientific community has to reach a consensus through debate on a number of points, such as when does a theory stand falsified? If the event under question goes contrary to the theory, does the competing theory explain the recalcitrant event? And so on.

The second position is anti-subjectivism, that is, the conviction that there are other minds which think differently. Popperian methodology assumes dissent and therefore criticism and debate. There must be minds which doubt the explanatory power of the ruling theory.

Third, the conviction that there is a world independent of consciousness and theory which can be used as a testing ground for hypotheses. Without such an objective reality, falsification of a theory is impossible.

Fourth, the methodology assumes the principle of convergence of theories. If domain D2 of the theory T2 was not a superset of domain D1 of the theory T1, there would be no ground to suppose that T2 is an improvement over T1. The theories would have been incommensurable.

In sum, Popper believes that the rationality of science is grounded in its methodology and that the methodology works only if the four hypotheses of rationalism, anti-subjectivism, realism and convergence are accepted.

The Nature of Mathematics

Before examining the applicability of Popperian rationality to mathematics, it might be worthwhile to discuss the nature of mathematics. In this exercise it will be profitable to follow Wittgenstein.

Mathematics begins with a mathematical technique which is drawn or conceptualised from several empirical practices. For example, for a primitive man, adding up two black and three white sheep, two yellow and three green bananas are different processes. They are conceptualised in the technique of adding 2 and 3. Thus the techniques begin from a chosen, coherent experience which Wittgenstein calls the form of life (Wittgenstein 1972, p. 241).

In the beginning, a technique is inextricably connected with the empirical processes. But soon this relationship becomes progressively tenuous. In the first stage the technique is detached from experience[1] (Wittgenstein 1964, p. 160). Next, it is made timeless and universal (Wittgenstein 1979, pp. 74, 100). Finally, it becomes a rule which is incorrigible in the face of adverse experience[2] (Wittgenstein 1964, p. 14).

Once a technique is detached from experience, various rules in it cohere to form a unity in the sense that they become responsible to one another (Wittgenstein 1979, p. 126). It is not possible to tamper with one without disturbing the whole. Further, each result in the technique ceases to refer to external objects; thus, instead of being a synthetic statement, it becomes an internal relation (Wittgenstein 1979, p. 75).

Thus a technique is grounded in experience and is sanctioned by a universal and extempore consensus. This consensus is made possible by a shared form of life which is manifest in empirical processes. But then a technique separates itself from experience in two senses. First, it does not properly address itself to any experience in its entirety and, second, it does not stand refuted in any circumstances. Thus it becomes a set of rules which operate on symbols. And yet we find that mathematical techniques do seem to describe what actually happens (Wittgenstein 1979, p. 51).

A mathematical symbol does not represent any physical object or a class of objects. It represents a concept[3] (Wittgenstein 1964, p. 186). Which does not capture any physical objects in their totality; rather, it contains their essence or meaning (Wittgenstein 1964, p. 12; 1979, p. 205). Thus a mathematical concept does not exhibit itself. It is the human mind which recognises a physical object as falling under a particular concept. The concept of a triangle, for instance, does not refer to this or that triangle but to all entities which have so far been recognised as triangles. Not only this, but in the future we may recognise as entity as a triangle with total freedom in the sense that what we today call a circle may also fall under the concept of 'triangle' if some future consensus so decrees. In this sense, a concept is never fully determined. What we have recognised so far as triangles may guide us in arriving at a future consensus but that will not determine our decision.

A mathematical technique has a threefold relationship with reality. It describes the reality; it acts as a standard for reality and it is independent of reality in the sense that no experience falsifies the technique. Similarly, a mathematical concept has a threefold relationship with reality. It acts as a description of certain objects in reality which may be said to form its domain; it is a standard for those objects and yet the objects do not completely determine its domain.

Every new development in mathematics is either an invention of a new concept or a novel extension of an old one. For instance, the Bolyai Lobachevsky Geometry (BLG) was a novel extension of the existing concepts of 'geometry', 'line' and 'triangle'. Similarly, Lebesgue Integration required the invention of two new concepts, namely, the measurable set and the measurable function. Thus we recognise it (i.e. concept formation) as the essential thing about a great part of mathematics (Wittgenstein 1964, p. 180).

Every time an old concept is extended or a new one introduced, mathematicians have to face the question of whether the new development is legitimate. The answer mainly depends upon the instrumental value of the extended or invented concept. In a large number of cases the extension or invention is a response to a specific question and therefore the instrumental value of the concept is manifest. Not much doubt is raised about its legitimacy. For instance, the concepts of measurable sets and functions were accepted without any debate about their legitimacy because these two concepts allowed a new formulation of the notion of integration which made many of the functions integrable which were not so under the previous notion.

However, sometimes, the extension or invention of a concept is forced upon mathematics through its autonomous growth. The discovery of irrational numbers was a consequence of Pythagoras' theorem and it is clearly a case of autonomous growth. It is well known that this discovery generated a profound debate within the Pythagorean school which formed the mathematical community at that time. A relatively recent debate of the same kind centred upon the discovery of transfinite numbers by Cantor.[4] It is this kind of development which is significant to our discussion since it raises deeper questions about the rationality and legitimacy of mathematics.

Rationality

The question of rationality, as we have seen, arises when either the extension of an existing concept or the innovation of a new concept is called for. Both occur in response to two sorts of situations: when the solution to a problem is being sought or when autonomous growth takes place.

In the first kind of development, fallibilist methodology and rationality are applicable in a modified form. Let us consider the extension of the concept of integration from the Riemannian one to the Lebesguian one.[5] In the last quarter of the nineteenth century, Riemann's theory ruled the field. According to this theory, every function defined and bounded on a closed interval was integrable provided its discontinuities were not 'too many'. Owing to an autonomous growth of the concept of a function, a new function called Dirichlet's function[6] appeared on the scene. This function was discontinuous at every point and hence not Riemann integrable. At this point, roughly speaking, the first stage in Popper's methodology is over.

However, it must be noted that Popper's methodology applies here in a significantly modified form. Firstly, Dirichlet's function is not an objective fact in the sense that bending of light rays is; indeed, at the time whether Dirichlet's function could be regarded as a function at all, was a hotly debated question. Dirichlet's function is a theory-bound fact. Secondly, Dirichlet's function could not be regarded as a falsifier for Riemann's theory since the latter never professed to make every function, especially so pathological a one as Dirichlet's, integrable. Therefore, at best, Dirichlet's function merely underlined the inadequacy of Riemann's theory. Thirdly, the overthrow of a theory is possible only if there is another competing theory in the field. But there was none when Dirichlet's function was noticed. Indeed, the supersession of Riemann's theory had to wait for more than two decades.

The next stage finds no place in Popper's scheme, namely, the construction of a new theory to solve a specific problem: the integrability of Dirichlet's functions. This was achieved by innovating two new concepts, namely, the measurability of a set and of a function.

Here we find that the principle of convergence which Popper did not emphasise, plays a crucial role. Though measurability of a function was undoubtedly a new concept, it was at the same time shown that all functions integrable in Riemann's sense are also measurable. Therefore, all Riemann integrable functions remained integrable in Lebesgue's sense also. Moreover, one could demarcate the domain of Riemann's integrability in such precise terms that it excluded Dirichlet's function. The two definitions of integrability thus became commensurate and Lebesgue's definition proved to be an improvement over Riemann's. Since the development of the theory of integration from Riemann to Lebesgue did carry the mark of Popperian rationality, although in a modified form, the progress was not controversial.

A more serious question regarding the rationality of a mathematical construction arises when autonomous growth generates an entity which cannot be accommodated within existing mathematics. It took mathematicians a long time to accept Dirichlet's function as a function; and this acceptance only came about after the established view of regarding something as a function was totally revised. A second case was the discovery of the irrational numbers as a consequence of Pythagoras' theorem. A proper way of viewing irrational numbers was developed only after 2,000 years. A third and more controversial problem centred on the transfinite numbers. It generated a widespread debate not only within the mathematical community but in philosophy and logic as well. Most of Wittgenstein's philosophical writings on mathematics are focused on this debate.

Rather than go into a detailed discussion of these cases I shall just note some of their common features. There are well-established concepts and processes in all branches of mathematics. A synthesis of them forces into existence an entity which claims to be mathematical by displaying a certain relationship with another well-established entity. Thus, for example, the transfinite numbers claimed relationship with natural numbers. BLG expected to be bracketed with Euclidean Geometry and so on. Thus the insistence of these entities on finding a place in mathematics could not be easily brushed aside. At the same time, these entities appeared to have no instrumental value since they were not discovered to solve any specific problem.

The incorporation of new entities into mathematics takes place through the introduction of new concepts which are then shown to include some existing concepts. Thus the new concept is later seen as an extension of an old one. Irrational numbers, for instance, were incorporated by introducing the concept of a real number. Both rational and the irrational numbers were then seen as instances of the concept of real numbers. To legitimise the transfinite numbers, a new concept called cardinal numbers was created. Then natural numbers were seen as cardinal numbers. The concept of manifolds with constant curvature was formulated and both BLG and Euclidean Geometry were displayed as particular instances of it. We call this procedure of incorporation 'retroactive revision'.

The crucial issue here is not how the entity produced through autonomous growth is later rationalised and brought into the fold of mathematics, though that does throw light on how it must have come into existence. What is important is why the construction of such an entity is allowed in the first place. For though the established theories hold out the possibility of such a development, the entity is discovered by a human agency.

In each of the examples just discussed, this point was raised by the opposite camp of mathematicians. The intuitionists described the development of non-Euclidean geometries, transfinite numbers and of continuous but non-differentiable functions as an embarrassment for mathematics. Apart from later rationalisation, on what grounds can we justify this autonomous development? Can we control such growth?

Control

From time immemorial, the autonomous growth of mathematics is sought to be justified on grounds of some version or the other of a mathematical reality. A modern version of mathematical reality is given by Hardy. He says, 'For me, and I suppose for most mathematicians, there is another reality which I call "mathematical reality" . . . . I believe that mathematical reality lies outside us, that our function is to discover it and to observe it' (Hardy 1967, pp. 123–24). On this view, transfinite numbers, for example, already existed in mathematical reality and Cantor only observed them. They were already a part of mathematics and therefore needed no justification.

A less exotic version of mathematical reality was given by Frege. For Frege the meaning of a term is an object and therefore there is a realm of meanings. Mathematics is a region in this world of meanings. Its rules constitute the physics of mathematical objects (Wittgenstein 1979, p. 240).

The concept of a mathematical reality has been questioned by many scholars. The three geometries, for instance, are based on different and contradictory descriptions of space. Hence we must assume that there are different mathematical spaces corresponding to the geometries. Wittgenstein observes, 'similarly there will be a reality for which 25 × 25 = 625 and another for which 25 × 25 = 624 . . . The whole thing crumbles down because you are making the assumption that when you are in the right world you will find out' (Wittgenstein 1979, p. 145).

If we abandon the notion of a mathematical reality, can we think of any other authority that would exercise control on the autonomous growth of mathematics? Here Wittgenstein refers to the historical development of mathematics. As mentioned earlier, mathematics is a collection of techniques that were drawn from primitive empirical processes. Indeed, the techniques became possible because there was a consensus on their rules. This consensus was not arrived at through any referendum. It was extempore and universal because it came from a shared form of life. Thus the roots of mathematics go back into the natural history of mankind.

This embedding of mathematics in wider fields of human endeavour is most strikingly seen in the use of mathematical terms in ordinary language (Wittgenstein 1979, p. 14). A term in mathematics must have a consistent interpretation in language. This is a responsibility which mathematics owes to its relationship with extra-mathematical reality. If mathematics closes upon itself it will become uninteresting; it has remained worthwhile because it is homogeneous with the totality of human culture.

Even though a certain autonomous growth appears inexorable within mathematics, thinkers from outside can seriously question its rationality. Wittgenstein did precisely this in the case of transfinite numbers. Hilbert (1862–1943) had proclaimed that mathematicians would not leave the paradise created by Cantor. To this

Wittgenstein's rejoinder (1979, p. 103) was: 'I would try to show that it is not a paradise so that you will leave on your own accord.'

Conclusion

It is now history that despite serious objections by thinkers like Wittgenstein, mathematicians went ahead with the Cantorian transfinite. They created a technique or arithmetic of cardinals which, when applied to finite cardinals, i.e. real numbers, coincided with ordinary arithmetic. Wittgenstein (1979, p. 159) expresses his dissatisfaction with cardinal arithmetic on the ground that it is artificial.

Does this mean that the development of the transfinite is irrational? Here we must observe that rationality is after all a function of tradition. Though Popper emphasises that his methodology is independent of the history of science, he also insists that on careful reading it will be seen that history bears out that methodology. On the other hand, thinkers who criticise Popperian methodology and rationality base their criticism on history. Therefore, it is seen that rationality is not something which can be legislated from outside a field of knowledge. It is very much a part of its tradition.

Coming to mathematics, we find that though mathematics has its origin in primitive empirical processes, the dominant trend from Plato and Pythagoras has been de-factualisation or going away from experience. That is why Plato regarded the contemplation of mathematical truths as worthy of the Deity (Russell 1959, pp. 59–60). According to this tradition, conceptualisation of the transfinite is surely a chosen task of mathematics.

In this purely contemplative domain, what is the scope of rationality? It is the principle of convergence and of logical consistency. Transfinite numbers are rationally constructed, for a logically consistent view can be adopted according to which both natural numbers and transfinite numbers are instances of a new concept called 'cardinal numbers' and these latter are mathematical entities, for an arithmetic of them has been developed. The empirical referrent of a transfinite number is a issue foreign to this conception of rationality.

Wittgenstein's understanding of mathematics, its history and its place in human culture is in sharp conflict with the understanding of mathematics cherished by this dominant tradition which is aptly summed up by Russell in the following words:

> Remote from human passions, remote even from the pitiful facts of nature, the generations have gradually created an ordered cosmos, where pure thought can dwell in its natural home; and where one, at least, one of our nobler impulses can escape from the dreary exile of the actual world (1959, p. 61).

Notes

1. 'A rule qua rule is detached, it stands, as it were, alone in its glory; although what gives it importance is the facts of daily experience.'
2. See remark no. 37.
3. 'A number is a mark of a mathematical concept.'
4. For Cantor's own version of this discovery see Cantor (1955).
5. See Lakatos (1976) for how Popper's methodology of fallibilism applies to Euler's conjecture. See Bhave (1993) for the application of fallibilism to Lebesgue Integration.
6. The function f(x) = 1 when x is rational; and f(x) = o when x is irrational, is called Dirichlet's function. Dirichlet's function was a product of the autonomous growth in the theory of functions; it was a fact and hence a problem for Riemann's theory.
7. According to Wittgenstein this responsibility was violated in three directions when transfinite numbers were discovered. Firstly, mathematics began to treat numbers as objects; secondly, in applying the procedure of counting to number-objects the old technique of counting is abused; thirdly, the picture of transfinite number which emerges is not consistent with its use in ordinary language.

References

BELL, E.T. 1940. *The Development of Mathematics*. London: McGraw Hill.

BHAVE, S.M. 1993. 'Proof Analysis and Lebesgue Integration'. Mathematics Student. *Journal of Indian Mathematical Society* (to be published).

HARDY, G.H. 1967. *A Mathematician's Apology*. Cambridge: Cambridge University Press.

LAKATOS, IMRE' 1976. *Proofs and Refutations*. London: Cambridge University Press.

POPPER, K.R. 1972. *The Logic of Scientific Discovery*. London: Hutchinson & Co. (Publishers).

RUSSELL, B. 1959. *Mysticism and Logic*. London: George Allen & Unwin Ltd.

WITTGENSTEIN, L. 1964. *Remarks on the Foundations of Mathematics*. Oxford: Basil Blackwell.

———. 1972. *Philosophical Investigations*. Oxford: Basil Blackwell & Mott Ltd.

———. 1979. *Lectures on Foundations of Mathematics*. Sussex: The Harvester Press Ltd.

IV

Rationality, Language and Logic

Preface

As several contributors remind us, the roots of rationality as a concept are intertwined with those of logic and of language. It is therefore entirely unsurprising that throughout this volume, logical and linguistic issues arise at almost every step of the discussion, however varied the routes chosen to approach the central theme itself.

What may be in need of justification is the grouping together in one section of a number of papers devoted to, or exhibiting concern for, logic and/or language. The classificatory principle however is straightforward enough: there appears to be a basic level at which the question of the very intelligibility of the concept at hand may be raised—the transcendental level, some would say, the level where conditions of possibility are deployed: for there to 'be' rationality, there must 'be', in some way or other, language and logic. These conditions may be contrasted with properties, varieties, degrees of rationality. In other words, there is at least an appearance of a dichotomy: the study of the most general concept of rationality involves logic and language in a pure mode, as opposed to the various ways in which rationality is applied to specific dimensions or realms of thought. So while the logic–language duo entertains with rationality crucial relations along every dimension, the pure analysis involves nothing besides, whereas once the concept of rationality is constituted, its connection with the duo tends to involve some third term: science, morality, action etc.

This section is therefore concerned with the primordial link between rationality, language and logic. How early in the Indian tradition that link was uncovered and thematised is brought out in *V.N. Jha's* contribution, which also and more specifically shows that philosophy of language and the metaphysics of truth opened the way for Indian thinkers of the past just as it did in the West, millennia later, for the founders of analytic philosophy and precursors of cognitive science. *John Vattanky's* paper explains the role played by grammar, according to the *Nyaya* philosophy of language, in the identification of denotation—a theme familiar to readers of Chomsky but treated by the Naiyãyikas in a different spirit: the grammatical constituents are those of the word, rather than the sentence, so that the denotation raises deep and original ontological questions which are settled at the price of subtle analyses moving from language to world and back.

Two contributions then lead the reader into the world of contemporary philosophy of language. *Kalyan Shankar Basu* tackles head on the arduous question of the relation between truth and meaning as posed by Donald Davidson. He argues against a full reduction of meaning to truth conditions, while accepting Davidson's view that interpretation as it occurs in the intercourse between hearer and speaker, involving as it does the structural dimension of language, rests entirely on the recursive Tarski-style clauses known as conditions (*T*) which jointly constitute a theory of truth for the language under consideration. That there is more to language than this structural dimension, which by itself would make language no more than a code, is shown, according to Basu, by the possibility of liar-type paradoxes. They point to another function of language, to wit naming, which calls for a different kind of interpretation, unshackled to communication, and bound instead to the projecting activity of the single individual engaged in meaning-giving play. Rationality requires precisely this duality of functions: naming the constituents in order to give sense to the world *vs*. constructing and transforming adequate structural descriptions of the world in order to correctly appraise complex situations.

François Recanati engages another, more specific problem arising in the understanding of the communication of certain kinds of thoughts. As first pointed out by Frege, so-called 'first-person' thoughts seem incommunicable in a sense, contrary to what experience shows: how can I transmit the thought that I am in pain by uttering the sentence 'I am in pain' if the hearer must choose between the obviously inadequate interpretation of 'I' as what 'I' would refer to if she uttered the same sentence (i.e. me, the hearer), and the more plausible yet still unsatisfactory interpretation of 'I' as the speaker? Recanati's solution to the paradox rests on a distinction between linguistic and psychological modes of presentation. The communication situation provides the hearer with the means of correctly identifying the referent of 'I' from its linguistic mode of presentation in the utterance at hand, while linking the fact of being in pain now to her psychological mode of presentation of the speaker, a process metaphorically described by Recanati as an appropriate addition to her 'dossier' on the speaker. Now what deserves emphasising here is that rationality requires correct identification of referents across modes of presentation: my belief that someone is in pain leads me to a response which is rational only if under normal conditions I am able to infer from my cognitive environment who the individual in question is—ignoring real-life complexities, taking an aspirin is appropriate in exactly those cases where saying 'Take an aspirin!' is not. (This problem is approached from a different angle in Pierre Jacob's contribution to section I.)

Up to this point, logic in the strict sense has had not much role to play. The three remaining contributions in the present section, on the other hand, focus on the fit between forms of rationality and systems of logic. *Amitabha Gupta* and *Nilesh Sutaria* set the stage by posing a dilemma: if we reject, for whatever reason, the traditional picture of epistemology as a normative enterprise ruled by the canon of classical logic, we seem to be reduced to the Quinean view of naturalised epistemology: our cognitive procedures are what they are, and psychology alone is in charge of saying precisely what they are. Gupta and Sutaria reject both horns of the dilemma and seek a third way. There is no reason, they argue, to give up the normative dimension of rationality simply because classical logic fails to provide a plausible and sufficiently rich set of rules. One should instead look for a realistic, cognitively sound, ecologically valid logic. This search has been underway for a good thirty years now, and many proposals have been put forward. Gupta and Sutaria critically yet sympathetically examine some of the most important ones; then they attempt to salvage the part of truth in each of them by incorporating their basic ideas into a new logical system, whose very name, Defeasible Relevant Minimal Logic (DRML), points to those ideas: defeasibility (the focus of non-monotonic logics), relevance, and feasibility (a constructive use of the finitude of all real cognitive systems).

Daniel Kayser adopts a somewhat similar tack, but takes as his starting point the results of Artificial Intelligence (AI) rather than philosophy. The poverty of logic (as traditionally construed) should not, according to him, lead to a total rejection of logic, as has often been advocated in AI; in fact, as he shows, there is a contradiction between pursuing the goals of AI and giving up on logic altogether, but logic must be taken in a reasonably broad sense. The challenge is to discover the real scope of logic as it operates in computational systems, and seek in parallel the principles of this sought after, computationally valid, logic. Kayser has been pursuing for a number of years one particular proposal for meeting the challenge; he calls it variable-depth reasoning. The problem it seeks to resolve is that of the change of ontology in the course of reasoning—in a sense, it can be seen as a computational echo of Basu's naming/(de)coding dichotomy. Non-monotonicity comes in as a crucial ingredient of the proposed system: as a contradiction is encountered, a new level of ontology is tacked on in an attempt to overcome it; cautious revision is then called for to maintain overall consistency.

How to compose and maintain a coherent representation of objects, states of affairs or processes built from heterogeneous elements is a recurring theme in computational models of reasoning—like philosophers, cognitive systems must have a way of determining 'how it all hangs together'. *Som Prakash Bandyopadhyay* attempts to spell out conditions of coherence in texts comprising both text and pictures. He argues that coherence is deployed on three levels, syntactic, semantic and pragmatic. His analysis brings to light the rich repertoire of 'logical' relations which obviously enter in the understanding process, where 'logical' is taken in the commonsense construal, and his proposal is to incorporate such relations in a logical system in the strict, albeit liberalised, sense.

Both the language and logic-oriented parts of this section attempt to remedy grave insufficiencies in the classical conception of rationality without rejecting it wholesale and in fact, preserving much of its spirit. Refusing to yield to the sirens of romanticism, constructivism or hermeneuticism, all the contributions reviewed here fit squarely in a rationalistic tradition. They abundantly show that the rationalist view of cognition is in need of both firmer foundations and wider scope, and point to ways of meeting those demands. The conception of rationality which will eventually emerge from the present efforts may well turn out to incorporate, to 'co-opt', many of the insights of the romantics, just as tomorrow's logic or logics will manifestly include dimensions ruled by the founding fathers out of the province of logic.

DANIEL ANDLER

Language and Cognition: An Ancient Indian Perspective

V.N. JHA

There is hardly any other country in the world apart from India that has paid so much attention in its past to language, its nature and function, and its relation with the universe and with man. This paper highlights some of these aspects as reflected in Sanskrit philosophical and systematic literature.

Human Behaviour

Human behaviour can be categorised into three types. A person can move forward (*pravṛtti*); or he can move backward (*nivṛtti*); or he can remain indifferent (*audāsīnya*). To illustrate, when a person goes to collect silver, his behaviour belongs to the first category. When he runs away on seeing a snake, his behaviour fits into the second category. And when a *Sanyāsi* neither moves forward to collect the silver nor runs away after seeing a serpent, his behaviour belongs to the third category.

Further, human behaviour may be consistent or inconsistent. Consistent behaviour is that by which a person gets the same object that his knowledge tells him he wants. If the knowledge shows that there is silver and if he goes and gets the silver and can prepare ornaments for which he wanted it, his behaviour is considered to be consistent (*saphala-pravṛtti*). But when he sees silver and goes to collect it and is frustrated in the attempt, his behaviour will be said to be inconsistent (*viphala-pravṛtti*). Here, the person does not get the object shown by his cognition.

The important point is that both types of behaviour—consistent or inconsistent—are prompted by one's cognition. Further, true cognition prompts consistent behaviour while false cognition prompts inconsistent behaviour, because the truthfulness or falsity of cognition depends on the consistency or inconsistency of the respective behaviour. Another important point is that cognition that prompts behaviour must be capable of being verbalised, because only verbalised cognition can clearly reveal the content of the cognition, giving it a form and name. Unless the content of cognition is revealed in this way, behaviour cannot take place.

The Universe

One behaves with the universe in which one lives. When we say that it is cognition that prompts behaviour, we mean that it is cognition that prompts one to behave with its content. Thus, one behaves with the content of the cognition that prompts one's behaviour. To be more precise, one behaves consistently with the object of true cognition that prompts one's consistent behaviour and one behaves inconsistently with the object of false cognition that prompts one's inconsistent behaviour. This object of cognition is the universe around us. It is our experience that we behave both consistently and inconsistently.

The question therefore arises: does this object of cognition i.e. the universe with which we behave, have any ontological reality? Philosophers are divided on this issue. Those who belong to the realistic school such as the Nyāya-Vaiśeṣika and Mīmāṁsā, hold that the world with which we behave is real, while idealists like the Buddhists and Advaitins hold that the world of our behaviour is a mental projection and so the universe has no ontological reality.

This raises the issue of 'what is reality?' In other words, how to decide whether *x* is real or unreal. The logicians attempted definitions of reality and falsity in terms of true and false cognitions. If a cognition is true, its content is true; if it is false, its content is also false. The problem is how to decide whether a cognition is true or false. They provided a test: if a cognition is subsequently contradicted, it is false and if it is not contradicted, it is a true cognition. For example, if I know 'It is *x*' and subsequently if, again, I know 'it is not *x*' then it is a case of contradiction and the cognition 'it is *x*' in this case is said to be false. Thus, an object of a cognition which is not contradicted by any subsequent cognition is said to be real, whereas an object of a cognition that is contradicted by a subsequent cognition is said to be false (*abādhitajñāna-viṣayaḥ satyaḥ* and *bādhita-jñānaviṣayaḥ asatyaḥ*).

This was the realists' position. The idealists, on the other hand, believed that the universe has no ontological reality. However, they suggested degrees of reality for the sake of explaining human behaviour. For idealists like the Buddhists, however, nothingness or voidness (*sunyatā*) may alone be the reality. For the Vedāntins, the ultimate reality is the Brahman, and really speaking there is no other reality in true sense of the term. But since human beings behave consistently with this world, they had to posit a reality called *vyāvahārika* or mundane reality. Since human beings also behave inconsistently with the object of a false cognition, they had to postulate a third type of reality called *prātibhāṣika* or mental reality. Thus, for the Advaitin, there are three types of reality (*a*) *pāramārthika* (ultimate), (*b*) *vyāvahārika* (mundane), and (*c*) *prātibhasika* (mental or false). The silver with which one makes ornaments is *vyāvahārika* reality, the silver seen in the glittering conch shell is *prātibhāsika* reality and the *Brahman* is *pāramārthika* reality. The silver seen in the conch shell has existence (*sattā*) only up to the time one discovers that it is not silver but only a conch shell. This *sattā* is called *prātibhāsika sattā*.

The relationship between these realities is quite interesting. The existence of the silver seen in a conch shell has a duration equal to that of the cognition of the silver till it is known that it is not silver. That is why, this silver is described as *jñanaikasattāka* (of the duration of the cognition itself). Thus, if the cognition exists for, say, one hour, the mistaken silver exists for one hour. In other words, it has reality in some sense during that time and it disappears immediately after one acquires the knowledge that it is a conch shell. It is like the silver in a dream which disappears immediately after awakening.

Likewise, the Advaitins argue, the silver with which we behave consistently is going to disappear after one realises the ultimate truth, the Brahman. From the point of view of ultimate reality there is no difference between a *prātibhāsika-sattā* and *vyāvahārika-sattā* and if at all there is difference, it is only in terms of the duration of time. The existence of silver seen in the conch shell has a shorter duration, while that of the silver which is used for ornaments has a longer duration; but ultimately, when the contradictory knowledge (*bādhakajñāna*) comes to light, even the *vyāvahārika-rajata* disappears. Just as the knowledge of the conch shell dispels *rajata*, the knowledge of the truth will dispel what we normally consider to be real i.e. the mundane world.

The Sources of a True Cognition

Since the object of a true cognition is true, and since true cognition alone prompts our consistent behaviour with this world, it follows that the world is true. Now, the question arises: how does true cognition come into existence? In other words, what are the instruments or sources of true cognition?

Cognition varies in type, depending upon the difference in its source. The process involved in generating a perceptual cognition is different from that involved in generating an inferential or verbal or analogical cognition. Since the processes differ, the resultant cognitions also differ. This process or source of a cognition is also called *pramāṇa*.

Indian philosophical systems differed in their conception about the number of sources through which cognition can be derived, ranging from one to eight. The Cārvākas accepted only one *pramāṇa* i.e. perception or *pratyakṣa*. The Vaiśeṣikas and the Buddhists posited two, perception and inference (*anumāna*), while the Samkhya philosophers added *Śabda* (verbal testimony) and accepted three *pramāṇas—pratyakṣa, anumāna* and *śabda* (verbal testimony). The Naiyāyikas accepted four—*pratyakṣa, anumāna, śabda* and *upamāna* (analogy). The Prabhakaras added *arthāpatti* to this list and made it five. The Bhāṭṭa school posited six by accepting *anupalabdhi* or *abhāvapramāṇa*. The *paurāṇika* tradition accepted eight by adding two more to the list, namely, *aitihya* and *sambhava*.

The Indian tradition has examined these proposals thoroughly through the centuries. The Nyāya-Vaiśeṣika argues that it is only necessary to accept the first four *pramāṇas*; the remaining four could be included in these four. Thus, *pratyakṣa* (perception), *anumāna* (inference), *upamāna* (analogy) and *śabda* (verbal testimony) are required to be accepted as being sources of valid cognitions. *Arthāpatti* (implication) can be included in *anumāna; anupalabadhi* (non-availability) can be accepted only as an auxiliary cause for generating the knowledge of absence, and *aitihya* (narration) and *sambhava* (inclusion) can be included in *śabda* and *anumāna* respectively.

Language as Source of a Cognition

It is clear from this that *śabda* or language is one source of cognition. But on this issue, too, the Indian philosophers were not entirely in agreement. The Buddhists held that language does not speak the truth and it has nothing to do with reality. For them, languages do speak of the world, but this world of our behaviour is unreal and simply the projection of our mind. The Advaita Vedānta also believed that while language undoubtedly refers to this world of our behaviour, this world is ultimately untrue. Thus, language cannot refer to absolute reality, which is beyond the reach of *all language* (*avāṅ-manaso gocaraḥ*). But there is a difference between the views of the Buddhists and of the Advaitins. While the former did not accept anything positive as real and consistently advocated *sunyatā* (nothingness) the Advaitins accepted the Brahman as the only reality. Consequently, they had to accept the temporary reality of this mundane world which is referred to by language. Thus, though a language may speak of this mundane world, it cannot speak of the ultimate reality, the Brahman.

In contrast, the Indian logicians (Naiyāyikas) took the stand that a language always speaks of reality. Thus, what language refers to is real. The entire world, therefore, is real and true. For them, the world of our behaviour is a set of entities (*padārtha* or 'referent of a term').

Bhartṛhari, the grammarian philosopher, took another stand. For him, the entire world was a manifestation of *śabda*, the supreme reality. We cannot conceive of this world without that supreme reality. Not even thinking about it is possible. He believed that there can be no knowledge of this world without the interference of *śabda* or language. According to him, any knowledge which is capable of being verbalised is a transformation of that *supreme language*, reality itself. Therefore, for him there is no knowledge which cannot be verbalised. Both right and wrong knowledge is capable of being verbalised. Thus, language can speak both the truth and fiction. Language can speak of a flower in the garden and also a flower in the sky, which is fictitious; it can speak of the son of a person and also of the son of a barren lady; it can describe a person taking a bath in the Ganges and also in a mirage. Since the whole universe is a modification and transformation of *śabda* (the supreme reality), language has reference to each and everything in this world, whether real or fictitious.

As against this, the Indian logicians (Naiyāyikas) and all those who believed in the reality of the mundane world around us, believed that language speaks both truth and untruth. Thus it can be the source of a true cognition as well as of a false cognition. Language which is the source of a true cognition is called *Śabda-pramāṇa* and that of a false cognition is known as *pramaṇābhāsa*. According to this school of thought, a language encodes the knowledge which is to be verbalised. The speaker or writer expresses that encoded language and the hearer or reader decodes that language and acquires the knowledge which was encoded by the speaker or writer. Language is the medium through which the speaker or writer transmits his or her knowledge to the hearer or reader. Thus, the speaker or writer wants to take the hearer or reader to the same world, the knowledge of which he has encoded in his sentence. If he succeeds in taking them there, there is said to be *samvāda* (agreement) between a speaker and a hearer or between a writer and a reader. It is against this background that one should understand why the Naiyāyikas call the universe *padārtha* (the referent of language).

This is an extreme view in the sense that it presupposes that language always speaks of reality; and when it appears that language has expressed absurdity, it has simply verbalised a false cognition. If one understands 'sky-flower' as an entity that has a real existence, one is mistaken because there is no entity called sky-flower in this world. Thus language always speaks the truth and it faithfully verbalises a false cognition also. Sentences like 'the circle is a square' will generate only false cognition since their content never correspond to reality. This leads one to think about the process by which a sentence generates understanding.

The Process of Verbal Understanding

The following steps have been postulated in generating verbal understanding:

Step I: The hearer acquires knowledge of the morphemic constituents of a sentence (i.e. *padajñāna*).

Step II: He will remember the referents from each of the respective constituents (*Padārtha-Smaraṇa*).

Step III: He will come to know the relationships among the remembered referents.

Step III is the resultant verbal understanding. It may be noted that for proceeding to Step II from Step I, the presupposition is that the hearer or reader already has the knowledge of the relationship between a *pada* (morpheme) and its referent (*artha*).

Again, to reach the Step III from Step II, it is necessary to know the intention of the speaker or writer. Unless that is known, the hearer or the reader will fail to decide with what intention the speaker has uttered or the writer written the sentence and without that knowledge, it will remain doubtful whether the hearer or reader has understood exactly what the speaker or the writer wanted him to understand. Thus, *tātparyajñāna* or the knowledge of the intention of the speaker or writer, plays an important role in generating verbal understanding.

To concretise the entire process, let us assume that a + b + c + d is a sentence consisting of four morphemic units, and a′, b′, c′ and d′ are the respective referents of a, b, c, and d. The vertical line is the relation between a, b, c, and d, and their respective referents, and the horizontal line represents the relationship among the referents. Thus, the entire process may be represented as follows:

a + b + c + d

a′ — b′ — c′ — d′

When one knows these horizontal lines, one is said to have understood the sentence-meaning. Thus these horizontal lines i.e. the relations among the meanings, represent the sentence-meaning.

Theories of Verbal Understanding

It should be noted that the horizontal lines i.e. the sentence-meanings are not referents of any linguistic element of the sentence a + b + c + d, although they are the content of the verbal understanding. Thus nothing is allowed to appear in the verbal understanding other than the relations among the referents. In other words, it is only the sentence-meaning that can appear in the verbal understanding even if it is not a referent of any linguistic unit of the sentence (*apadārthaḥ api vakyārthaḥ śābdabodhe bhāsate*). This is the stand taken by the Nyāya-Vaiśeṣika system.

However, this issue has given rise to several theories of verbal understanding. The question has been raised is that if a *padārtha* (a', b' etc.) is expressed by *pada*, what expresses *vakyārtha* (horizontal lines)? The answers to this question have led to various theories, such as *Abhihitānvaya-vāda, Anvitabhidhana-vāda, Tātparya-vāda,* and *Saṁsarga-maryāda-vāda*.

Abhihitānvaya-vāda

This theory originates from the answer to the above question given by the Bhāṭṭa School of *Pūrvamīmāṁsā*. According to it, the *padas* express their *arthas* and by *lakṣaṇā* i.e. by the *śakti* of the *arthas* the sentence-meaning is understood. Thus, the relationships among the expressed senses are established after the meanings are expressed by the individual *padas* (*abhihitānām arthānām anvayaḥ*).

Anvitabhidhana-vāda

This is the contribution of the Prābhākara School of *Pūrvamīmāṁsā*. According to this theory, the sentence-meaning is also expressed by the constituents of the sentence. The Prābhākaras believe that the *padas* not only express their own meanings but they also express the relationship with other meanings (*anvitānām padārthānām abhidhānam*).

Tātparya-vāda

This theory holds that the relationship among word-meanings will be established by the force of the intention of the speaker. Jayantabhaṭṭa, the author of *Nyāyamañjarī*, considers *tātparya* to be a *vṛtti* (function) that should account for the sentence-meaning.

Saṁsarga-maryāda-vāda

A contribution of the Navya Nyāya, according to this system of Indian logic, sentence-meaning is obtained because of the juxtaposition of meanings. In verbal understanding, sentence-meaning appears in the capacity of a relation. The meanings presented to the mind get connected in accordance with their capacity and compatibility. This theory is very close to that of the Bhāṭṭas in the sense that the meanings themselves have the capacity to connect themselves.

Structure of a Śābdabodha

There is another point that needs to be mentioned here. The content of a verbal understanding reflects a structure. The structure is of the type of a qualificand, qualified by a qualifier. Any acceptable sentence gives rise to an understanding which shows its content and the content always appears arranged in a particular form or structure, which can be divided into two parts—qualifier (*viśeṣaṇa*) and the qualificand (*viśeṣya*). Analysis reveals that the division observes the principles of immediate constituent analysis which can be transformed into a tree diagram. To illustrate, if the content has the element a', b', c' and d', one may analyse it as follows:

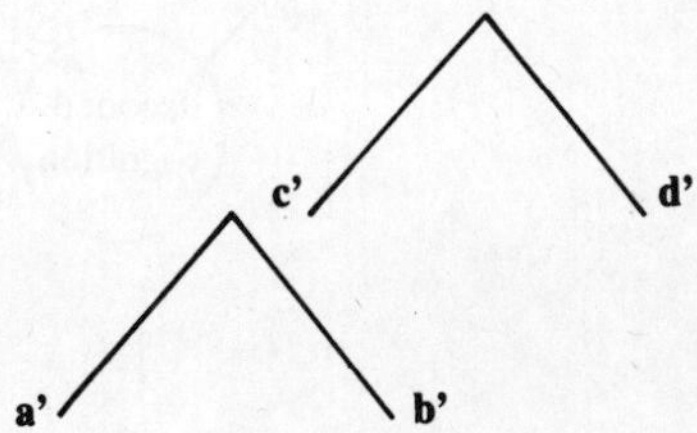

Conclusion

It is clear that the ancient Indian philosophers had paid great attention to the analysis of verbal behaviour. Their entire attempt to analyse language has been from the listener's or hearer's point of view. That is why we find meticulous analysis of the process of verbal understanding i.e. the epistemology of cognition that arises from a sentence i.e. the *śābdabodha*. As philosophers of language, they took divergent stands—from saying that language does not speak the truth to saying that it cannot but speak the truth, to saying that it speaks both, truth and untruth. To my mind, these divergent views paved the way for the evolution of an analytical philosophy, that is Indian philosophy. The Nyāya Vaiśesika School contributed a great deal to this development.

The speaker cognises this universe; he encodes this cognition in a language and speaks it out; the hearer receives that language, decodes it and gets to the same cognition as that of the speaker, and through that cognition to the same world of the speaker.

When this happens, it is said that there is agreement (*saṁvāda*); when there is absence of such a circle, it is said to be a case of disagreement (*vi-saṁvāda*).

To illustrate:

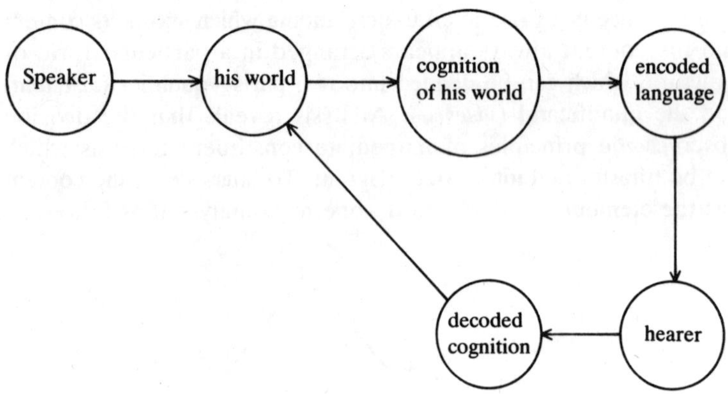

References

JHA, V.N. 1985. *Studies in Language, Logic and Epistemology*. New Delhi: Pratibha Prakashan.

———. 1987. *The Philosophy of Injunction*. New Delhi: Pratibha Prakashan.

———. 1987. *The Logic of Intermediate Causal Link*. New Delhi: Indian Books Centre.

———. 1987. *Viṣayatāvāda* (Introduction only). Poona: Centre of Advanced Study in Sanskrit, University of Poona.

———. 1990. *The Philosophy of Relations*. New Delhi: Indian Books Centre.

———. (ed.) 1992. *Relations in Indian Philosophy*. New Delhi: Indian Books Centre.

MATILAL, B.K. 1971. *Language, Logic and Grammar*. The Hague: Mouton.

———. 1986. *Language and Reality*. New Delhi: Motilal Banarasidas.

Grammar as a Means of Knowing Denotative Function

JOHN VATTANKY

An important topic in the *Nyāya* philosophy of language relates to the means of cognising the denotative function (*śakti*). After discussing the nature of the denotative function, *Muktāvalī, Dinakara*, and *Rāmarudra* (hereafter abbreviated as M., D., R., respectively) present detailed discussions on the means of knowing the denotative function. Quoting an early source, M. points out that there are eight means of knowing the denotative function. They are grammar, comparison, dictionary, sentence uttered by an authoritative person, action brought about by usage, the rest of a sentence, explanation, and proximity of a well-known word. These means are discussed by the texts, but the discussions centring around grammar are long and intricate. In this essay, an attempt is made to examine the discussions of M., D. and R. on grammar as a means of knowing the denotative function.[1]

First of all, M. explains that grammar is a means of cognising the denotative function in the sense that the denotative function of roots, stems and suffixes etc. can be discerned from grammar. We know from grammar that the root *bhū* means to be, that '*laṭ*' denotes the present tense and so on. While *Dinakara* does not explain the text, Rāmarudra goes into detailed explanation. He first explains the meaning of the expression 'roots, stems, suffixes etc.' (*dhātuprakṛtipratyādi*) in the following way: 'root' (*dhātu*) means verbal root. Stem (*prakṛti*) means the base, i.e. a noun in its uninflected state (*prātipadika*). Suffixes (*pratyayāḥ*) mean cases of nouns (*vibhakti*); affixes forming nouns from other nouns (*tadhita*), tense suffix (*ākhyāta*), affixes used to form derivatives like nouns and adjectives etc., from roots (*kṛt*). The word 'etc.' in the expression 'roots, stems, suffixes etc.' stands for compounds (*samāsa*), since the grammarians also accept the denotative function for compounds. Statements in grammar such as 'the root *bhū* means to be' help us to know the denotative function of roots. The knowledge of the denotative function of derivative nouns in their uninflected state (*prātipadika*) such as *pācaka*, is also gained from grammar. This is because grammar teaches us that the root *pac* means cooking (*pākah*); the *kṛt* suffix *aka* after the root *pac* means the agent of cooking. It is to be noted here that the knowledge of the denotative function of independent nouns in their uninflected state such as *go* (cow), can only be gained from the dictionary.

According to Rāmarudra, there is a difficulty in this explanation. He points out that the word 'stem' (*prakṛti*) stands for the nouns in their uninflected state (*prātipadika*); he also asserts that the word 'etc.' in the expression 'roots, stems, suffixes etc.' stands for compounds, which are also nouns in their uninflected state (*prātipadika*). This is not correct since 'stem' (*prakṛti*) means *prātipadika* in which compounds are also included. Therefore, Rāmarudra interprets the expression 'roots, stems, suffixes etc.' in another way: root-stem (*dhātuprakṛti*) means stem in the form of a verbal root (*dhāturūpaprakṛti*). Hence it is possible to interpret 'etc.' in the sense of compound and other *prātipadikas*. In order to differentiate between verbal roots and suffixes, verbal roots are referred to as stems (*prakṛti*).

A statement in grammar like *vartamāne jaṭ*, i.e. *laṭ* tense (which refers to the present) is the means of knowing the denotative function of the *laṭ* suffix. While in general these rules are accepted by the Naiyāyikas, they do not accept them in cases where there is an opposing factor. For instance, the grammarians hold that the denotative function of the verbal suffix is in the agent. Take the sentence *caitraḥ pacati* (Caitra cooks). Here grammarians assert that the verbal knowledge arising from the sentence is through the relation of non-difference of Caitra with the agent, i.e. the verbal suffix *ti* means agent and Caitra is related to this agent by the relation of non-difference. The Naiyāyikas reject this on the ground that it is cumbersome. They point out that the verbal suffix *ti* has its denotative function in effort (*kṛti*) because of simplicity and that effort is a qualifier of the meaning of a word ending in the nominative case. Dinakara explains the reason for this simplicity: the universal 'effortness' (*kṛtitvajāti*) is the determinant of 'the-state-of-being-denoted-meaning' and this is simpler than efforts which are many and which are such determinants according to the grammarians.

Dinakara and Rāmarudra brilliantly summarise the long drawn-out arguments between the grammarians and the Naiyāyikas on these points. It has already been mentioned that the basic position of the Naiyāyikas is that the denotative function of the verbal suffix is in effort (*kṛti*). The argument in support of this view is as follows: the universal of effortness (*kṛtitvajātiḥ*) is the determinant of the-state-of-being-denoted-meaning (*sakyatāvacchedaka*) because such a determinant is simple in comparison with efforts that are taken as determinants by the grammarians. The reason is that according to grammarians, the denotative function of the verbal suffix is in the agent (*kartā*). And an agent means one who has effort (*kṛtimān*). Therefore, the determinant of the-state-of-being-denoted-meaning is the-state-of-having-effort (*kṛtimattvam*), which means just effort (*kṛtiḥ*) and efforts are many. Hence if, as the Naiyāyikas say, effort is taken as the denoted meaning of the verbal suffix, then effortness (*kṛtitvaṃ*), which is the determinant of the state-of-being-denoted-meaning and which being one, is simple.

Rāmarudra explains in detail how effortness (*kṛtitvam*) is simpler. The general rule of the Naiyāyikas is that objects other than universal, and indivisible imposed property (*akhaṇḍopādhiḥ*) become objects of knowledge with some distinctive property (*jātyakhaṇḍopādhtyatiriktānāṃ kiṃcidrupeṇaiva bhanaṃ iti niyamaḥ*). If an individual becomes the object of knowledge, the universal residing in it also becomes the object of knowledge but without any further distinctive characteristic in the universal itself becoming the object of knowledge. For example, when a cow is the object of knowledge, cowness (*gotvaṃ*) also becomes the object of that knowledge, by itself, without any further characteristic, i.e. cownessness (*gotvatvaṃ*) also becoming the object. A universal designated as universal becomes the object of knowledge only with some distinctive characteristic (*jātyullikhitapratītau jātiḥ kiṃcidrupeṇaiva bhāṣate iti niyamaḥ*). For instance, 'cowness' in the sentence 'there is cowness' becomes the object of knowledge only with the characteristic of cownessness (*gotvatvaṃ*).[2] Therefore, effort (*kṛti*) can never become an object of knowledge without some distinctive characteristic, i.e. effortness (*kṛtitva*). It follows, then, that effort by itself (*svarūpataḥ*) cannot become the determinant of the-state-of-being-denoted-meaning.

In order to avoid this difficulty of cumbersomeness (*gauravaṃ*), modern grammarians hold that the suffixes of active and passive verbs have their denotative function in mere substratum (of actions from volitional effort up to bodily action—*kṛtiparyanto vyāpāraḥ dhātvarthaḥ*). It should be noted that effort, action and result are denoted by the roots themselves.[3] The-state-of-being-substratum (*āśrayatva*) is an indivisible imposed property which can be an object of knowledge by itself, without any further distinctive characteristic. And so there is no cumbersomeness (*gauravaṃ*).

Naiyāyikas, however, point out that such an interpretation is against the Pāṇini sūtra, *Jaḥ karmaṇi ca bhāve cākarmakebhyaḥ* (3.4.69). The full meaning of the sūtra as given by Bhaṭṭojidīkśita is: *lakārāḥ sakarmakebhyaḥ karmaṇikartari ca syuḥ akarmakebhyo bhāve kartari ca*, i.e. after the roots of transitive verbs, the tense suffixes (*lakārāḥ*) are enjoined in the meaning of the agent and object of action (substratum of result which is produced by action) and after the roots of intransitive verbs, the tense suffixes (*lakārāḥ*) are enjoined in the meaning of *bhāva* (*dhātvartha*, meaning of root) and in the meaning of the agent. Hence the tense suffixes do not have the meaning of mere substratum (*āśraya*). Nor should the word 'agent' (*kartā*) be interpreted as substratum (*āśrayaḥ*). Such an interpretation would render another Pāṇini sūtra useless viz. *svatantraḥ kartā* (1.4.54), according to which the agent is one who is independent, and independence is defined as the-state-of-having-an-action-not-depending-on-the-actions-of-other-kārakās (*kārakāntaravyāpārānadhīnavyāpāravattvaṃ*. Therefore, if the active verbal suffix denotes the meaning 'agent', then the-state-of-being-agent (*kartṛtva*), which is defined as independence (*svātantryaṃ*), is the determinant of the-state-of-being-denoted-meaning (*sakyatāvacchedakaṃ*). As has just been explained, such a determinant is indeed very cumbersome. Similarly, Bhaṭṭojidīkśita's statement[4] that root denotes action as well as result and verbal suffixes denote substratum, is also against the Pāṇini sūtra, *laḥ karmaṇi ca bhāve ca akarmakebhyaḥ* (3.4.69) referred to earlier.

Another problem encountered in taking 'substratum' (*āśrayaḥ*) as the meaning of the verbal suffix is that since Pāṇini took the meaning of the verbal suffix to be the agent (i.e. one who has effort, *kṛtimān*), modern grammarians can take the meaning of verbal suffix as substratum only by implication (*lakśaṇā*). If this is so, then it would be easier to take effort (*kṛti*) as the implied meaning since effortness (*kṛtitva*) is the determinant of the-state-of-being-the-meaning-of-verbal-suffix (*ākhyātārthatāvacchedakam*). And this is indeed simpler, for the-state-of-being-substratum cannot be taken as an indivisible imposed property (*akhaṇḍopādhi*) because there is no reason to do so. Hence the-state-of-being-substratum is different according to the different substratums because the-state-of-being-substratum is the relation in the form of substratum (*adhāratārūpasayā āśrayatvasya svarūpasaṃbandharū patvāt*). For all these reasons the denotative function of the verbal suffix is in effort (*kṛti*).

Now the grammarians raise an objection against this position of the Naiyāyikas. One of Pāṇini's rules (2.3.18) enjoins that the instrumental case should be used to express the agent.[5] This rule occurs under the governing rule *anabhihite* (2.3.1) according to which, when the agent is not expressed by *tiṅ* and so on, the instrumental case should be used to denote it. Now if, according to the Naiyāyikas, in such statements as *Caitrastandulaṃ pacati*, the verbal suffix denotes only effort, then *ti* does not denote agent and so the instrumental case-ending should be used after the word *Caitra* to denote the agent. This is the objection.

The Naiyāyikas answer this objection by asserting that the reason for using the instrumental case-ending after the word *Caitra* is that the verbal suffix does not denote the number of the agent. This means that in statements like *Caitrastandulam pacati*, the verbal suffix denotes the number, i.e. singular number of the agent. The instrumental case is therefore not used after the word Caitra, to signify the agent. Here the verbal suffix *ti* denotes only the number of *Caitra* and not the number of *taṇdula* (rice). For this there must be some definite reason; otherwise, in statements like *Caitreṇa pacyate taṇdulaḥ* also there would be the possibility of the verbal suffix denoting the number of *Caitra*. The instrumental case thus cannot be used after the word *Caitra* since, according to the Naiyāyikas only when the verbal suffix does not denote the number of the agent in question can the instrumental case be used after it. The question therefore arises: why does the verbal suffix denote the number of the agent? M. tells the reason: that-which-is-not-connected-with-the-state-of-being-the-object-(of action) etc.[6] and that-which-is-to-be-known-by-the-word-ending-in-the-nominative-case, is capable of having its number expressed by the verbal suffix. Dinakara and Rāmarudra explain the two qualifications of this reason in detail.

First of all, Dinakara explains why the expression that-which-is-not-connected-with-being-the-object-(of action) (*karmatvādya-navaruddha*) is to be used. If this expression is not used, then *prathamāntapadopasthāpya* alone would be the cause for the verbal suffix to denote number. In that case, in statements like *Caitraḥ pacati taṇḍulaḥ*, the word *taṇḍula* would be in the nominative case which would convey the meaning-of-being-the-object-of-action by implication (*lakṣṇayā*). It is then possible that the verbal suffix *ti* may denote the number of *taṇḍula*. When, however, it is added that the verbal suffix can denote the number of only-that-which-is-not-connected-with-being-the-object-of-action, then the verbal suffix *ti* cannot denote the number of *taṇḍula* since it is connected-with-being-the-object-of-action (i.e. it is *karmatvādyavaruddha* and not *karmatvādyanavaruddha*).

Now an objection can be raised against this position. The expression *karmatvādyanavaruddhatva* can be explained in two ways: first, not being a qualifier of the meaning of the *kārakavibhakti* like the accusative case etc. And *kārakavibhakti* means a case of nouns which (i.e. case) denotes a *kāraka*, and a *kāraka* is the meaning of a case which (i.e. the meaning) brings about the action which is the meaning of the root (*dvitīyādikārakavibhaktyarthāviś-eṣaṇatvam*). Second, not-being-related-to-the-state-of-being-the-object-of-action (*karmatvādyananvitatvam*). If we take the first meaning in sentences like 'Maitra goes like Caitra', since Caitra is not the qualifier of the-state-of-being-the-object-of-action and since he is only the qualifier of similarity which is the meaning of the word *iva*, there could be the connection-of-number which is the meaning of the verbal suffix *ti* with Caitra. If we take the second meaning, then in such statements as 'cooked rice is eaten' (*pak-vamannam bhujyate*), since rice is connected with the-state-of-being-the-object-of-the-action of cooking, there cannot be the connection-of-number which is the meaning of the verbal suffix *te* with rice (*anna*). Therefore the expression *karmatvādyānavarud-dhatvaṃ* is defined in another way, viz. not being the object of intention as a qualifier for another meaning (*itaraviśeṣaṇatvena tātparyaviṣayatvaṃ*).

The question is: how can the expression *karmatvādyānavaruddhatva* be understood in the meaning of 'not-being-the-object-of-intention-as-a-qualifier' for another meaning. Dinakara points out that such a meaning is obtained by understanding the expression *karmatvādi*, in the sense of 'another' (*itara*) and by interpreting the expression *anavaruddhatva* in the sense of 'not-being-the-object-of-intention-as-a-qualifier'. But there could be an objection against this interpretation. If the expression *anavaruddhatva* means 'not-being-the-object-of-intention-as-a-qualifier', then the expression *karmatvādi* in the sense of 'another' is useless because a qualifier always applies to another. Therefore, it is pointed out that the whole expression *karmatvādyanavaruddhatvaṃ* means not-being-intended-as-a-qualifier (*viśeṣaṇatvena tātparyāvisayatvaṃ*).

Dinakara raises a couple of objections on this point. In statements like *Caitraḥ taṇdulam pacati*, the number which is the meaning of the verbal suffix is connected with Caitra. The reason for this is that Caitra is not the object of intention as qualifier for anything (*viśeṣaṇatvenatātparya viṣayatvaṃ*) but it is enough to say that the reason is merely 'the-state-of-not-being-qualifier' (*aviśesaṇatva-mātraṃ*). In the above sentence Caitra is the qualificand. He is not a qualifier for anything and therefore, a connection with number takes place with regard to Caitra. The reason for the connection with number is thus simply the-absence-of-being-a-qualifier (*aviśeṣaṇatvaṃ*).

But this cannot be admitted because of the following reasons. First of all, what is 'the-absence-of-being-a-qualifier?' It is the-state-of-not-having-qualifier-ness. But qualifier-ness (*prakāratā*) is always associated with an instance of knowledge. Verbal knowledge (*śābdabodha*) is an instance of knowledge and so before the verbal knowledge from a sentence such as 'like Caitra, Maitra goes', takes place, there is the absence of qualifier-ness (*prakāratā*) in Caitra. Therefore, there could be a connection-of-number (which is the meaning of the verbal suffix *ti*) with Caitra. This objection is answered by pointing out that in the sentence 'like Caitra, Maitra goes', the qualifier-ness in Caitra is indicated by the qualificand-ness in similarity (*śādṛsya*) and the expression 'like Caitra' is a part of the same sentence. Now, it is impossible to have the verbal knowledge of the meaning of the whole sentence without having knowledge of the meaning of the part of the sentence prior to it. So how can there be the-absence-of-qualifier-ness in Caitra?

Therefore, it is impossible to have the connection-of-number with Caitra.

This argument cannot be admitted. In the *Nyāya* system there are two opinions about how verbal knowledge arises from sentences. According to one, the prior verbal knowledge arising from parts of a sentence is the cause of the verbal knowledge arising from the whole sentence; it is on this basis that the objection is raised. But the other opinion is that there is no need of the verbal knowledge arising from parts of a sentence for the verbal knowledge arising from the whole sentence; the recollection of the meanings of all the words of a sentence simultaneously brings about the verbal knowledge of the meaning of the whole sentence. This kind of verbal knowledge is produced following the maxim of 'pigeons in the threshing floor' (*khale kapota nyāyena*). When the verbal knowledge arising from a whole sentence is produced in this way, there is no prior verbal knowledge arising from parts of a sentence in such sentences as 'like Caitra, Maitra goes'. Therefore, since Caitra has the absence-of-qualifier-ness, and it is possible for Caitra to have the connection-of-number, the objection stands. Hence, it is said that the cause of connection with number is not-being-intended-as-qualifier (*viśeṣṣaṇatvena tātparyāviṣayatvaṃ*). Since in 'like Caitra, Maitra goes', Caitra is intended by the speaker as a qualifier of similarity denoted by the word *iva* (like), there is no absence 'of-being-intended-as-qualifier' in Caitra which makes it impossible for Caitra to have a connection with number.

But now another objection arises: In a statement like *Caitra eva pacati* (Caitra alone cooks), Caitra is a qualifier of 'another' which is the meaning of *eva*. Therefore, in Caitra, there is an absence of absence-of-being-intended-as-qualifier, which means that it cannot have a connection with number; but actually there is a connection with number. This is the objection.[7] It can be answered by pointing out that *karmatvādyanavaruddhatva* means not-being-intended-as-being-a-qualifier-to-anything-other-than-the-meaning-of-*eva* (another). In the present instance, Caitra is a qualifier only to the meaning of *eva* (another) and so Caitra could have a connection with number.[8]

But then another difficulty arises: in sentences like 'Caitra sees Caitra', Caitra is indeed intended as a qualifier to-the-state-of-being-the-object-of-action and such a state is different from the meaning of *eva*. Hence, Caitra could not have any connection with number for Caitra. This is the objection. The answer to it is: in the example given, for Caitra to have a connection with number, it is enough that Caitra is not intended as a mere qualifier because it is now asserted that the cause of connection with number is not-being-intended-as-a-mere-qualifier (*viśeṣaṇatvamatreṇatātparyā-viṣayatvaṃ*).

Rāmarudra clarifies exactly what is meant by not-being-intended-as-a-mere-qualifier. What brings about the connection with number is the absence of the-state-of-not-being-intended-as-being-qualifier. In the instances cited of 'Caitra alone cooks' and 'Caitra sees Caitra', although the subject Caitra is a qualifier of the meaning of *eva* (another) in the former, and a qualifier of the meaning of being-the-object-of-action in the latter, in both instances the subject Caitra is intended also as the principal qualificand. There is thus no difficulty in Caitra having a connection with number.

But another objection may be raised against the statement that what brings about the connection with number is not-being-intended-as-a-qualifier. According to the *pūrvapakṣin*, knowledge of intention cannot be regarded as a cause of verbal knowledge in general. Where words like *saindhava* (salt, horse), which have many meanings, are used, the word is sometimes, understood in the sense of horse, sometimes in the sense of salt. We can assert that the knowledge of intention is a cause of verbal knowledge only in instances where words have many meanings. When words have only one meaning, knowledge of intention cannot be attributed to verbal knowledge. Therefore, in such instances 'like Caitra, Maitra goes', the words have only one meaning; knowledge of intention is thus not a cause of verbal knowledge. Hence Caitra is not intended-as-a-qualifier; so there could be a connection of number with Caitra.

The objection can be met by the following argument: the absence-of-the-state-of-being-qualifier means the state-of-not-being-denoted-by-the-respective-words which end in the nominative case and the knowledge of which brings about the knowledge of the state-of-being-mere-qualifier. In 'like Caitra, Maitra goes', the world Caitra, which ends in the nominative case, brings about the knowledge of the state-of-being-mere-qualifier for Caitra with reference to similarity which is the meaning of *iva*. Therefore, there is no connection of number with Caitra. In 'Caitra alone cooks', Caitra is both the principal qualificand and a qualifier of 'another' which is the meaning of *eva*. And so, since Caitra is not merely a qualifier, there is no difficulty in having a connection of number with Caitra.

But it may be objected that in 'like Caitra, Maitra goes', before the verbal knowledge, there is the recollection of Caitra, by means of the word Caitra. In that knowledge (in the form of recollection), Caitra is the object of knowledge as principal qualificand, and so the word is not a mere qualifier. Hence, there could be a connection of number with Caitra. This can be refuted by asserting that the-state-of-being-mere-qualifier as understood here is with reference to the verbal knowledge (*śābdabodhanirūpitaviśeṣaṇatvamātraṃ*) and not with reference to recollection. Caitra is only a qualifier and not a qualificand. Therefore, there can be no connection of number with Caitra.

It has been stated that the cause of connection with number is the-state-of-not-being-denoted-by-the-respective-words which end in the nominative case and the instances of knowledge they provide bring about the knowledge of the-state-of-being-mere-qualifier. What is the purpose of words ending in the nominative case which are mentioned in the explanation of 'the-state-of-not-being-a-qualifier' (*aviśeṣaṇatvaṃ*)? In a sentence like 'Caitra sees Caitra', (i.e. himself), the second Caitra is in the accusative case. Hence, Caitra denoted by the word ending in the accusative case is only a qualifier to 'the-state-of-being-the-object-of-action' (*karmatva*). Now this Caitra is the same as the Caitra denoted by the first Caitra which ends in the nominative case. Therefore, it would be impossible to have a connection of number with Caitra. But when the words ending-in-the-nominative-case are added, this difficulty is avoided since Caitra is denoted by the second Caitra which ends in the accusative case and not in the nominative case.

But this leads to another difficulty. In a sentence like *Caitraḥ pacati Caitra iva Maitraḥ sundaraḥ* (Caitra cooks; like Caitra, Maitra is handsome), where there is conjunct verbal knowledge (*samūhālaṃbana*) brought about by these two sentences, there cannot be a connection-of-number which is the meaning of the tense suffix *ti* in the sentence *Caitraḥ pacati*, with Caitra. This is because the knowledge of Caitra brought about by the word Caitra in the sentence *Caitra iva Maitraḥ sundaraḥ* which ends in the nominative case and which is the cause of the knowledge of the-state-of-being-mere-qualifier. And since the Caitra mentioned in both sentences is the same, there is the absence of the-absence-of-the-state-of-being denoted by the word ending in the nominative case and which is the cause of the knowledge of the-state-of-being-mere-qualifier (*saṃkhyānvayaniyāmakasya viśeṣaṇatvamātraprayojaka prathamāṅtapadānupasthāpyatvasyābhāvāt*).

The answer to this objection is as follows: the cause of the connection with number (which is the meaning of the verbal suffix *ti*) is the-absence-of-the-state-of-being which the object of knowledge brings about by that particular word ending in the nominative case. This particular word is the cause of the knowledge of the state-of-being-mere-qualifier. The word ending in the nominative case should be understood as that particular word ending in the nominative case which, together with the tense suffix, produces the verbal knowledge in which the-states-of-being-object (i.e. *ākhyātārthaniṣṭaprakāratā* and *prathamāntārthaniṣṭaviśesyatā*) which are brought about by both (i.e. that particular word ending in the nominative case and the tense suffix) mutually indicate.

Thus, for example, a sentence like *Caitraḥ pacati* produces the verbal knowledge in which the-states-of-being-object (i.e. *ākhyātārthaniṣṭaprakāratā* and *prathamāntacaitrapādārthaniṣṭaviśesyatā*) mutually indicate. And since this word Caitra, ending in the nominative case, is not *viśesaṇatāmātraprayojaka*, there is no difficulty in having the connection-of-number with Caitra. In the instance of the conjunct knowledge mentioned above, such a verbal knowledge is not produced by the verbal suffix *ti* in the sentence *Caitraḥ pacati* and by the word Caitra ending in the nominative case, occurring in the sentence *Caitra iva Maitraḥ sundaraḥ*. Therefore, although in Caitra, there is the-state-of-being-object-of-knowledge brought about by the word Caitra ending in the nominative case which is the cause of the knowledge being mere qualifier and which occurs in a different sentence, there is no harm in having a connection-of-number with Caitra in the instance *Caitra pacati, Caitra iva Maitraḥ sundaraḥ*.

It has already been stated that the cause of connection with number is *karmatvadyānavaruddhatva* and *prathamānthapadopasthāpyatva*. So far, all the discussions have centred on the first part of the expression describing the cause for the connection with number viz. *karmatvadyānavaruddhatva*. Now it is time to turn to the purpose of the second part of the expression describing the cause for the connection with number, viz. being-known-by-the-word-ending-in-the-nominative-case is being discussed.

M. begins by pointing out that in some instances as *taṇdulaṃ pacati*, i.e. 'rice cooks', when the speaker does not intend rice as a qualifier of the-state-of-being-the-object-of-action, then rice is not the object of intention as a qualifier of another. Hence *taṇḍula* could have a connection-of-number. In order to avoid such a possibility, in mentioning the cause of the connection with number being-known-by-the-word-ending-in-the-nominative case is also added. In the present example, the word *taṇḍula* is used in the accusative case and so in *taṇḍula*, there is the absence of being denoted by the word ending in the nominative case; therefore, there is no possibility of *taṇḍula* having a connection with number.

Dinakara points out a possible objection to this answer: If in *taṇḍulaṃ pacati* (rice cooks), *taṇḍula* is not intended as a qualifier for the-state-of-being-the-object-of-action, then the word *taṇḍula* cannot be put in the accusative case. This is because of *karmaṇi dvitīya* (Pāṇini sūtra, 2.3.2) according to which the accusative case is enjoined only when the meaning of the stem is intended as a qualifier for the-state-of-being-the-object-of-action.

Dinakara answers this objection by asserting that M.'s words, 'not-intended-as-a-qualifier' mean 'not-intended-as-mere-qualifier' (*na viśeṣaṇatvamātre tātparyaṃ*). Here taṇḍula is intended as a qualifier and as the principal qualificand in the following way: rice is having action producing the act of cooking the object of which is rice (*tāṇḍulakarmakapākānukūlavyāpāravāṃstandulaḥ*). Thus there can be a connection with number here since not-intended-as-a-qualifier-for-another-(meaning) is interpreted as not-intended-as-mere-qualifier. In order to avoid this fault the words being-known-by-the-word-ending-in-the-nominative-case is added to the cause of connection with number. In *taṇḍulaṃ pacati*, since rice is denoted by the word *taṇḍula* ending in the accusative case, it is not denoted by the word *taṇḍula* in the nominative case. Hence, *taṇḍula* cannot have a connection with number.

If someone says that in sentences like *taṇḍulaṃ pacati* (rice cooks), when there is intention of having the above mentioned verbal knowledge (i.e. rice is having action producing the act of cooking the object of which is rice), then it is desired that there should be a connection-of-number with rice. Accordingly, the expression being-known-by-the-word-in-the-nominative-case, as a part of the description of the cause of the connection with number, would be useless in sentences like *taṇḍulaṃ pacati*, and so another sentence where such an expression (i.e. *prathamānta-padopasthāpyatvam*) would be needed in order to have a connection with number is pointed out by Dinakara.

In a sentence like 'there is sleeping of Caitra', which is the verbal knowledge arising from the sentence *Caitreṇa supyate*, in the impersonal voice (*bhāve prayogah*), the real meaning of the suffix *te* is the meaning of the root, i.e. sleeping (Pāṇini sūtra: *laḥ karmaṇi ca bḥave cakarmakebhyaḥ*, 3.4.69). Actually 'sleeping', which is the meaning of the root, is conveyed by the root itself; therefore the suffix *te* is used in order that the sentence may be correct, i.e., in Sanskrit the root itself is only used with an appropriate suffix, never alone. Therefore, the meaning of the root which, here, is 'sleeping', has the cause of connection with number, viz. not-intended-as-qualifier-of-another (*itaraviśeṣaṇatvena tātparyāviṣayatvaṃ*). Therefore 'sleeping' could have a connection with number. In order to avoid this fault, the expression that-which-is-to-be-known-by-the-word-ending-in-the-nominative-case (*prathamāntapadopasthāphyatvaṃ*) is also added to the description of the cause of connection with number. And when this is done, sleeping will not have a connection with number since it will not be denoted by a word ending in the nominative case but by the root *svap*.

Now the meaning of the first part of the description of the cause of the connection with number is given in another way in order to explain the purpose of the second part in a different manner. Thus the meaning of the first part is said to be not-being-the-qualifier-for-a-meaning-which-is-other-than-the-meaning-of-the-root. Therefore, there is no connection of number with Caitra in sentences like *Caitra iva Maitro gacchati*, since Caitra is a qualifier for similarity which is the meaning of *iva*, and which is other than the meaning of the root. The second part, viz. 'that-which-is-to-be-known-by-the-word-in-the-nominative-case' is added in the description of cause for the connection with number in order to avoid the connection of number with the meaning of *stokam* in sentences like *stokam pacati* ('he cooks a little').[9] The reason for this is that the adverbial meaning is denoted by a word which is not in the nominative case, viz, *stokam,* little. In should be noted here that according to Vaiyākaraṇas, 'adverbs have accusative case endings and neuter gender'.[10]

If this is the meaning of the first part of the description of the cause of the connection with number (i.e. if *karmatvādyānavaruddhatvaṃ* means *dhātvarthātiriktāviśeṣaṇatvaṃ*), then in a sentence like 'Caitra alone cooks' (*Caitra eva pacati*), there would be no connection-of-number with Caitra, because Caitra is a qualifier of 'another' which is the meaning of alone (*eva*). And so in Caitra, there is the-absence-of-not-being-a-qualifier-of-a-meaning-which-is-other-than-the-meaning-of-the-root.

In order to avoid this difficulty, the meaning of 'not-being-' the-qualifier-for-a-meaning-which-is-other-than-the-meaning-of-the-root, is described as not being intended as mere qualifier for a meaning which is other than the meaning of the root (*dhātvarthātiriktāviśeṣaṇatvamātreṇa tātparyāviṣayatvaṃ*). Here, however, Caitra is not intended as a mere qualifier of 'another' which is the meaning of *eva*, but also intended as the principal qualificand of the meaning of the tense-suffix. More technically, this could be expressed as follows: that which is not intended as the principal qualificand, while at the same time, intended as a qualifier of something which is other than the meaning of the root; the absence of this is what brings about the connection with number. Therefore, there is no difficulty in having a connection-of-number with Caitra.

I shall now turn to the opinion of the author of *dīdhiti*. What follows is a summary of Raghunātha Siromaṇi's *Akhyātaśaktivāda*,[11] particularly about what exactly is the effect brought about by that which is known by the word in the nominative case. The state of being denoted by the word in the nominative case is not independently or directly the cause of bringing about the knowledge of the connection with number. This is because the tense suffix invariably brings about the knowledge of the connection with *bhāvanā*, i.e. the-state-of-being-agent or the-state-of-being-the-object-of-action; but it does not invariably bring about the knowledge of the connection with number, because the tense suffix cannot do this without at the same time, also having *bhāvanā* as its object.[12]

In other words, the-state-of-being-knowledge-of-connection-with-the-state-of-being-an-agent or the-state-of-being-knowledge-of-connection-with-the-state-of-being-an-object-of-action (*bhāvanānvayabodhatvaṃ*) is the pervader. And the-state-of-being-knowledge-of-connection-with-number (*sāṃkhyanvayabodhatvaṃ*) is the pervaded. So the cause of the-knowledge-of-connection-with-the-state-of-being-an-agent or the-state-of-being-an-object-of-action is the-state-of-being-denoted-by-the-word-in-the-nominative-case. Therefore, because of the absence of the-state-of-being-denoted-by-the-word-in-the-nominative-case, which is the cause of the knowledge of connection with the *bhāvanā* (i.e. the-state-of-being-an-agent or the-state-of-being-the-object-of-action), there cannot be the connection-of-number with something which is not denoted by the word in the nominative case.

Someone may object here that there is no argument to establish whether the knowledge of connection with number or the knowledge of connection with *bhāvanā* is the effect of the-state-of-being-denoted by the word in the nominative case. The answer to this objection is as follows: in such sentences as *dārāḥ santi*, there is wife, the plural tense suffix brings about the knowledge of connection with the state of being the agent without also having the connection-of-number as its object. Here, the tense suffix is in the plural; so it conveys the meaning of a plural number. But although the word *dārāḥ* is in the plural, this is contradicted by the meaning of the word *dārāḥ*, since there is only one wife here. It should be noted that though the word *dāra* is always used in the plural, in meaning it is singular. Hence the-state-of-being-denoted-by-the-word-in-the-nominative-case brings about the knowledge of connection with *bhāvanā* (the-state-of-being-agent or *kartṛtva* or the-state-of-being-the-object-of-action or *karmatva*) and not the knowledge of connection with number.

Here another objection may be raised: in such sentences as *dārāḥ santi*, the tense suffix does not bring about the notion of connection with number since in the meaning of the word *dāra* (i.e. wife), the connection with a plural number is contradicted. It has been pointed out earlier that there could be an instrumental case after the word denoting agent, in case the tense suffix does not denote the number of the agent. Therefore, in the present example, since the tense suffix does not denote the number of the agent, there could be an instrumental case after the word *dāra*, i.e. *dārāih santi*. The answer to this objection is that when the tense suffix brings about the notion of agency, the nominative case is needed; this being so in this example, the nominative case is used after the word *dāra*.

So far, I have concentrated on controversies between the Naiyāyikas and Vaiyākaraṇas about precisely what the denotative function of the tense suffix is. According to the Naiyāyikas, it is in the effort (*kritau śaktiḥ*); according to the Vaiyākaraṇas, it is in the agent (*kartari śaktiḥ*). Now the texts discuss and reject the opinion of the Mīmāṃsakas on this point, according to which, the denotative function of the tense suffixes is in action (*vyāpāra*). The Naiyāyikas hold that the denotative function of the tense suffix is in effort (*kṛti*) where the agent is animate as, for example, in the sentence

Caitraḥ gacchati. Here, the tense suffix *ti* denotes effort since Caitra is animate. In sentences like *ratho gacchati*, the chariot goes, the tense suffix *ti* denotes by implication (*lakṣaṇā*) movement of action or the-state-of-being-the-substratum (*āśrayatvaṃ*) of the act of going (*gamana*) which is the meaning of the root *gam*. Since 'chariot' is inanimate the tense suffix *ti* cannot denote effort (*kṛti*); instead, it denotes movement or action by implication (*lakṣaṇā*). The Mīmāṃsakas reject this on the ground that it would be better to accept the denotative function of tense suffix in movement or action (*vyāpāra*) which is common to both animate and inanimate agents. And so when the agent is inanimate, the denotative function of the tense suffix *ti* can be said to be in movement or action (*vyāpāra*) without recourse to implication.

The Naiyāyikas reject this on grounds of cumbersomeness. This is because while action (*vyāpāra*) being a produce of that, is the producer of that which is produced by that (*tajjanyatve sati, tajjanya janakaḥ vyāpāraḥ*). In the present context of *ratho gacchati*, the movement of the chariot is *vyāpāra*; this movement which is produced by the chariot produces the connection of the chariot with the village and this conjunction is produced by the chariot (*ratha-janyatve sati rathajanya uttaradeśasaṃyogajanakaḥ*). In the definition of *vyāpāra*, *tajjanyatve sati tajjanyajanakatvam*, the first *tajjanya* denotes the movement produced by the chariot. The second *taj-janya* denotes the final result of conjunction with the village, which is brought about by the chariot. Here, therefore, the *vyāpāra* is the movement of the chariot. Such a *vyāpāra* is defined as *tajjanyatve sati tajjanyajanaka*.

Similarly, in the sentence *Caitrah taṇḍulaṃ pacati* (Caitra cooks rice), that which is denoted by that (*tad*) is *Caitrah*; and that which is produced by that (*tadjanyaṃ*, i.e. *Caitrajanyaṃ*) is action (*vyāpāra*) and the second *tajjanya* (*Caitrajanya*) is cooking (*pāka*). If such an action is taken as the meaning denoted by the tense suffix, the-state-of-being-action (*vyāpāratvaṃ*) is the determinant of-the-state-of-being-denoted (*śakyatāvacchedakaṃ*). If effort is the meaning of the tense suffix, then the state-of-being-effort (*kṛtitvaṃ*) is the determinant of the-state-of-being-denoted (*śakyatāvacche-dakaṃ*). Of these two, the latter is simple in comparison with the former which is cumbersome because it has to be defined as *tajjanyatve sati tajjanyajanakatvam*.

So the difference between the Mīmāṃsaka position and the

Nyāya position is that according to the former the tense suffix means action by denotative function; according to the latter it means action by implication. The Naiyāyikas hold that the tense suffix means effort by denotative function and wherever this meaning cannot be accepted because of contradiction, as in the sentence 'the chariot goes', the tense suffix is taken as meaning action by implication. It should, however, be clarified that this opinion is held by the old Naiyāyikas.

The stem *gam* means section, i.e. that which brings about connection with a place that is further. The tense suffix means action in the form of the-state-of-having-connection with a rope which is tied to the horses and so on, which brings about the action which is the meaning of the root *gam*. Thus according to the old Naiyāyikas, the full verbal knowledge of the sentence 'the chariot goes', is as follows: the chariot is having action which brings about movement which brings about connection with a place which is further (*uttaradeśasaṃyogānukūlakriyānukūlavyāpāravan rathaḥ*).

But the modern Naiyāyikas believe that the verbal knowledge from the sentence 'the chariot goes' is: the chariot is the substratum of the action which brings about connection with a place which is further (*uttaradeśasaṃyogānukūlakriyāśrayaḥ rathaḥ*). Therefore, the moderns say that the meaning of the verbal suffix in sentences like 'the chariot goes' is 'being-substratum' (*āśrayatvaṃ*) by implication and not action (*vyāpāra*)—as the old Naiyāyikas hold.

In sentences like 'Caitra knows', the tense suffix means the-state-of-being-substratum (*āśrayatvaṃ*) (of knowledge) by implication, because the Naiyāyikas do not admit that knowledge is produced by effort. So the verbal suffix in this instance cannot mean effort by denotative function. But an objection may be raised against this position: in such instances as 'knows' (*jānāti*) one cannot claim that the tense suffix invariably means being-substratum by implication, because in a sentence like *cakṣurjānāti* (the eye knows), the eye is not the substratum of knowledge. The tense suffix here means action (*vyāpāra*) or being-the-cause (*janakatva*) by implication, i.e. eye has action that produces knowledge; or the eye is the cause of knowledge.

The answer to this objection is that in M.'s expression 'knows etc.' (*jānātītyādau*), the word 'etc.' (*ādi*) can also be taken to mean desires, endeavours (*icchati, yatate*). Here, the tense suffix invariably means the-state-of-being-substratum because there is no effort to produce desire or endeavour. Thus the verbal knowledge in sentences like *Caitraḥ icchati* or *Caitraḥ yatate*, is that Caitraḥ is the substratum of desire or effort (*icchāśrayaḥ* or *kṛtyāśrayaḥ*) since 'to desire' and 'to endeavour' can be used only with sentient beings. Therefore, in sentences which contain *icchati* and *yatate*, implication is necessarily to be accepted in the state-of-being-substratum.

In such instances as it perishes (*naśyati*) the tense suffix implies the state-of-being-counterpositive by means of well-established implication. Here, it may well be asked why the-state-of-being-substratum is not considered as the meaning by implication. The answer is that after destruction, the thing itself does not exist and it therefore cannot be the substratum of destruction. Therefore, the verbal knowledge of the sentence 'the pot perishes' is 'the pot is the counterpositive of destruction' (*ghaṭo nāśapratiyogī*).

I shall now describe the actual form of verbal knowledge from some sentences. What is the verbal knowledge from the sentence *Caitrastaṇḍulaṃ pacati* (Caitra cooks rice)? The meaning of the accusative case is the-state-of-being-the-object-of-action (*karmatvaṃ*). This means the result (produced by action); mere action is root meaning. Rice is connected with this result, which is the meaning of the accusative case by the relation of being-superstratum.[13] This simply means that the result (of the action of cooking rice, i.e. softness of rice, *viklitthi*) resides in rice. The result is related to cooking, which is the meaning of the root *pac* by the relation of causality. This cooking is related to effort, which is the meaning of the tense suffix by the relation of causality. This effort is related to Caitra by the relation of being-substratum. Hence the full form of the verbal knowledge is as follows: Caitra has the effort which produces the action of cooking which produces the result in the form of the softness which resides in rice (*taṇḍulaniṣṭa viklittirūpaphalajanakapākakriyānukūlakṛtimāṃścaitraḥ*). This, as Dinakara points out, is in accordance with the old school of Naiyāyikas.

The modern Naiyāyikas, however, believe that the verbal knowledge arising from the sentence *Caitrastaṇḍulaṃ pacati* is as follows: the meaning of the root *pac* is result and action (*phala* and *vyāpāra*, i.e. *viklithi* and *pākakriyā*). The meaning of the accusative case in the expression *taṇḍulam* is the-state-of-being-superstratum (*ādheyatvaṃ*). Rice is connected with the-state-of-being-superstratum by the relation of being-indicated (*nirūpitatvaṃ*). The-state-of-being-superstratum is connected with the result by the relation of being-substratum (i.e. *āśrayatā*). The result is connected with the action (of cooking, i.e. *vyāpāra*) by the relation of being-producer (*janakatā*). This *vyāpāra* is connected with effort, which is the meaning of the tense suffix *ti* by the relation of being-producer (*janakatā*). This effort is connected with Caitra by the relation of being-substratum (*āśrayatā*). Hence the full form of the verbal knowledge is that Caitra has effort which produces the action which produces the result which is the substratum of the state of being superstratum which is indicated by rice (*taṇḍulanirūpitādheyatāśrayaphalajanakavyāpārajanakakṛtimāṃścaitraḥ*).[14]

In sentences 'like the rice is being cooked by Caitra' (*Caitreṇa pacyate taṇḍulaḥ*), the verbal knowledge is as follows: the instrumental case after the word Caitra means agency (*kartṛtvaṃ*). Agency is the-state-of-having-effort (*kṛtimattvaṃ*), i.e. effort. So Caitra is related to effort by the relation of being-superstratum.[15] Thus, we arrive at the meaning, the effort of Caitra, i.e. *Caitraniṣṭakṛtiḥ*. This is related to the act of cooking by the relation of being-produced (*janyatāsaṃbandhena*), since cooking is the result of the effort. This cooking is connected with the result (i.e. the softness of the rice, *viklitthiḥ*) which is the meaning of the suffix of middle voice (i.e. *te*) by the relation of being-produced (*janyatā*) since the softness of the rice is caused by cooking. This result, i.e. the softness of rice, is connected with rice by the relation of being-substratum.[16] Hence the full form of the verbal knowledge of the sentence, *Caitreṇa pacyate taṇḍulaḥ* is: the rice is the substratum of the result (i.e. softness of rice) which is produced by the cooking which is produced by the effort which is in Caitra (*Caitravṛttikṛtijanyapākajanyaphalāśrayaḥ taṇḍulaḥ*).

I shall now discuss the meaning of the *kṛt* suffixes. It is said that the denotative function of *kṛt* suffixes is in the agent or in the object of action (i.e. *karma*, or kriyājanyaphalāśrayaḥ). The reason for this is that in sentences like *Caitraḥ paktā*, Caitra is the agent of cooking, and Caitra and the agent of cooking are related by the relation of non-difference. Similarly, in the sentence *taṇḍulaḥ pakvah* (rice is cooked), rice and what is cooked are related by the relation of non-difference. Thus that which is signified by the substantive ending in the *kṛt* suffix (i.e. *paktā* or *pakva*) and the meaning denoted by the other substantive are respectively experienced as identical. Therefore, *kṛt* suffixes have a denotative function in agent or the object of action.

But there is an objection raised against this position. Caitra and the agent of cooking and rice and what is cooked need not be related by the relation of non-difference. The *kṛt* suffix can have a denotative function in agency (*kartṛtva*) or the state-of-being-the-object-of-action (*karmatva*). These could be related to the meanings of the respective other substantives by a relation other than non-difference, i.e. by the relation of being-substratum (*āśrayatā*). According to this explanation, the verbal knowledge of *Caitraḥ paktā* and *taṇḍulaḥ pakvaḥ* is respectively that Caitra is the substratum of the agency of cooking (*pākakartṛtvāśrayaḥ*) and rice is the substratum of the-state-of-being-the-object-of-cooking (*pākakarmatvāśrayaḥ taṇḍulaḥ*). Therefore, the meaning of the *kṛt* suffix is the-state-of-being-agent (*kartṛtvaṃ*) or the-state-of-being-the-object-of-action (*karmatvaṃ*).

One can refute this objection with the argument that if the connection of the meanings of two substantives by a relation other than non-difference is admitted, then from the words 'floor pot' (*bhūtalaṃ ghaṭaḥ*) there could be the verbal knowledge that the pot is on the floor. This is because floor can be connected with pot by the relation of difference, i.e. the-state-of-being-superstratum (*ādheyetā*). The connection of meanings of two substantives by a relation other than non-difference is therefore not admitted, because when the meanings of two substantives which have the same case are connected by a relation other than non-difference, there is no expectancy (*ākāṅṣā*) which is the cause of the verbal knowledge. In *bhūtalaṃ ghaṭaḥ* (floor pot) there is no expectancy which would bring about the verbal knowledge that the pot is on the floor. In order to have such a verbal knowledge, there should be the seventh case after the word *bhūtala*.

But this position too lends itself to an objection. There is the word Caitra after the word *paktā*; the knowledge of these words in this order is the cause of the verbal knowledge in the form that Caitra is connected with agency of cooking by a relation other than non-difference, i.e. by the relation of being-substratum (*āśrayatā*). Similarly, there is the word *taṇḍula* after the word *pakva*; the knowledge of these words in this order is the cause of the verbal knowledge in the form of 'rice is connected with the-state-of-being-the-object-of-action' (*karmatva*), i.e. the result of cooking (i.e. *phala*, here *viklithiḥ*, softness of rice), by a relation other than that of non-difference, i.e. by the relation of the-state-of-being-substratum (*āśrayatā*). This kind of particular instance of expectancy is the cause of the verbal knowledge of the connection of the meanings of two substantives by a relation other than non-difference. In the expressions *bhūtalaṃ ghaṭaḥ* (floor pot) there is no such expectancy and therefore, there is no verbal knowledge of the connection of meanings of the two substantives by a relation other than non-difference, i.e. the pot is on the floor. Therefore, the *kṛt* suffix has a denotative function in agency (*kartṛtva*) and the-state-of-being-the-object-of-action (*karmatva*).

The answer to this objection is that if we accept the causality of verbal knowledge as has been described in the previous paragraph, then we will have to take the countless number of words ending in *kṛt* suffixes and the words following them and then attribute causality to them for bringing about verbal knowledge of the meanings of two substantives by a relation other than non-difference. Therefore, the Naiyāyikas hold that expectancy is cause only for the verbal knowledge of connections of meanings of substantive and suffix by a relation other than non-difference, and not for the verbal knowledge of the connection of meanings of two substantives by a relation other than non-difference. Since in our experience, the knowledge of connection of the meanings of two substantives is by the relation of non-difference, Naiyāyikas hold the view that *kṛt* suffixes have a denotative function in agent and the object of action.[17]

Notes

1. The relevant section dealing with grammar as a means of cognising the denotative function can be seen in *Kārikāvalī* of Visvanāthapancānana, with the commentary M., D. & R. edited with footnotes by Atmārāṃ Nārayan Jere. Krişnadās Academy, Varanasi, 1988, pp. 359–66.
2. Cfr. the discussion on R., p. 376, II, 29–30. (*Nirdharmitāvacchedakakaviśiṣṭabudherbādhabudhyapratibadhyatayā*)
3. The point of the modern grammarians is this: in the example *Caitraḥ taṇdulaṃ pacati*, the active verbal suffix *ti* denotes directly the substratum of effort and actions. But the volitional effort (*kṛti*) and bodily actions (*vyāpāra*) are denoted by

the root itself viz. *pac*. Therefore, the meaning of verbal suffix is mere substratum (*āśrayaḥ eva*). In the example *Caitreṇa pacyate taṇdulaḥ*, the verbal suffix *te* also means mere substratum (of result). The final result of the act of cooking viz. softness of rice (*viklitti*) is denoted by the root *pac*.

4. *Phalavyāpārayordhāturāśraye tu tiṅaḥ smṛtāḥ*, Cfr. Bhaṭṭojidīkṣita, *Vaiyākaraṇabhūṣaṇasāra*, section on *dhatvarthanirṇayaḥ*, Kārikā, no. 2.
5. Here, the full śūtra is: *kartṛkaraṇostritīyā* 2.3.18. and the governing sūtra (*adhikāra*) is *anabhihite* (2.3.1). Therefore, the whole sūtra means that when the agent or instrument is not expressed (by *tiṅ* and so on) the instrumental case ending is to be used in the meaning of agent or instrument.
6. I have translated *karmatvaṃ* as the-state-of-being-the-object-of-action, although the full meaning of *karmatva* is, *kriyājanyaphalāśrayatvaṃ*, the-state-of-being-the-substratum-of-result-which-is-produced-by-action.
7. Here Rāmarudra goes into a very interesting discussion of the meaning of *eva*, which could be summarised as follows: The meaning of *eva* (alone) is threefold: absence of relation with another (*anyayogavyavacchedaḥ*); absence of non-relation (*ayogavyavacchedaḥ*); absence of absolute non-relation (*atyantāyogavyavacchedaḥ*). Absence of relation with another applies when *eva* is associated with the word which denotes the qualificand (*viśeṣyasaṃgataivakārasya anyayogavyavacchedofrthaḥ*); for example, in such sentences as 'Caitra alone cooks' (*Caitra eva pacati*) the relation of the agency-of-cooking (*pākakartṛtva*) with persons other than Caitra is denied. Now an objection could be raised: relation is a universal positive (*kevalānvayī*) and so absence of relation is unestablished. Hence the denotative function of the word *eva* cannot be cognised in absence of relation. Therefore, the denotative function of the word *eva* is cognised separately in another (*anyasmin*) in relation (*yoge*) and in absence (*vyavaccheda*). Thus the word *eva* has three separate denotative functions. The meaning 'another' becomes the object of knowledge as qualificand of Caitra. This is the meaning of *Caitrānya*, i.e. a person other than *Caitraḥ*. The connection of effort which (i.e. effort) is the meaning of the verbal suffix *ti* with relation which is one of the meanings of *eva* is by means of its (i.e. of relation) being-an-adjunct (*pratiyogitaya*, i.e. =*nirūpakatayā*). The connection of such a relation with absence which is one of the meanings of the word *eva* is by means of its being a counterpositive. This absence is related to 'another' (i.e. the person other than Caitra) which is one of the meanings of *eva* as a qualifier (*viśeṣaṇavidhayā*). Thus the whole form of the verbal knowledge is in the following form: Caitra has the effort which brings about the cooking; any person other than Caitra has the absence of the relation of effort which brings about cooking (*Caitraḥ pākānukūlakṛtimān, Caitrānyaḥ pākānukūlakṛtisaṃbandhābhāvavān*).

   The meaning of *eva* associated with a word which denotes the qualifier (*viśeṣaṇasaṃgataivakārasya ayogavyavacchedofrthaḥ*) is absence of non-relation (*ayogavyavacchedaḥ*). As for instance, when it is said, 'conch is only white' (*śankhaḥ pāndara eva*), the meaning is that the absence of the relation of whiteness is denied with reference to conch.

   The meaning of *eva* associated with the verb is the absence of absolute non-relation (*kriyasaṃgataivakārasya atyantāyogavyavacchedofrthaḥ*). For example in such sentences as 'there is indeed a blue lotus' (*nīlaṃ sarojaṃ bhavatyeva*), the absolute non-relation of blueness with lotus is denied since in some cases the lotus is not related with blueness, i.e. lotuses can be white, red etc. Now an objection can be raised against this explanation. Here absoluteness in non-relation (*ayoge*) means the state-of-being-pervader of lotusness (*ayoge atyantatvaṃ sarojatvavyāpakatvaṃ*). And so since a lotus can also be blue, it would be impossible to deny the relation of blueness with lotus because there is no state-of-being-pervader of lotusness (*sarojatvavyāpakatvaṃ*) in the absence of the relation of blueness (*nīlatvasaṃbandhābhāve*); yet it is possible to deny absoluteness (*atyantatvaṃ* i.e. *sarojatvavyāpakatvaṃ*) in non-relation (of blueness). This is what is called absence of absolute non-relation. And so the aforesaid objection does not arise.

8. In M. *karmatvādyanavaruddhatvaṃ* is *itaraviśeṣaṇatvenatātparyāviṣayatvaṃ*. Dinakara pointed out that the word *itara* is not necessary. Now according to the present explanation of Rāmarudra, the word *itara* becomes useful because *itara viśeṣaṇatvaṃ* now means *evakārārthādatirikteviśeṣaṇatvaṃ*.
9. Note that here a little, *stokaṃ*, is used as a qualifier of the action of cooking (*pākakriyāyāṃ viśeṣaṇaṃ*), and not in the sense of cooking a small quantity of rice.
10. Cfr. Rāmarudra: *kriyāviśeṣaṇānāṃ karmatvam napuṃsakatā ca*. He ascribes this to *varttikaṃ*; for further discussion cfr. S. Subramania Sāstri's *Sabdataraṅgaṇi*, pp. 205–6.
11. Cfr. S. 185, L. 3ff and also *Tattvacintāmaṇi* by Gaṅgeśa Upādhyāya *Sabdakhaṇda*, pp. 867–1009, edited by Pandit Kāmākhyanātha, *ṭarkavāgiśa*, Calcutta, 1897.
12. *Bhāvanatvaṃ kartṛtvakarmatvānyataratvaṃ* cfr. R., p. 365, LL. 15–16.
13. *Prakṛtipratyayau sahārthaṃ brūtaḥ; tayostu pratyayaḥ-prādhānyena*. This means that the stem and suffix together convey the meaning. Between the two, the suffix is the principal part, i.e. the meaning of the suffix is the qualificand (*viśeṣyaṃ*) and the meaning of the stem is the qualifier (*viśeṣaṇaṃ*). So in the present instance of *taṇdulaṃ*, the meaning of the accusative case, result (of the action) is the qualificand and the meaning of the stem, i.e. rice is the qualifier. Hence in verbal knowledge, we say that rice is connected with the result and not that the result is connected with rice.
14. Cfr. S., p. 185, L. 18ff.
15. *Adhāraḥ adheye adheyatāsaṃbandhena anveti*, i.e. *svaniṣṭādhāratānirūpitādheyatāsaṃbandhena anveti ityarthaḥ*. Here *svam* is *ādhāraḥ*. Similarly, *adheyaḥ ādhāre ādhāratāsaṃbandhena anveti svaniṣṭādheyatānirūpitādhāratāsaṃbandhena anveti ityarthaḥ; svaṃ adheyaṃ*.
16. Here the softness of rice is the *adheya* and rice is *adhāra*. Softness is therefore connected with rice by the relation of *adhāratā* (*āśrayatā*), i.e. being-substratum, cfr. note no. 12.
17. Parīkṣit Thampurān adds another powerful argument to show that the denotative function of *kṛt* suffixes is in *kartā* (agent) and not in *kartṛtva* (i.e. *kṛti*, i.e. effort). In such sentences as *paktā paśyati* i.e. 'the cook sees', if the *kṛt* suffix means agency (*kartṛtvaṃ*), then the connection of meaning (i.e. being-substratum, *āśrayatvaṃ*) of the tense suffix with the meaning of the *kṛt* suffix (i.e. agency) is not possible; in ordinary language the usage is that the agent of cooking is seeing and not the effort of cooking is seeing; hence the meaning of the *kṛt* suffix is the agent or the object of action (cfr. S. p. 185, L. 21ff).

Reference and Interpretation

KALYAN S. BASU

The claim that the question of meaning suffers not a bit in being reframed as the question of truth, has received its most sophisticated defense in the writings of Professor D. Davidson. The arguments have been structured around an axis that is central to Davidson's philosophical concern—the notion of interpretation, especially that of an interpretative practice. Viewed along this axis these arguments are, to my understanding, insightful and convincing. Yet there is, in my opinion, a certain tension between the concept of truth and that of interpretation, that inhabits the margins of Davidson's discourse and promises much if only seriously articulated. The tension is of this nature: the concept of truth must, in order to conserve our intuition of its material adequacy, admit to a transparency across contexts of reference, relative to its domain, which is sentences; yet, the concept of interpretation admits,—which is the very reason for its universality as a fact of language usage, as opposed to an unmediated access to the world and the intentions of the other—a certain inscrutability at the roots of our language. How do we square one against the other? Or is the very question ill-posed since the relevant phenomena are at incommensurable conceptual levels? I do not think that this last is really the case: for, there is indeed a point where they, or at least their effects converge—in paradox. This essay attempts to see these questions in a clearer light and in this quest, paradox is the principal point of entry.

In the landmark essay 'Truth and Meaning', Davidson (1967) argues that an adequate theory of meaning for a language must entail all sentences of the form: *s* means that *p*, where *s* is replaced by a structural description of a sentence of the language, and *p* by a sentence. Unfortunately the phrase 'means that' carries a threat of non-extensionality. To counter this, Davidson proposes a radical alternative. Replace the schema with: *s* is *T* if and only if *p*. We shall refer to this as schema (*T*). It is not necessary, at this point to give a more explicit characterisation of the predicate 'is *T*'; we would merely require of our theory of meaning that it place enough restrictions on the predicate to entail all sentences got from the schema when *s* is replaced by the structural description of a sentence and *p* by that sentence. From this it is just a step to argue that the sentences to which the predicate 'is *T*' applies are the only true sentences of our language! Hence the conclusion that a theory of meaning is adequate for a language *L* if it contains a recursive definition of truth-in-*L*.

While this move seems to propose a mere schematic shift—we are indebted for our extensionality to the grace of 'if and only if'—the rest of Davidson's essay and much of his subsequent writing persuades us otherwise. This particular schema is strong enough to bear all the burden that we would like any theory of meaning to support. This is so because the giving of necessary and sufficient conditions for the truth of every sentence of a language is a way of giving the meanings of those sentences. The following statement from 'Truth and Method' sums it up: 'To know the semantic concept of truth for a language is to know what it is for a sentence—any sentence—to be true, and this amounts, in one good sense we can give to the phrase, to understanding the language.'

The weight of the argument, as can clearly be seen, is borne by the form of the schema (*T*). Since this will play a significant role in the sequel, let me state it more perspicuously. Consider any sentence *s* of a language *L*, and let *X* be a singular referring expression which refers to *s*. Then an equivalence of the form (*T*) is any legitimate instantiation of the following schema: *X* is true if and only if *s*. We shall refer to this schema as the condition (*T*), for it is nothing but the form of A. Tarski's condition for the material adequacy of a definition of truth. Tarski intended the schema to constrain the class of theories of truth for any specified language. A truth definition according to any such theory would be considered as adequate from the material point of view if and only if all equivalences of the form (*T*) could be asserted and be shown to follow from it. That this does indeed square with our intuitive grasp of the notion of truth, is, I think, a matter beyond any serious contention.

However, as Tarski himself went on to elaborate, unrestricted application of this seemingly innocent principle leads to paradoxes of various kinds: the infamous antinomy of the liar is an example in point, and Tarski's work on the semantic conception of truth, was an attempt to circumvent this and related paradoxes. The diagnosis of this pathology is well known, as is its therapy. Roughly speaking, languages for which these paradoxes could be generated are too promiscuous in some sense. The predicate 'is true' is type-promiscuous, for it applies to objects, which at least on an analysis *a posteriori*, are of different type levels. There is also a degree of referential promiscuity, for the language can generate names for its sentences which, as referential contexts, can function as both transparent and opaque. We shall make these remarks more precise in the sequel; for now it suffices to say that it is possible to avert this problem by stratifying the domain of our theory into an *object-language* for which the truth definition is sought, and a *meta-language* in which the definition is given. Since the *T*-equivalences are to be asserted in the meta-language, we require that the object-language be contained in the meta-language (or at least be capable of being translated into it). Most importantly, we have to impose a formal type discipline on the two languages in order to prevent the sort of promiscuities which led to the paradoxes.

Formal details of the stratification apart, the point to be noted is that our preconceptual grasp of the notion truth—which we seek to conserve through the condition (*T*)—is in some significant sense *underdetermined*, and this underdetermination can always be instanced in the process of pushing it into a formal contradiction. The situation is in some sense parallel to the pathology of the notion of a set. Perfectly universal at the intuitive level, it can easily be pushed into a formal contradiction. I prefer to view such instances of underdetermination as concepts bearing, in some sense, an informational overload. In order to equilibriate them as formally correct concepts, we have to push some of this information into their domains of application: for instance, by ramifying the domains through a type discipline. Russell's creation of the predicative hierarchy in set theory is paralleled precisely by Tarski's construction of a hierarchy of typed languages. But in the absence of a formal theory of information of this sort, these remarks would have to remain imprecise, or at best suggestive.

However, in this pathology we catch the glimmer of an aporia: for, if the concept of truth indeed admits to underdetermination, are we justified in foisting this underdetermination on to the concept of meaning—by claiming that a theory of the former captures all that we seek in the latter? Material adequacy, formal correctness, antinomies, stratification and underdetermination, are significant landmarks in the story of truth: surely they form no part of the story of meaning. Nor did they of the story of truth at first sight; the problem upon refinement led to their emergence. Perhaps the problem of meaning too appears deceptively innocent prior to analysis.

The vital question here is—paradox. Paradox sullies the manifold of truth—but how salient is it to the notion of meaning? To my understanding, not at all. For the simple reason that a paradox is meaningful enough. The judgement that a paradox is indeed a paradox, is parasitic upon the extraction of its meaning. And in any case, if grammaticality is the condition of meaningfulness, then paradoxes, being grammatical, are meaningful enough. But then, do they have a *meaning*? Not if their meaning is to be framed within, or elicited on the basis of truth conditions. But then how seriously do we take a theory that understands something as meaningful, but denies them a meaning within its framework?

Of course, Davidson has a clear understanding of this problem and addresses it in some detail later on in his essay. Tarski himself was pessimistic about the applicability of a theory of truth to natural languages. Davidson identifies two themes in this pessimism: the universality of natural languages (which leads to paradox) and their amorphousness (which resists formal methods). As he correctly observes, the second is not a problem in principle: it is not difficult to imagine a history of methodological refinement that would lose this difficulty in craft. On the other hand, he confesses that he does not have a definite answer to the first. There is a suggestion that perhaps the problem could be tackled by reconfiguring quantifiers and their scopes. I do not find this too relevant for the problem at hand *viz.* the antinomy of the liar. I am not clear about what part quantification theory plays in this; it appears to be in essence a problem of reference, which is why a type discipline can quite saliently circumvent it. I shall have occasion to elaborate on this after a few paragraphs.

Actually it is not as if the two—universalism and amorphousness—are distinct problems: they are related aspects of the *same* problem—the problem of formal correctness of a certain theory. A solution of one usually entails a solution of the other. A decently categorematic grammar built around a formal type discipline takes care of both at one shot. However, since amorphousness is at least not a problem in principle, I would claim that neither is universalism: in principle, neither is a problem to a theory of meaning. They are problems of craft thrown up by the need to formally structure an underdetermined notion like truth. A particularly perspicuous instance of the perfectly healthy divergence between the formal order of truth and the practical order of (meaningful) performance, occurs in Computer Science. Typed programming often deploys with healthy unconcern the logically blasphemous principle of a type of all types. It is a logical antinomy and the paradox of Burali–Forti could be derived on its basis. Yet the system is computationally functional and even has a non-trivial semantics.

Yet, even if truth were to have a vulnerable core with regard to formal structuration, it would be facile to disregard it on just this account. Davidson's later writings, and especially his 1989 Dewey Lectures on the *Structure and Content of Truth* (1990) builds an extremely sophisticated and cogent case for truth on the grounds of translation, interpretation and practice. I must say that I find the substantial theses perfectly convincing. But then the problem remains: what it is precisely in truth that allows it to function as the condition of translation and of an interpretative practice, and yet at the critical juncture of formalisation, pushes it into incoherence? The answer, I propose, is to be found in the notion of reference.

We can gain a foothold on this concept and its anomaly by looking at the antinomy of the liar more closely. Let me express it schematically. Let *X* be a singular term that refers to a certain sentence of a language *L*. Consider the sentence '*X* is false', and let it be, that contingently, *X* refers to just this sentence, i.e. 'X is false'. Then we have by the condition (*T*): '*X* is false' is true if and only if *X* is false. But note that the occurrence of the second *X* in this sentence is in a non-opaque context. Hence, substituting it with its reference, we have: '*X* is false' is true if and only if 'X is false' is false.

Actually, the trick is done by the mention of the singular term *X* twice in the same sentence—in syntactically identical but referentially contrary contexts. The first context viz. 'X is false' is a name, and hence referentially opaque; the second context viz. *X* is false—is referentially transparent and conserves the truth of the equivalence (*T*) under co-referential substitutions (in this case, by the name of the referred object). These are two points which ought not to distract us later on in the analysis. In the generation of this paradox, any property salient to the predicate 'is true', other than its extensional formulation as in the *T*-condition, is quite irrelevant for our purpose. Second, any valorisation or otherwise of disquotationality in this context is quite irrelevant for our purpose, unless viewed as a purely formal device to turn a (referentially) opaque context into a syntactically identical but transparent one.

In this context, the significant effect of this transformation is to endow the overtly unstructured string 'X is false' with the sentential structure of: *X* is false—leaving it morphologically untouched. The point is, *if there is any fact to the matter of this transformation, it is not any fact of structure* per se. The generation of the paradox is, in fact, dependent on the essential condition that the restructuration leaves the structure itself absolutely intact. Hence the device of signalling the effect through a device that forms no part of morphology—the quotes (and their removal), which I would view more or less as a diacritic. In fact this diacritic has no phonological realisation either.

Contemporary syntactic theory (after the theory of government and binding) has pointed out the need to consider elements in the structural matrix of a phrase, that have neither a morphological nor a phonological realisation. Such elements have significant structural effects, and considerable explanatory and predictive potential. In this case however, we have a diacritic which effects a perlocution of the form 'elide structure!' I would strongly argue that this effect cannot be captured as an effect internal to the domain of structure—that it is properly *orthogonal* to structure and cannot be given an account or any site within an economy of structural effects. In quite an analogous sense, predications of provability or of decidability are orthogonal to the domains respectively of proofs and of algorithms, and attempts to draw them into the respective domains leads to inconsistencies. It may be recalled that classical proofs of incompleteness and of undecidability exploit precisely this phenomenon.

It is important to understand the limits of structure. Structure operates universally and transparently over its domain. A particular string is always endowed with the same structural description under a specified structural syntax in every context that does not have systematic structural effects on embedded sub-strings *specifiable within the structural syntax*. It has already been argued that the perlocution of the quotes is not such an effect.

This understanding of the transparency and the universality of the application of structure has significant implications for a theory of reference that is parasitic upon a theory of structure. Reference, by definition, is extensional and the reference of expressions depends systematically upon the references and effects thereupon of their constituents. This dependence is determined by and along the structures through which the constituents constitute the whole. In other words, reference is hypotactic. Extensionality and hypotaxis together imply that co-referential substitutions of structural sub-constituents leaves the reference of the whole unaltered. More significantly, in contexts not explicitly marked as non-extensional, not only is reference determined by structure, but the latter in fact drives the process of interpretation towards referential primitives viz. names. Structurally complex referential terms carry intensional information and interpretation processes this information—unless otherwise directed—towards extensions. In view of this, the sentence: *X* is false, of the antinomy, has already an inherent referential instability. The above principle, along with the idea that structure-eliding effects are orthogonal to the structural domain—and hence to reference, which is parasitic upon it—creates a potential that drives the sentence inherently towards the regressive form: "*. . ." is false" is false" is false*. Of course, what arrests and stabilizes this, is the structure-eliding force of the quotation: the process halts on the first step and yields: '*X* is false' is false. In the case of the paradox, the disquotation restores elided structure and we 'evaluate' the reference of the second *X*, to yield the contradiction.

To take stock, what we have been arguing towards thus far, is this: the concept of truth effects, by virtue of the condition (*T*), an adequation between incommensurable elements of the interpretative economy: structure, on the one hand, and opacity (or the elision of structure) on the other. This elemental dichotomy manifests itself on the semantic level as the dichotomy between referring and naming. The concept of truth, which by virtue of its extensional formulation, operates transparently through an economy of structural effects, can indeed be made incoherent by the force of opacity that is orthogonal to this economy. Names have no structure, and are to be understood as interpretatively opaque; every other structure (and that includes singular referring terms) speaks. Devices—diacritic or otherwise—that elide or revive structure, leave an easy access to formal antinomies, in contexts such as condition (*T*) which assert an adequation between structured and opaque entities.

A somewhat different way of characterising this condition may be illuminating. One understands by the term *code* a system of signs, for which name and structure coincide. In other words, the name of any piece of code *is* its structure, and vice versa. Then one would maintain that in so far as the concept of truth, through the condition (*T*), admits an adequation between names and structures, it countenances the view of any language as a *code*, and of interpretation as a process of *decoding*.

An interesting connection emerges from this account: the presence of structure is deeply coupled with the possibility of interpretation. Structure *speaks*; it triggers interpretation and delineates the path that it shall trace. As interpretation steps through structure, it culls its content through hypotaxis. On the other hand, anything that admits of an interpretation, must have a structure. In fact, I would maintain that as far as the knowledge of interpretative practice goes, only that part of it is, in principle, publicly and evidentially accessible, that is in the form of a knowledge of structure, and of constitutive and evaluative precepts that recurse along structure. Interpretative knowledge shared between speaker and hearer, which is also the knowledge that permits evidencing of claims, and the attributions of intentions, is limited to such structural knowledge. Yet if we are to have such knowledge, we cannot do better than a theory of truth for the language: for truth, as we have seen, aligns its field with and effaces it into structure; opacity leaves not the slightest residue. Hence, the salience of a truth theory to a theory of interpretation viewed as a social practice.

The process of interpretation can, however, be punctuated, as it steps through structure, by forces that operate orthogonally to it and through parataxis. Of these, the force that elides structure to make contexts opaque is perhaps the most incorrigible within its economy. The knowledge of these forces and their operation forms no *determinate* part of that shared and publicly accessible knowledge that sustains the social practice of interpretation. In actual practice, they always exist as disruptive potential and confer on it a measure of inscrutability and indeterminacy. Certain empirical adjustments do tend to bring their effects into line with approximate truth conditions, but a gap always remains potentially. Such adjustive tendencies fore-structure the practice, and goes in Davidson under the name of charity, and in Plato, perhaps, as the good will to understand.

In my view, the interpretation-as-social-practice perspective underemphasises the fact that interpretation is also an individual and intentional act; the intention to interpret and the intention to make meaningful, both pre-form the practice and make it inscrutable at its roots. It is this intention which secures the deployment of those forces that operate orthogonally to structure, and which sustains the norm of charity in interpretation. The operation of this intention can clearly be seen within certain typical strategies internal to the practice. I can mention three that are particularly prominent. First, the strategy by which references are not taken, so to speak, all at once to the world, but suspended or held at varying shades of opacity. It is precisely such a strategy that allows us to grasp as interpretable certain self-referential expressions that entail a referential regress. Second, the strategy that allows the negotiation of contexts, so that references may be taken coherently and conservatively with regard to them. This is a frequently followed strategy, especially in novel or unfamiliar contexts. Third, the attribution of intentions to agents and revisions thereof, with a view to maximise understanding.

The point in all of these—the suspension of references, the negotiation of contexts and the attribution of intentions—is not, as in Davidson, that they are done under the norm and on the basis of a theory of truth—internalised or explicit. I agree with this point of Davidson's completely. What interests and intrigues me is that while the concept of truth, under the condition (*T*), excurses the field of structure in a transparent way, in the case of the strategies mentioned above, the operation of structure itself is interrupted, suspended or elided, and through forces beyond its ambit; in such instances, the concourse of truth becomes indeterminate. We have already seen an example of how in such a situation, an insistence on determinate truth conditions leads to a paradox.

In my understanding, this indicates an aporia in the discourse on truth and interpretation. This aporia is not something that we must rush to repair: in itself, it calls us to deepen our understanding of the field of interpretation. It indicates, in my view, a discontinuity between a theoretical perspective that views interpretation as a social practice with truth as its norm and ground, and a perspective that views interpretation as an agentive and intentional projection of signifying clusters into the sphere of the meaningful. Within the first perspective, truth is gridded on to structure without residue, and so too interpretation through its truth. Within the second, the intention to interpret sustains, at the interstices of structure, a play between structure and the world; this interstitial play is not determinately accessible as public evidence, nor decidable as truth condition. Between this play and that determination, the game of interpretation is viably sustained: and in order to play it, the concept of truth is the clearest and the most basic illusion we must hold.

References

CARDELLI, L. (1986). *A Polymorphic λ - Calculus with Type: Type*. DEC Systems Research Centre Report, No. 10.

DAVIDSON, D. (1967). 'Meaning and Truth'. *Synthese* 17.

———. (1985). *Inquiries into Truth and Interpretation*. Oxford: Clarendon Press.

———. (1990). 'The Structure and Content of Truth'. *Journal of Philosophy* LXXXVII(6).

TARSKI, A. (1944). 'The Semantic Conception of Truth and the Foundations of Semantics'. *Philosophy and Phenomenological Research*, Vol. 4.

———. (1983). *Logic, Semantics, Metamathematics. Papers from 1923 to 1938*. 2nd edition. Translated by J.H. Woodger. Indianapolis, U.S.A.: Hackett Publishing Company.

The Paradox of the First Person

FRANÇOIS RECANATI

Modes of Presentation and the Rationality Constraint

Frege made a famous distinction between the reference of a singular term—that which it represents or denotes—and the mode of presentation, i.e. the way the reference is presented. Modes of presentation are needed essentially to make sense of our propositional attitudes, e.g. to make sense of the fact that we believe 'Cicero is bald' while disbelieving 'Tully is bald'. The reference of 'Tully' and 'Cicero' being the same person, there must be something other than the reference which the terms 'Tully' and 'Cicero' contribute to the propositions respectively expressed by 'Tully is bald' and 'Cicero is bald', and which explains that one proposition can be believed while the other is disbelieved. In order to play this role, modes of presentation must obey a constraint which Schiffer (1978) refers to as 'Frege's Constraint' and which I will call, for reasons having to do with the topic of this symposium, the Rationality Constraint:

> Necessarily, if *m* is a mode of presentation under which a minimally rational person *x* believes a thing *y* to be F, then it is not the case that *x* also believes *y* not to be F under *m*. In other words, if *x* believes *y* to be F and also believes *y* not to be F, then there are distinct modes of presentation *m* and *m'* such that *x* believes *y* to be F under *m* and disbelieves *y* to be F under *m'*. Let us call this *Frege's Constraint: it is a constraint which any candidate must satisfy if it is to qualify as a mode of presentation* . . . . In effect, Frege's Constraint provides *the* motivation for the introduction of modes of presentation. (Schiffer 1978: 180)

From the Rationality Constraint plus other plausible assumptions we can derive a number of interesting consequences. The first one is quite obvious: I call it the Principle of Difference for Modes of Presentation (PDMP): Two modes of presentation m and m' are distinct if it is possible for a rational person not to realise that they are modes of presentation of one and the same object.

The second consequence is less obvious. It is this: we need to draw a distinction between two sorts of modes of presentation—linguistic and psychological. Psychological modes of presentation are modes of presentation in Frege's sense, that is, satisfiers of the Rationality Constraint. But there is another sort of mode of presentation, namely, linguistic modes of presentation. Such modes do not satisfy the Rationality Constraint. What I am calling the linguistic mode of presentation is closely related to what Kaplan calls the 'character' of a singular term. The character of a term is a linguistic rule which, given a context, determines or helps to determine the reference of that term in that context. Thus the character of 'I' is the rule that 'I' refers to the speaker, i.e. to the person who utters this token of 'I'. By virtue of this linguistic rule, the reference of 'I' is presented as being the speaker. The character of a word like 'I' can therefore be described as a certain mode of presentation of the reference—the *linguistic* mode of presentation. It has the following properties (Recanati 1990, 1993):

(*a*) It is conventionally determined by the rules of the language. The way the reference of 'I' is presented (viz., as being the speaker) is determined by the linguistic rule that 'I' refers to the speaker.[1]

(*b*) As a consequence of (*a*), the linguistic mode of presentation is constant and does not vary from context to context, contrary to the reference.[2]

Clearly, linguistic modes of presentation thus characterised do not have to satisfy the Rationality Constraint. Consider the demonstrative phrase 'this ship'. The linguistic meaning of this expression, and therefore the linguistic mode of presentation of the reference, is constant from occurrence to occurrence: the reference is always presented as being a salient ship. But the psychological mode of presentation is not constant. In fact, it cannot be. Consider the utterance 'This ship <pointing to a ship through one window> is a steamer but this ship <pointing to a ship through another window> is not a steamer' and suppose that, unbeknownst to the speaker, the same, very long ship is being demonstrated twice.[3] The speaker is not irrational even though she says of the same ship both that it is and that it is not a steamer. She is not irrational because she does not realise that there is only one ship; and she does not realise this because she thinks of the ship under two different (psychological) modes of presentation. In other words, we have to posit two different (psychological) modes of presentation of the reference, one corresponding to each token of the expression 'this ship', in order to make sense of the utterance. But there is only *one* linguistic mode of presentation, the same for both tokens of this expression.

In the case of 'I', it is clear—though for different reasons—that two different sorts of mode of presentation are involved. On the one hand the reference of 'I' is presented as being the speaker, by virtue of the linguistic rule that 'I' refers to the speaker. Yet 'I' also expresses the concept of *oneself*. The thought expressed by 'I am the speaker' is neither the thought that the speaker is the speaker, nor the thought that, say, Recanati is the speaker. (I could think 'I am the speaker' even though, suffering from amnesia, I did not remember that I was Recanati.) This shows that 'I' expresses a special mode of presentation which is not 'the speaker', but rather something like *self*.

That there is a distinction between the two modes of presentation ('the utterer' and *self*) both associated with the word 'I' is shown by a simple application of the PDMP, for someone may well not realise that *he himself* is the utterer. So we have two distinct modes of presentation, and one problem: how are they connected? What is the relation between the two modes of presentation in the case of 'I'? Frege had to face a very similar problem, for he saw the need for the distinction between the two modes of presentation in the case of 'I'.

The Communication of First Person Thoughts

In his article 'The Thought', Frege writes:

> Every one is presented to himself in a particular and primitive way, in which he is presented to no one else. So, when Dr.

> Lauben thinks that he has been wounded, he will probably take as a basis this primitive way in which he is presented to himself. And only Dr. Lauben himself can grasp thoughts determined in this way. But now he may want to communicate with others. He cannot communicate a thought which he alone can grasp. Therefore, if he now says 'I have been wounded', he must use the 'I' in a sense which can be grasped by others, perhaps in the sense of 'he who is speaking to you at this moment' . . . (Frege 1918–19: 25–26)

From this passage, two important ideas can be extracted. The first one is commonly accepted nowadays:

(*a*) First person thoughts concerning a person A can be grasped or entertained only by A. Another person, B, can entertain thoughts about A, but not first person thoughts about A: only A can think of himself in the first person. To be sure, B can also entertain first person thoughts; but these thoughts will be about B, not about A. (Even if B falsely believes that he *is* A, that would not make his first person thoughts, thoughts about A.)

Once we accept (*a*), there arises a problem which I will call 'the paradox of the first person'. First person thoughts are private, hence incommunicable; yet we do communicate them, by uttering first person sentences. How do we manage to do this? Frege's second idea is meant to solve the paradox:

(*a*) There are two sorts of senses or modes of presentation associated with the first person. Let us call the 'special and primitive' mode of presentation which occurs in first person thoughts *self* where 'x' stands for the name of the person thinking the thought (for example *self* $_{\text{Lauben}}$ in the case of first person thoughts about Lauben).[4] This mode of presentation must be distinguished from the mode of presentation associated with the word 'I' in communication ('he who is speaking to you at this moment'). The latter can be grasped by others, the former cannot.

The paradox is solved because, according to Frege, we do *not* communicate the original, incommunicable first person thought involving the mode of presentation *self*$_x$, but a different thought involving the other sort of mode of presentation.

One might think (and some have said) that the distinction between the two modes of presentation is *ad hoc* and designed only to solve a particular problem. I, on the contrary, think it is an essential distinction—one which lies at the very foundation of the theory of language use. It corresponds to the distinction I have just made between what I called 'linguistic' and 'psychological' modes of presentation—a very general distinction, which affects every indexical, not merely the first person, and which holds whether or not we like Frege's solution to the paradox of the first person (Recanati 1990, 1993).

Even though it is far from *ad hoc*, Frege's solution is not altogether satisfactory, for it is sketchy and incomplete. What is the relation between the two sorts of mode of presentation mentioned in (*b*)? What makes it possible for the intersubjective sense associated with 'I' in communication to *stand for* the private sense *self*$_x$ which cannot be directly communicated? These are important questions which Frege does not address, let alone answer.

There is an alternative solution to the paradox, which does not rely on the distinction between the two sorts of mode of presentation but on another distinction (Dummett 1981: 122–23). The speaker's first person thought is not, and cannot be 'communicated' because the hearer does not, and cannot, come to *entertain* that very thought as a result of the communication process. The speaker who says 'I have been wounded' expresses a first person thought, to the effect that he himself has been wounded; but the hearer, upon understanding the utterance, can only form a different thought: '*He* has been wounded'. Contrary to the speaker, the hearer does not think of the referent (i.e. the speaker) in a first person way. So the speaker's thought has not been 'communicated' in the strong sense of the term. Yet it has been communicated in a weaker sense, for the speaker's first person thought which is expressed by the utterance 'I have been wounded' can be *recognised* as such by the hearer. Even if the speaker's thought is unavailable to the hearer, the utterance may inform the hearer that the speaker entertains a certain type of thought, which he himself (the hearer) is unable to entertain. Thus Leo Peter knows which thought Lauben has expressed by saying 'I have been wounded', even though he himself, Peter, cannot think that very thought. Along these lines, it may be found unnecessary to distinguish between two sorts of thought, that which Lauben privately entertains and that which he communicates. To account for the communication of first person thoughts despite the 'incommunicability' of the latter, one need only draw a distinction between two forms of communication: an utterance can 'express' a thought (weak communication) even if that thought is not thereby made available to the hearer (strong communication).

My primary aim in what follows is to fill the gap in Frege's account so as to make it satisfactory. But I will start by considering the alternative account and raising an objection to it. To meet the objection I will suggest an improvement of the alternative account which makes it indistinguishable from Frege's. The problem of the relation between the two modes of presentation will be solved along the way.

Semantic Content and Thought

Even though it departs from Frege's own solution, the alternative account is Fregean in spirit. From a Fregean point of view, it has the great merit of allowing one to maintain the equation of thought and semantic content.

The semantic content of an utterance—that which the utterance expresses and which must be grasped for it to be correctly understood—is by definition an 'objective' property of that utterance which can be recognised by both speaker and hearer and which remains stable in the process of communication. But the required stability cannot be found at the level of thoughts, or so it seems. The first person thought which the speaker expresses by saying 'I have been wounded' differs from the hearer's thought formed upon understanding the utterance—they involve different modes of presentation of Lauben. The sentence means the same thing for speaker and hearer, and the statement that is made—to the effect that Lauben has been wounded—is also the same for both, but the associated thoughts change as communication proceeds from speaker to hearer. This is what makes the Russellian notion of a (singular) 'proposition' an arguably better candidate for the status of semantic content than the Fregean notion of a thought. For the proposition ('what is said', the 'statement' that is made) remains constant from one person to the next, in contrast to the thought.

As John Perry says, 'one reason we need singular propositions is to get at what we seek to preserve when we communicate with those who are in different contexts' (Perry 1988: 4).

The alternative account disposes of this objection to Frege's equation of thought and semantic content. Even though the speaker's thought is tied to his own point of view and cannot be *entertained* by someone else (e.g. the hearer), still it is this thought which is expressed by the utterance and can be *recognised* as such by the hearer. Its being publicly recognisable confers a sufficient objectivity to the speaker's thought, despite its essential subjectivity, to make it a plausible candidate for the status of semantic content.

Let us analyse the theoretical move at work here. Two points of view are involved in the communication process: that of the speaker and that of the hearer. In Frege's example the speaker's thought includes the mode of presentation $self_x$, while the hearer's thought, formed upon understanding the utterance, is a demonstrative, third person thought: '*He* has been wounded.' As long as the speaker's thought is seen as being on the same footing as the hearer's, it is tied to a particular point of view and lacks the sort of objectivity needed to equate it with the semantic content of the utterance. The move consists in *privileging* the speaker's thought and giving primacy to his point of view over the hearer's. On the 'alternative account' I have sketched (following Dummett), it is the speaker's first person thought rather than the hearer's which is objectively 'expressed' by the utterance and recognised as such by all participants in the speech episode.

One possible objection to this move is that it is somehow arbitrary. How do we choose the particular point of view to be privileged? On intuitive grounds it seems natural to select the point of view of the speaker, yet there are *also* reasons to select the hearer's point of view. As Evans emphasised (after Dummett), what matters, when we want to individuate semantic content, is what would count as a proper *understanding* of an utterance (Evans 1982: 143n, 171, etc.); now 'understanding' defines the task of the hearer. Thus it is the hearer's point of view which Evans privileges; as a result of that choice he is led to deny that the grammatical first person expresses the concept $self_x$: the grammatical first person expresses a demonstrative concept akin to that expressed by the demonstrative phrase 'that person', according to Evans.

To overcome the difficulty, one may try a slightly different route. Instead of *privileging* a particular point of view (that of the speaker or that of the hearer), we may decide to focus on what is *common* to both points of view. This more or less corresponds to the Russellian strategy, according to which what is common to the speaker's thought that he himself has been wounded and to the hearer's thought that that man, Lauben, has been wounded, is the singular proposition: <Lauben, the property of having been wounded>, that is, the state of affairs which both thoughts represent (their common 'incremental truth-conditions', in Perry's terminology [Perry 1990]). Now this commonalist strategy can also be opted for in a Fregean framework, for there is more that is common to both thoughts than merely the state of affairs they represent at the level of 'incremental truth-conditions'. In particular, there is more that is common to the modes of presentation under which Lauben and his addressee respectively think of Lauben than merely the reference, i.e. what these modes of presentation are modes of presentation of.

The Solution: Nondescriptive Modes of Presentation as Dossiers of Information

Most authors in the field, whether Russellian or Fregean, draw a distinction between descriptive and nondescriptive modes of presentation. Indexicals and proper names typically express nondescriptive modes of presentation, in contrast to (attributively used) definite descriptions, which express descriptive modes of presentation.[5]

What are nondescriptive modes of presentation? Like a number of contemporary authors, I construe 'nondescriptive' modes of presentation as *dossiers of information*. Thus *self*$_{\mathrm{Lauben}}$ is Lauben's dossier for whatever information he gains about himself. What is 'special and primitive' about this sort of dossier is that Lauben—like all of us—has a particular way of acquiring information about himself, such that (*i*) only Lauben can acquire information about Lauben in this way, and (*ii*) Lauben can acquire information in this way only about Lauben. A *self*-dossier serves as repository for information gained in this particular way (the first person way).

Qua dossier, a nondescriptive mode of presentation *contains* information about whatever the dossier concerns. This allows for the following possibility: two dossiers which differ by their global content and/or by the sort of dossier they are, may nevertheless have something in common, namely *part* of their content—some particular piece of information which they both contain. This is what happens in Frege's example. Both the thought of the speaker and that of the hearer include a mode of presentation which corresponds to their respective dossiers concerning Lauben. The modes of presentation in question are quite different from each other: the speaker's is a first person mode of presentation (i.e. it corresponds to a dossier based on the special way of acquiring information mentioned above) while the hearer's is a third person mode of presentation. Nor do they contain the same information: there are things which Lauben knows about Lauben which his hearer does not know, and the other way round. But there are also pieces of information which both dossiers contain—there are things which both Lauben and his hearer know about Lauben. The latter provide *identificatory facts* which Lauben and his hearer can appeal to in order to secure reference when communicating about Lauben. In particular, both Lauben's and his hearer's dossiers concerning Lauben include the information that Lauben is the utterer of this token of 'I have been wounded'. This is part of Lauben's current notion of himself as much as it is part of his hearer's current notion of Lauben: Lauben is conscious of being the utterer, and hearer also knows that Lauben is the utterer, the man speaking to him at this moment. The information is part of both dossiers, even though one is a first person dossier and the other a third person dossier. Now that specific aspect common to both the speaker's and the hearer's notion of the reference is, I suggest, what is expressed by the linguistic expression 'I'. The reference of 'I' is presented as being the utterer of this token (linguistic mode of presentation). That linguistic mode of presentation is intersubjective, unlike the psychological mode of presentation which is subjective (i.e. the notion of *himself*, on the speaker's side, or the notion of *that man*, on the hearer's side); but the former may be construed as an *aspect* or *part* of the latter, an aspect (or part) which is *common* to the speaker's and the hearer's point of view.

Note that the identificatory fact which Lauben appeals to in order to secure reference to himself in communication belongs to a special category of identificatory facts: the category of *communication-specific* identificatory facts. Those facts do not exist independently of communication but are created in the very process of communication (Benveniste 1956, 251–57). They are aspects of the speech situation, and as such they are automatically (and mutually) known to both speaker and hearer *qua* participants in that situation. Thus both the speaker and the hearer (in a normal conversational setting) know that the speaker—say Lauben—is the speaker, that the hearer—say, Leo Peter—is the hearer, and so forth. This enables the speaker to use these mutually manifest facts in referring to the speaker, the hearer and other aspects of the speech situation. Indexicals are conventional means of doing so: the linguistic modes of presentation conventionally expressed by indexicals such as 'I' or 'you' ('the utterer', 'the addressee') correspond to facts about their referents which are created by the speech situation itself and are therefore mutually manifest to participants in the speech situation.

Since the identificatory fact appealed to by virtue of the linguistic sense of the indexical is mutually known to the speaker and his hearer, it belongs to their respective dossiers concerning the reference—the speaker's first person dossier and the hearer's third person dossier. The linguistic sense of the indexical can therefore stand for both dossiers through a cognitive process of 'synecdoche' (Recanati 1993): the part stands for the whole. The linguistic sense of 'I' ('the utterer of this token') stands for the speaker's notion of himself because it corresponds to an aspect of that notion, to some information which the speaker's *self*-dossier contains (the information that he is the utterer of this token). Interpreting the utterance—the hearer's task—consists in going back from the part to the whole; but it is not the same whole at both ends of the communicative process. We start with the speaker's thought which involves his first person dossier. The latter does not go into the semantic content of the utterance, which contains only the linguistic sense of 'I', corresponding to an aspect or part of the original dossier. At the other end we find the hearer's thought, formed by 'interpreting' the semantic content of the utterance and somehow enriching it with the hearer's own dossier concerning the speaker, which also contains as a component part the identificatory fact appealed to by the indexical.

On this view Frege was right: that which is communicated—the semantic content of the utterance—is not quite the speaker's original thought. It involves a mode of presentation ('the utterer of this token') which is closely related to the linguistic meaning of 'I' and differs from the special and primitive mode of presentation $self_x$ which occurs in the speaker's first person thought that he himself has been wounded.

Notes

1. Kaplan (1989: 505): 'The character of an expression is set by linguistic conventions and, in turn, determines the content of the expression in every context. Because character is what is set by linguistic conventions, it is natural to think of it as meaning in the sense of what is known by the competent language user.'
2. Perry (1977: 479): 'When we understand a word like 'today', what we seem to know is a rule taking us from an occasion of utterance to a certain day. 'Today' takes us to the very day of utterance, 'yesterday' to the day before the day of utterance, 'I' to the speaker, and so forth. I shall call this the *role* of a demonstrative . . . . The object a demonstrative takes us in a given context, I shall call its value in that context or on that occasion of use. Clearly, we must grant 'today' a role, the same on both occasions of use. And we must, as clearly, give it different values on the two occasions.'
3. I borrow this example from Perry (1977: 483). See also Evans (1982: 84).
4. This notation is essentially Peacocke's. See Peacocke 1981, 1983.
5. The modes of presentation in question, whether descriptive or nondescriptive, are all 'psychological' modes of presentation in the sense of §1.

References

BENVENISTE, E. 1956. La nature des pronoms. Reprinted in his *Problèmes de linguistique générale*. Paris: Gallimard (1966), pp. 251–57.

DUMMETT, M. 1981. *The Interpretation of Frege's Philosophy*. London: Duckworth.

EVANS, G. 1982. *The Varieties of Reference*. (J. McDowell ed.). Oxford: Clarendon Press.

FREGE, G. 1918–19. The Thought: A Logical Enquiry. English translation by A. and M. Quinton in P. Strawson (ed.), *Philosophical Logic*, Oxford University Press (1967), pp. 17–38.

KAPLAN, D. 1989. Demonstratives. In J. Almog, H. Wettstein and J. Perry (eds.), *Themes from Kaplan*. New York: Oxford University Press, pp. 481–563.

PEACOCKE, C. 1981. Demonstrative Thoughts and Psychological Explanation. *Synthese* 49: 187–217.

———. 1983. *Sense and Content*. Oxford: Clarendon Press.

PERRY, J. 1977. Frege on Demonstratives. *Philosophical Review*, 86: 474–97.

———. 1988. Cognitive Significance and New Theories of Reference. *Noûs* 22: 1–18.

———. 1990. Individuals in Informational and Intentional Content. In E. Villanueva (ed.), *Information, Semantics and Epistemology*, Oxford: Basil Blackwell, pp. 172–89.

RECANATI, F. 1990. Direct Reference, Meaning, and Thought. *Noûs* 24: 697–722.

———. 1993. *Direct Reference: From Language to Thought*. Oxford: Basil Blackwell.

SCHIFFER, S. 1978. The Basis of Reference. *Erkenntnis* 13: 171–206.

Executable Justificational Rationality as Naturalised Epistemology

AMITABHA GUPTA
NILESH SUTARIA

Introduction

Naturalised Epistemology and Epistemic Rationality

The naturalistic approach to epistemology has posed a challenge to the concept of epistemic rationality. The rationality principles, in general, demand that the cogniser should be as logical and rational in justifying epistemic claims as he is in formal logic, i.e., that human beings be logically perfect in their day-to-day reasoning. But this rationality claim is in jeopardy while undertaking the enterprise of naturalised epistemology, for naturalised epistemology is based on a reasoning system which implies that human beings, instead of being logically perfect, more or less invoke the heuristic principle in inference with a minimal deductive ability.

Historical Background of the Epistemic Rationality

Some forms of so-called naturalised epistemology (Quine 1969) go to the extent of completely ignoring the rationality principle in epistemology. For example, Quine claims that epistemology should be replaced with psychology and psychology for him is behavioural. On the other hand, classical epistemology espouses an excessively idealised epistemic rationality principle (e.g. Hintikka 1962). It is true that the foundationalism in espistemology, propounded by

Descartes, does not hold promise any more, as it has been found that the *a priori* or armchair and speculative normative methods are futile and unproductive. Consequently, certain forms of the naturalised epistemology no longer look up to classical logic as a model of inquiry (Halpern, 1986).

Some Relevant Questions Regarding the Principles of Epistemic Rationality

At this point the following questions can be raised: Can rationality be totally banished from an enterprise which is conceptually a normative study? If not, to what extent should an agent be rational to be able to perform his epistemic activities? 'Naturalistic' epistemologists have tried to retain the rationality conception of an agent while naturalising epistemology. In any form of epistemology (naturalised or otherwise), it is always desirable that some form of rationality principle is to be preserved else the whole enterprise would be suspect. Hence the central question which we would like to address is: In what manner are rationality principles to be welded with naturalised epistemology?

Section 2 of this article examines classical epistemic logic and its corollary the K-K thesis, which is based on an ideal conception of rationality. It is found to be inadequate on the grounds that it (*a*) leads to the problem of logical omniscience, and (*b*) takes no cognisance of some of the fundamental limitations of human cognitive capabilities. This prompts us to opt for naturalised epistemology, which, however, has several forms.

In Section 3, instead of subscribing to any specific brand of naturalised epistemology, we explore the nature and requirements of an adequate epistemology and the underlying concept of epistemic rationality.

In Section 4 we look for an 'epistemically friendly' logic which may articulate the notion of rationality implicit in a naturalised epistemology. We critically evaluate Pollock's Defeasible reasoning, Garfield's claim that Relevant logic captures the pre-analytic intuition about inferences in the epistemic context, and Cherniak's notions of Minimal rationality and Feasible inference.

In Section 5 we give an outline of a logic called DRML (Defeasible Relevant Minimal Logic), based on the insights drawn from

Pollock, Garfield and Cherniak in a unified theory of justification of belief in the epistemic context.

Classical Epistemology and K-K Thesis: Their Normative and Idealistic Nature

Epistemology, the study of the origin, nature and the limits of knowledge has a long tradition in philosophy dating back to the Greeks. However, the first systematic treatment of epistemic logic, i.e., the formal logical analysis of reasoning about knowledge, known as the classical model for knowledge and belief, can be traced back to the work of Von Wright (1951) and Hintikka (1962).

Here, we demonstrate that classical epistemology is both normative and idealistic. Although the normativity of epistemology does not pose any problem, the latter characteristic is the source of some serious difficulties which will be enumerated a little later.

Normativity of Classical Epistemology: The Rationale for Epistemology being Normative

Epistemology is the normative or evaluative study of knowledge acquisition and knowledge management. It is concerned with justification or non-justification and warrantedness of beliefs. Mere true belief is not knowledge. In order to acquire the status of knowledge a true belief must be justified, hence knowledge is defined as 'justified true belief'.

Definition of Knowledge in Traditional Epistemology

Schematically, *S* knows that *P* if and only if

(*i*) *P* is true;
(*ii*) *S* believes that *P*; and
(*iii*) *S* has adequate evidence for believing that *P*.

Normativity of the Definition of Knowledge

One of the consequences of the claim that a belief is justified or warranted i.e. it meets condition (*iii*), is that the cogniser is permitted or licensed to believe in that doxastic state. The very act of justification, therefore, requires epistemic rules or principles which permit or license the cogniser to believe in one belief state from the preceding belief states, which are held to be justified. The formulation of these rules and principles is thus the job of epistemology. This rule-giving and principle-stating role of epistemology relegates it to the status of a normative study.

Idealistic Nature of the Classical Epistemology

The classical model, articulated in terms of a set of sound and complete axioms of knowledge put forward by Hintikka, is founded on the notion of possible world. We assume in this logic the notion of a world, or situation which is possible relative to an individual cogniser *i*. The cogniser *i* knows a fact ϕ if, and only if, ϕ is true in all the worlds which are possible for *i*.

The K-K Thesis and Idealistic Nature of the Classical Epistemology

Given the definition of knowledge as justified true belief, stated in terms of conditions *i* to *iii*, Hintikka (1962) goes on to interpret condition (*iii*) as tacitly embodying what is known as the K-K thesis. It is stated as follows:

(*iv*) *K-K thesis*: " *i* knows that ϕ " entails that
" *i* knows that he knows that ϕ "

$$Ki\phi \rightarrow KiKi\phi$$

We are not particularly concerned about the validity of Hintikka's claim that condition *iii* can be construed as *iv*. Assuming that Hintikka is correct, we would like to demonstrate that such a formal logical approach is basically idealistic in nature.

There are reasons to believe that the K-K thesis itself gives rise to the definition of knowledge as justified true belief. The most important reason for this is that if knowing something requires that one knows that he knows it, then one will only have knowledge if one knows that one's beliefs have been formed in a way that renders them likely to be true; put in a different way, if one possesses a justification for the said belief. This is the reason why the K-K thesis is often called the Positive Introspection Principle.

A bit of introspection will show that the process of arriving at a belief, however, can in fact be acceptable even though one may not know that it is so, as one may be totally unaware of what that process is. Hence, one must make a distinction between first-order knowledge, that is knowing something and second-order knowledge, that is, knowing that one knows. The conflation of these two levels leads to the K-K thesis and justificationalism (i.e. knowledge is justified true belief). Moreover, negatively speaking, the K-K thesis leads one to assert that one can not know that ϕ unless one knows that one's claim to know that ϕ is correct, i.e.

(*v*) It is not possible for *i* to have adequate evidence for believing that ϕ when ϕ is false.

The tenability of both *iv* and *v* rest on the claim that there exists actual and effective procedures for determining whether ϕ is true and whether *i* has adequate evidence for believing that ϕ, i.e. whether conditions *i* and *iii* respectively have actually been fulfilled. Although the definition of knowledge given earlier includes conditions *i* and *iii* it fails to specify the *actual procedures* that will satisfy these conditions regarding the determination of truth and adequate evidence.

Ideal Procedure vs. Executable Procedure

Some proponents of naturalised epistemology, such as Dretske (1969) and Goldman (1976, 1978), have expressed their dissatisfaction with classical epistemology on the ground that while conditions *i* and *iii* are treated as part of the definition of knowledge, no effort is made to spell out the actual and effective procedures in terms of which those conditions are to be met. In the absence of such procedures the definition of knowledge merely states certain ideal conditions without bothering about how these conditions are to be executed or implemented in practice.

The Problems Arising Out of the Idealistic Nature of Classical Epistemology and Epistemic Logic

The possible-worlds approach in epistemic logic compels us to regard our cogniser *i* as an 'ideal knower', not bound in terms of memory and the duration of the cognitive task performed by *i*. This leads to two problems: logical omniscience and the finitary predicament of the human cogniser because of cognitive infirmities.

The Problem of Logical Omniscience

The notion of ideal knower in epistemic logic commits us to the view that the cogniser's knowledge is closed under implication. From the possible world model of knowledge and belief, it follows that if cogniser i knows ϕ and also knows $\phi \rightarrow \psi$, then both ϕ and $\phi \rightarrow \psi$ are true in all worlds possible for i, and, therefore, ψ is true in all the worlds possible for i so that i knows ψ. In particular, the cogniser i knows all valid formulas ψ and all logical consequences of his knowledge. This principle known as the principle of logical omniscience, can be formulated in terms of the following formula:

$$(Ki\phi \wedge [Ki\{\phi \rightarrow \psi\} \rightarrow Ki\psi])$$

Quite clearly, this does not appear to be a realistic model for any actual human cogniser. It seems to be all the more unrealistic if one considers the adequacy of this model for a knowledge base which is bounded in terms of (computation) time and space in memory that the cogniser can use (Parikh 1987, Halpern 1986).

The Finitary Predicament

The limitations of the cognitive capacities impose certain constraints on the notion of human rationality (Goldman 1978; Cherniak 1981a, 1984). These constraints (which will be discussed later) are the fundamental facts of human existence and, therefore, must be taken into account in epistemology.

For the reasons mentioned above, epistemology should be divested of idealistic procedures and principles. This compels us to look for an epistemology that would be based on realistic rational procedures and principles which are executable and are such that human beings do actually conform to them. It would be a version of naturalised epistemology.

Forms of Naturalised Epistemology

Since Quine (1969), three distinct approaches to naturalised epistemology have emerged (Almeder 1990):

a) the replacement thesis: epistemology should be replaced by psychology and natural sciences.

b) epistemology should be transformed and supplemented by the methods and insights of psychology, biology and cognitive science.

c) epistemology should neither be replaced nor transformed but incorporate the methods of the natural sciences to acquire knowledge.

Instead of entering into this debate, it might be worthwhile to ask a different question: what would be the nature and requirements of an adequate epistemological theory?

The Nature of an Adequate Epistemological Theory

The specific form of naturalised epistemology we wish to defend aims at providing a normative account about what should count as knowledge and how best knowledge should be acquired by demonstrating that this account would be compatible with the available scientific theories of human cognition. Some of the important findings about human cognition that should be taken note of by such an epistemology are as follows:

Cognitive Infirmities

One of the fundamental facts of human existence is the finitude of the human cogniser (Goldman 1978; Cherniak 1984; 1981a; Halpern 1986; Parikh 1987). According to Cherniak the human cogniser is '. . . in the finitary predicament of having fixed limits on our cognitive resources . . .' (Cherniak 1984, p. 141).

The claim to these limits is derived also from the implications of the Complexity Theory. Although philosophers have paid considerable attention to the philosophical implications of the Incompleteness Theorem of Godel (Lucas 1961; Lewis 1969; Delong 1971, especially ch. 5) and the classical unsolvability results of Church, Turing and Post (Hao Wang 1974, chs. IX, X; Webb 1980), drawing out the philosophical implications of Complexity Theory have just begun (Mendelson 1990; Cherniak 1984).

One of the important consequences of Complexity Theory is that it enables us to distinguish a formally correct deductive rule or procedure from a 'quick but dirty' heuristics or rule of thumb. The former is often so slow as to yield computational intractability or paralysis because of being NP-type, i.e. exponential time complexity. The latter, however, avoids this problem as it belongs to P-type, i.e. polynomial time complexity.[1]

It follows from the above that a problem has a practical computational solution if there is a procedure with polynomial time complexity that solves it. Since the exponential time complexity makes the computational time extremely large, it defies practical solution although there might be a formally correct algorithm available to solve it.

In contrast, with formally elegant deductive rules and procedures that often require more memory capacity and computing time, human beings often resort to 'quick but dirty' heuristics, which fare much better in terms of the computing time taken.

The Travelling Salesman Problem

A salesman has a list of cities, each of which he must visit exactly once. There are direct roads between each pair of cities on the list. Find the route the salesman should follow so that he travels the shortest possible distance on a round trip, starting at any one of the cities and then returning there.

If there are N cities, then the time required to solve this problem is O (N!). A good heuristic to solve this problem is the 'nearest neighbour algorithm':

* Arbitrarily select a starting city.
* To select the next city, look at all cities not yet visited. Select the one closest to the current city. Go to it next.
* Repeat step 2 until all cities have been visited.

This procedure executes in time proportional to N squared, a significant improvement over N!.

The empirical evidence for the use of sub-optimal heuristic strategies, rather than formally correct procedures in solving problems, has been provided in terms of some psychological experiments on 'irrationality' by some cognitive scientists (Tversky and Kahneman 1974; Kahneman, Slovic and Tversky 1982; Nisbett and Ross 1980).

Although coming from two different strands of research, the evidences provided by both complexity theory and the psychological experiments converge on *the problem of characterising the human inferential mechanism in the face of the 'finitary predicament'*. Cherniak calls this *less-than-perfect-deductive-ability* of an epistemic agent 'minimal rationality' (Cherniak 1981a). He locates the minimal rationality condition by steering a middle course between ideal and null rationality conditions, both of which turn out to be unacceptable after a careful scrutiny of human cognition. The ideal rationality condition regards the epistemic agent as endowed with virtually unlimited cognitive resources and without any inferential infirmities. This condition is impossible to meet as it claims that the epistemic agent's belief-set is deductively closed, i.e. he believes all the consequences of his beliefs. The null rationality condition, on the other hand, is based on the assent theory of belief which denies the obvious role which inferences play in knowledge acquisition.

Impracticality of the Principle of Total Evidence Requirement

One of the hallowed principles of traditional epistemology is the *total evidence requirement*, i.e. the requirement that the epistemic credence given to a statement at a given time is a function of the total evidence then available (Carnap 1950, p. 211; Hempel 1965, p. 64). Goldman (1985, pp. 219–20) has argued that on psychological grounds it is unrealistic to expect that one would be able to bring the entire stock of information in forming a belief.

In view of this psychological limitation it would be necessary, Garfield (1990) argues, to look for inference strategies and rules through which the cogniser brings together only the *relevant* evidence from the mass of data available to him. The fact that both scientists and laymen alike manage to formulate a hypothesis by ignoring the principle of total evidence based on what they consider to be relevant from the mass of the data goes to show that (*a*) our concept of what is 'relevant' requires to be made more precise; and (*b*) the logic that is implicit in epistemology cannot be a merely content-independent formal theory as classical logic happens to be.

Fallibilism and Defeasibility of Knowledge

Since Popper, a common theme in the epistemology of science has been that knowledge is fallible and unstable, rather than representing a continual accumulation and monotonic increase in the past stock of knowledge. We often arrive at a belief based on some reasoning and subsequently, in the light of additional information that undercuts the belief, we give it up. This shows that because we live in an *espistemically hostile environment* (Simon 1981; Garfield 1990, p. 99), our knowledge is non-monotonic.

Perception provides us with a clear example of non-monotonicity. '. . . something looking red to me may justify me in believing that it is red, but if I subsequently learn that the object is illuminated by red lights and I know that that can make things look red when they are not, then I cease to be justified in believing that the object is red.' (Pollock 1987, p. 481)

Memory exhibits the same features of non-monotonicity. Psychologists working on human memory organisation and efficiency recall, claim that belief and its reasons are stored separately so that when we recall a belief we may be unable to recall the justification for arriving at that belief (Nisbett and Ross 1980). Given this fact of human frailty, Pollock says:

> If one originally used *P* in reasoning to *Q*, but can no longer remember the fact, then he is not being irrational in retaining *Q* even though he later rejects *P*. We must regard his continued belief in *Q* as justified until he discovers that it was originally based upon reasoning that he would now reject. In other words, memory itself provides defeasible justification for remembered beliefs (1987, p. 487).

Moreover, we often entertain false and even inconsistent beliefs, some of which may be well-justified since the environment we inhabit often deceives us, or at least 'does not go out of its way to deliver the truth to us' (Garfield 1990).

Finally, non-monotonicity is central to the much vexing 'frame problem' (Pylyshyn 1987). In any intelligent system a chain of inferences based on facts as they are known at the moment may be undermined by information about changes to the world. A chain of reasoning based on the assumption of the form '*P* is true now, therefore *P* will be true later' might be falsified as a result of a change in the world.

The inference mechanism for belief fixation, extension and revision must not, therefore, be merely concerned with truth-preservation as classical logic is. An epistemically adequate logic must recognise and be sensitive to these facts about the non-monotonic character of the growth of our knowledge and should recommend rules of inference that are not only truth-preserving but also epistemically prudent. The search for such an 'epistemically friendly' logic has been one of the main objectives of some of the non-standard logics that have been developed in the recent past.

Search for an *Epistemically Friendly* Logic

Because of its idealistic nature, classical logic is claimed to be inappropriate to a theory of rationality and inadequate to meet the epistemological requirements outlined in the preceding paragraphs. There have been attempts to look for a substitute for classical logic either by replacing it altogether with an alternative logic or by putting restrictions on it. In what follows we will be concerned only with three such attempts viz. those of Pollock (1987), Cherniak (1981a, 1981b, 1984) and Garfield (1990). Although the suggestions contained in these attempts for an epistemically friendly logic are certainly improvements over classical logic, none of them are by themselves adequate.

We will consider each one of these attempts separately. Further, we will explore whether the limitations of each one of these approaches can be overcome by combining them in a logic that will be sensitive to the requirements of a naturalised epistemology. Needless to say, we will only be able to provide an outline of this suggested logic.

John Pollock's *Defeasible Reasoning*

Pollock wishes to develop an epistemically friendly theory of reasoning based on the building blocks of the concepts of conclusive reasons, prima facie reasons, rebutting defeaters and undercutting defeaters (Pollock 1987, p. 485). He calls such a theory of reasoning 'Defeasible Reasoning', which is roughly the same as non-monotonic reasoning in Artificial Intelligence. Defeasible reasoning, however, is earlier in origin and owes itself to the work of Chisholm (1957) and Pollock (1987).

One of the distinctive features of defeasible reasoning is the notion of 'undercutting defeaters', although the more familiar 'rebutting defeaters' form the basis for AI work on non-monotonic reasoning. Pollock is not merely concerned with developing a theory of reasoning based on these notions but wishes to make it consistent with the facts of defeasibility evident in certain knowledge-acquisition processes, e.g. perception and memory.

Pollock's main objective is to define the most important notion of 'warrant'. To this end he first construes the current perceptual and memory states as forming the 'epistemic basis'. The arguments of the various types mentioned earlier are meant to enable an epistemic agent to move from the basis to a warranted or justified belief. However, Pollock recognises certain *practical* difficulties in arriving at a warranted belief. A belief is warranted, according to Pollock, when it is ultimately undefeated by *all possible* arguments available to an epistemic agent at a single time (Pollock 1987, p. 504), i.e. it survives *all possible* defeaters.

Since the notion of an 'ultimately undefeated argument' involves 'the comparison of infinitely many arguments and, in the case of collective defeat, the infinite cycling of arguments through defeat and reinstatement' (Pollock 1987, p. 504), it smacks of idealism and impracticality. Pollock acknowledges the idealistic nature of his definition:

> I glossed 'warrant' as 'what an ideal reasoner would believe'. An ideal reasoner is one unconstrained by a finite memory or processing capacity. Warrant is an ideal to which 'real' epistemic agents aspire. But we cannot expect real epistemic agents to believe only warranted propositions. Warrant is a 'global' concept defined in terms of the set of all possible arguments available to an epistemic agent at a single time. No one can actually survey that infinite totality and decide what to believe by applying the definition of 'warrant' to it . . . . Actual rules for reasoning must appeal exclusively to 'local' considerations—readily accessible features of the epistemic situation' (1987, p. 504).

Instead Pollock suggests the more epistemically prudent notion of 'justified belief', which he wants to define based on 'local' considerations and readily accessible features of the epistemic situation. These 'local' considerations enable the epistemic agent to form only those limited number of beliefs which are 'appropriately related' (Pollock 1987, p. 504). The assumption here is that the number of such appropriately related beliefs would be limited and hence can be compared and subjected to the cycling of arguments through defeat and reinstatement. If the belief in question survives all such defeaters, it will be justified.

There are two important aspects of Pollock's notion of justified belief:

i) For Pollock the distinction between warranted and justified belief rests only on the size of the set of belief; otherwise both sets would undergo the same process of defeat and reinstatement to see whether the belief in question survives the ordeal.

ii) In the case of justified belief, however, the belief set is restricted only by considering the 'appropriately related beliefs'. Pollock defines this concept in terms of three types of production rules, viz., adoption rules, defeat rules and reinstatement rules.

The most basic production rule, according to Pollock, is the adoption rule which relates one set of belief to a given belief in terms of the psychological state of 'interest' or 'caring'. For example, Pollock's belief formation rule 6.3 states:

> . . . if you believe $P_1, \ldots, P_n$ and if you have no defeating beliefs and you care whether Q, then you should believe Q (Pollock 1987, p. 506).

However, Pollock, takes the intuitive understanding of 'care' as the ground or basis for an epistemic agent to 'care whether *Q*'. We feel that in order to define 'appropriately related belief' it is unnecessary to invoke such intuitive notions as 'caring' or 'interest', even if they result from and are evident in the local considerations of the epistemic situation. The introduction of these concepts, we believe, will create undue complications as they are notoriously subjective and relative.

We suggest that the relationship between beliefs, especially Pollock's notion of 'appropriately related beliefs', can be understood in terms of the notion of 'relevance' as defined in relevant logic. This is all the more justified because local considerations of the epistemic situation will surely demand that we avoid the gratuitous introduction of any random belief, i.e. there must be some 'intensional connection' if the beliefs are to be related.

Thus the problem of limiting the size of the belief set could be solved in terms of considering only 'relevant' beliefs. Accordingly one can reformulate the notion of justified and warranted beliefs in the framework of relevant logic. For instance, Pollock's rule 6.3 can be reformulated as follows:

> If you believe $P_1, \ldots, P_n$ and you have no defeating beliefs and $P_1, \ldots, P_n$ are relevant to Q then you should believe Q.

By incorporating relevant logic, Pollock's theory of defeasible reasoning could only be strengthened as it would avoid certain undesirable features based on intuitive and subjective concepts like 'caring' and 'interest'.

Garfield on Relevance Logic and Epistemology

Often the choice of a logic is based on the philosophical arguments justifying the appropriateness of that logic for a given purpose. The arguments claiming appropriateness appeal to pre-analytic facts in a given context and try to demonstrate that a given logic fits such facts more satisfactorily than any rival system of logic.

Both Barker (1989) and Garfield (1990) claim that the pre-analytic intuitions about inferences in the epistemic context fit more closely with relevant logic and hence it '. . . provides a better model for rational inference . . .' (Garfield 1990, p. 97) than classical logic.

> . . . it is rational to reason relevantly, but irrational to reason classically. The argument (for the justification for Relevant Logic) will be primarily epistemological, rather than logical, but this is because rationality is, after all, an epistemological notion. (Garfield 1990, pp. 97–98)

The argument between Barker and Garfield, however, ends with this. Barker suggests certain modifications in the Anderson and Belnap system, the most important of which is the admission of the Disjunctive Syllogism (the inference rule abbreviated as DS) with the truth-functional interpretation of 'or' as valid in epistemic contexts.

The admissibility of DS was urged by Meyer and Dunn way back in 1969, not on epistemological grounds but on purely logical grounds. Meyer and Dunn argue that if $\sim A$ and $A \vee B$ are theorems, so will be B. Such stipulations might save DS and work for the system of 'pure logic', but would fail to justify the retention of DS, in spite of Barker's protestation, in the epistemic context as Garfield convincingly argues.

Garfield's main arguments for Relevant Logic as more appropriate for epistemic purposes are based on the following:

(*i*) Garfield confines his discussions of relevant logic to the rejection of DS and the associated logical connective 'or'. Relevant logicians makes the distinction between the intensional sense of 'or' (i.e. the relevance/entailment of the truth-value of the other) and the familiar truth-functional definition. Based on this distinction Garfield claims that these two senses of 'or' exhibit that they are 'epistemically dissimilar' (1990, p. 106): while the intensional 'or' imputes a 'connection' between the disjuncts such that one disjunct is entailed by the family of the other, the truth-functional interpretation requires no such connection. The epistemic situation in which the former holds is different from that in which the latter does (as would be evident from (*iii*) below).

(*ii*) Garfield claims that one of the inescapable facts of the epistemic situation we live in is one of 'hostility' rather than 'cooperation'. This is because (*a*) the epistemic agent's 'belief acquisition mechanism' and 'inference engine' are both flawed, and (*b*) the environment/world is not obliging in terms of readily delivering the truth.

(*iii*) The main function of the inference rules in classical logic, including DS, is the preservation of truth. In an epistemically hospitable and cooperative world such rules would have sufficed. However, in an epistemically hostile world, Garfield argues:

> Since some of our beliefs are false, preservation of truth is not enough . . . In it, we need not only to preserve truth, but to avoid gratuitous introduction into our belief sets of wild falsehood . . . In such circumstances . . . admitting the disjunctive syllogism would lead to the gratuitous insertion of irrelevant falsehood into our belief set (1990, p. 107).

We are in general agreement with Garfield on this last point. However similar arguments need to be advanced for the other logical connectives, such as 'and' and the notorious 'if . . . then', and for the associated inference rules. For each of these connectives we must identify the appropriate intensional relationships or meaning-connections that hold between the constituents. This is done, for example, by replacing the extensional disjunction by intensional disjunction (or *fission*, symbolised by +), extensional conjunction by intensional conjunction (or *fusion*, symbolised by ×) and the material conditional by relevant implication (symbolised by →).

Moreover, Garfield is conspicuously silent about deductions or long chain of (immediate) arguments in epistemology and the necessity of appropriately defining them. Relevant logic is important here as it imposes restrictions on classical deduction techniques in order to ensure inferential connection, to keep track of propositional dependencies and to block deductions involving the introduction of irrelevant propositions. For example, in the relevant account of entailment *fusion* is allowed to play an important role, because

> fusion binds two propositions together in such a way that one is assured of their mutual relevance in deriving any consequence from the fused conjunction, as in the Deduction Theorem or in the Relevant Account of validity. Whatever $A \times B$ entails needs both A and B in its derivation. (Read 1988, p. 134).

Thus, for the claim that relevant logic provides a better model for inferences in epistemic context it is necessary to explicate the idea of logical 'connection' in general and not merely with reference to 'or'. Relevant logic does this by specifying

i) appropriate meaning-connection holding between the constituents linked by the various logical connectives; and *ii*) appropriate inferential-connection by imposing restrictions on classical deduction techniques.

Finally, Garfield's account or relevant logic needs to be supplemented by Defeasible Reasoning in order to define such key epistemic concepts as 'justified' or 'warranted' beliefs. The primary concern, however, of the systems of relevant logic developed so far, such as *E*, *R*, $R^{\square}$ or *DW*, is to define the fundamental logical notion of *entailment* or *R-validity*. In order to justify the use of relevant logic in the epistemic context one must demonstrate the role it can play in defining the concepts mentioned above.

Cherniak on Complexity Theory, Minimal Rationality and the Theory of Feasible Inference

Cherniak explores the possibility of a *predictive intentional theory* based on the concept of *less-than-perfect deductive ability* that suggests a plausible way of reconciling human imperfection with rationality.

The early proponents of intentional systems (Davidson 1963, 1974, 1976; Dennett 1971) endorsed the claim that 'the behaviour (of such a system) can be (at least sometimes) explained and predicted by relying on ascriptions to the system of beliefs and desires (Dennett 1971, p. 87) because a belief could be regarded as part of an agent's *reason* for a decision to act in a particular way. (Davidson 1976).

Unlike the much discussed theory of 'rule following' (Wittgenstein 1958; Kripke 1982) which explains the predictive and explanatory capacity of an intentional theory in terms of the behaviour conforming to rules or conventions, cognitivists like Davidson and Dennett, assign a central role to the notion of *rationality*. Davidson says:

> . . . when we use the concept of belief, desire, and the rest, we must stand prepared, as the evidence accumulates, to adjust our theory in the light of considerations of overall cogency: the constitutive ideal of rationality partly controls each phase in the evolution of what must be an evading theory (1974, p. 223).

Davidson founds this notion of rationality on the naturalistic approach to choice, based on psychology, such as decision theory

(Davidson 1976) for prediction and explanation of behaviour. Similarly, some other philosophers have invoked game-theory.

Both Davidson and Dennett, however, acknowledge that accurate predictions are not possible in terms of intentional theories (Davidson 1963, 1974) unless one descends to the design or physical stance (Dennett 1971) and physical domain in which alone 'strict' or 'serious' deterministic laws can be found.

Cherniak claims that the inability of Davidson and Dennett to construct a predictive intentional theory is due to the persistent attempts by philosophers to define rationality in such idealised terms that '. . . it cannot apply in an interesting way to actual human beings' (Cherniak 1981a, p. 163). Moreover, the admission of the impossibility of a predictive intentional theory leads to the acceptance of an instrumentalist rather than a realist account of mental entities by Davidson and Dennett because '. . . beliefs are so idealised (by Davidson and Dennett) as to be inapplicable to humans, then any attributions of beliefs to humans cannot really be true, the attributed entities are at most useful myths' (Cherniak 1981a, p. 163).

The complexity theory claims that rational methods need not be idealistic logical methods, which make unrealistic assumptions about virtually unlimited cognitive resources in terms of the assumptions of perfect information, unlimited memory capacity and inferential capabilities. On the contrary, the complexity theory has provided a basis for a hierarchy of qualitative distinctions between unfeasible and feasible tasks. The latter are performed in terms of heuristic or less-than-perfect deductive ability which constitutes 'minimal rationality' and reflect, more realistically, the ways of men.

Cherniak thus locates minimal rationality or less-than-ideal deductive ability by exhaustion of a trichomy, i.e. between the ideal and the null rationality condition. The null rationality condition is based on the assent theory of belief in which no inference is required, (Russell 1971, p. 311). The delineation of the middle ground as minimal rationality is based on the satisfaction of two conditions, viz. the minimal inferential condition[2] and minimal consistency condition,[3] which essentially attempt to specify under what circumstances an epistemic agent will draw which inference or eliminate which inconsistencies.

Cherniak assigns such tasks of specifications to his theory of *feasible inference*. Given the cognitive infirmities, the finitary predicament and the current belief-desire set, the theory of feasible inference specifies a minimally rational epistemic agent about which deductive inferences from the belief-set are most likely to occur. This would be achieved in terms of ordering of inferences with respect to their relative difficulty or easyness. A theory of feasible inference would incorporate such an ordering of inferences. In other words, for any minimally rational agent and a given connective there will be a ranking in the order of difficulty of a given argument schemata involving that connective.

Cherniak calls the easy inferences *constitutive inferences*. These are the primitives in one's inferential repertoire. A difficult inference might consist of a chain of constitutive inferences. The possibility of an alternative feasibility ordering has been left open by Cherniak since he believes that certain connectives might appear simple to some than to other people.

According to Cherniak, the criterion for weighting of deductive tasks with respect to their feasibility for the reasoner depends on the considerations of computing space and time or alternatively, on the number of steps required in terms of constitutive inferences. It is here that he wishes, perhaps, to connect his theory of feasible inference with the complexity theory.

> A theory of feasibility will include . . . ordering. Thus, other things being equal, inferring $\sim q \rightarrow \sim p$ from $p \rightarrow q$ typically is easier (and hence more likely) than inferring $(\forall x)\ Fx \rightarrow (\forall x) Gx$ *from* $(\exists x)(\forall y)(Fx \rightarrow Gy)$, and the latter task is much easier than one as difficult as determining that the axioms of choice are independent of the other axioms of set theory. *An even more difficult inference would be one which required more space and time* than is available in the galaxy before heat-death (Cherniak 1981b, pp. 250–51, our emphasis).

One notable feature of Cherniak's minimal rationality is that he avoids the two extremes of anti-psychologism on the one hand, and a logical theory based on descriptive psychology on the other. Frege (1983) clearly separated logic from psychology and developed formal models of valid inference by paying attention only to the logical form of the sentences involved in the inferences. However, recent empirical work on reasoning aims to show that psychological considerations bear upon principles of logic (Goldman 1985; Stich and Nisbett 1980; Thagard 1982). Such studies demonstrate the deviation of human reasoning from the logicians (Stich 1984; Tversky and Kahneman 1974; Johnson-Laird 1980, 1983, 1990). According to such an approach, which is based on empirical evidence on how human beings reason, many time-honoured principles of logic go by the board, e.g., the principle that a person's justified belief should be closed under *modus ponens* or Bayer's theorem.

Cherniak, however, belongs to neither of these camps. He falls back on the complexity theory, which is a formal mathematical theory of decidability concerned with the problem of the limits of computation based on certain constraints. He observes certain similarities between computational constraints and human inferential infirmities and retroductively ascribes a particular inferential mechanism on the human mind, viz. the theory of feasible inferences.

Based on the assumption as to what inferential mechanism (i.e. which primitive constitutive inferences and deduction techniques) is 'wired in' to one's mind, one might find that certain inferences require fewer steps than others despite their formal equivalence. This explains the simplicity or difficulty of certain inferences. (We owe this point to Garfield.)

The main thrust of Cherniak's approach seems to be to explain inferential behaviour in the light of information concerning human limitations based on an abstract mathematical theory about the algorithmic nature of cognitive tasks. The complexity theory assumes that cognitive tasks are formulable as effective procedures—some of which are feasible and others not. Based only on such assumptions, Cherniak imposes certain limitations, e.g. computing time on human inferential behaviour.

Moreover, such an analysis of human inferential behaviour does not pay any attention to the grounds of choice of a particular inferential mechanism, i.e. the goals which a specific set of primitive constitutive inference and deduction technique serve. Otherwise, the choice of a particular inferential mechanism might appear to be quite arbitrary. Epistemology and psychology can play a role in helping logic define these goals. This is one reason why the need to explore the close connection between logic and epistemology has been felt by naturalised epistemology.

Goldman (1986) acknowledges this need and claims that the necessary condition for justification of a belief is that it is *reliable*.

An inferential mechanism, according to this view, could be regarded as part of a belief-making mechanism method or process, whose aim would be to produce only reliable beliefs. In the absence of such goals, Cherniak's notion of minimal rationality fails to discriminate between those constitutive inferences (or inference rules) which are reliable and those which are not (e.g. Disjunctive syllogism) in justifying our beliefs.

> . . . while Cherniak's model is laudably motivated by a concern for the omnipresence of epistemic hostility, Cherniak fails to consider the possibility that epistemic hostility may make certain possible constitutive inferences bad moves (Garfield 1990, p. 101).

The importance of goals for a theory of rationality is admitted by most cognitive scientists. It is important, however, to realise that following a well-formed sequence of deductive steps is *not* sufficient ground to declare that the agent who performs the steps acts rationally, unless the goals of that agent are known and the steps are seen to form part of some proof of the goal. Here again relevant logic would play an important role, simply because it would refuse to accept the behaviour of merely following a well-formed sequence of deductive steps as rational, and it would reject all spurious steps of playing any role in the derivation of the result.

An Outline of an *Epistemically Friendly* Logic

From the discussion in the previous section it can be seen that classical logic is not sensitive to the available facts about human cognition. Moreover, the alternative or 'deviant' logics, like those of Pollock, Garfield and Cherniak, are not adequate by themselves, although they are certainly an improvement over classical logic. From a critical evaluation of these logics we can identify certain features which should be included in the formulation of a more satisfactory and epistemically friendly logic. They are:

(*i*) The logic must be defeasible, because some of the basic sources of human knowledge, e.g. perception, memory etc.

exhibit this feature. The core epistemic concepts like 'warrant' and 'justification' should be defined in terms of defeasible reason. Pollock (1987) has developed such a logic. However, in this logic we need to take into account the cognitive infirmities and the finitary predicament of the epistemic agent.

(*ii*) The logic should keep track of the propositional dependencies and block the gratuitous introduction of irrelevancies and unwanted deductions that are allowed in classical logic. Relevant logic is concerned precisely with these problems.

(*iii*) The logic should model inferential capabilities of the epistemic agent that are both feasible and reliable. Cherniak's notions of 'minimal rationality' and 'feasible inferences' are attempts to model inferential capabilities, although he ignores the problem of reliability of the inferential rules.

The logic we suggest, viz. Defeasible Relevant Minimal Logic (DRML) will combine the features of each of its antecedants and the deficiencies of each would be made up by the strength of the other features.

DRML

While considering any logic, one needs to spell out two of its important aspects, i.e. its syntax and semantics.

The syntax of a logic includes a set of formation rules, a set of rules of inference and the definition of the deducibility relation or an argument proof. The semantics of a logic is specified in terms of (*a*) the interpretation of the formal language in which the specific components are assigned specific objects from the domain of interpretation; and (*b*) the truth conditions under the interpretation that make the sentences (or wffs) of the language true or false. It also states the conditions under which an argument would be valid. It is customary for a logic to have a meta-theory in order to ensure that it has the two important properties of soundness and completeness. The meta-theorems which establish that a given logic possesses these properties aim to connect the syntactic notion of deducibility with the semantic notion of validity.

Syntax of DRML

We assume the standard formation rules with the restrictions imposed on V, →, Λ, ~ as connectives and ∀ and ∃ as quantifiers in relevant logic (Read 1988, pp. 52–55). We will assume that this formalism is adequate for the representation of the cognitive states. The set of inference rules will have the same restrictions as in relevant logic.

DRML will be based on a Gentzen-type natural deduction system with the restrictions imposed on it in relevant logic. The main concern of DRML is to capture features *i* to *iii* (referred at the beginning of this section) that are involved in cognitive-state transition, e.g., transition from a given cognitive state (say, a belief state) to another. The natural deduction format with appropriate restrictions seems to be eminently suited to capture this feature.

In classical epistemic logic, its semantics is given in terms of possible worlds (Hintikka 1962). Because of the idealistic nature of such a semantics, we will refrain from using it. DRML has been developed mainly with the derivational problems in epistemology in mind and consequently does not (yet) have its semantics worked out completely. However, we would assume the standard semantics already provided for certain aspects of DRML, e.g. the semantics for relevant logic. For example, the meaning of *implies*: one statement *implies* another only if one is *connected* with the other; only if the statements have enough to do with one another i.e. if $A \rightarrow B$ then B is connected with A. The connection may be based on a causal relationship.

DRML as an Epistemic Logic

Goldman (1985, pp. 310–17), in his attempt to define 'warranted' or 'justified' belief has argued that one of the concerns of epistemology is to suggest appropriate rules for cognitive-state-transition. He takes these rules to be psychological. Since we do not endorse a Fregean cleavage between logic and psychology, it would be our endeavour to ensure that DRML reflects the inferential behaviour involved in the process of knowledge acquisition.

Foundation State

To meet the demands of epistemology, DRML should model those arguments which, starting with some perceptual or memory state—called *foundation state*—forming the *epistemic basis*, moves on to beliefs and from these beliefs to new beliefs, and so on. (Pollock 1987, pp. 484, 490).

Forms of Epistemic Argument

This above consideration provides the most obvious form of epistemic argument, i.e. simply being in a particular perception or memory state (or any finite set of foundational states) would constitute an *adequate reason* for a belief. This may be called an *initial argument*.

However, one of the interesting features of the arguments in the epistemology is that the claims to the entailment of a belief-state by a foundational state is often retracted in the light of new or additional information, called defeaters. In such a situation the alleged foundational state provides merely a *prima facie* reason for a given belief. The notions of 'prima facie reason'[4] and 'defeater'[5] together constitute the defeasible or non-monotonic nature of epistemic arguments, e.g. arguments involving perception or memory.

Interaction between Arguments

A defeater and an initial argument, involving *prima facie* reason, may interact in two different ways. A defeater R of a prima facie reason P for a belief Q may lead to or be a reason for:

(*i*) the belief $\sim Q$ or
(*ii*) the denial of the entailment between the prima facie reason P and the belief Q i.e. $\sim (P \rightarrow Q)$.

The former is called the *rebutting defeater* and the latter the *undercutting defeater*. It is possible that the defeater itself is the conclusion or the legitimate line of another argument.

Once the interaction between an initial argument and a defeater is acknowledged, the possibility of a third argument defeating the defeater and *reinstating* the initiai argument cannot be ruled out. This exhibits the second type of interaction that may take place between arguments. Thus, a given argument may go through a cycle of several defeats and reinstatements and come out unscathed, vindicated and ultimately undefeated. The interaction of this type among arguments makes it possible to define some of the most crucial epistemic concepts, such as 'warrant' or 'justification'.

Strong and Weak Beliefs

It is conceivable that starting with certain foundation states, the defeasible epistemic reasoning could lead to several solutions through a cycle of defeats and reinstatements. This plurality of solutions gives rise to two possibilities: first, those ultimately undefeated beliefs which form the intersection of solutions through a cycle of defeats and reinstatements. These beliefs would constitute 'strong beliefs'. Secondly, the set of beliefs contained in at least one solution through a cycle of defeats and reinstatements (hence, their union). This set would constitute the set of 'weak beliefs'. With the strong and weak beliefs at the two ends of the continuum, one can think of beliefs with possible intermediate degrees.

Propositional Dependencies and Restriction on Irrelevancies

One of the objectives of DRML is to keep track of the propositional dependencies and prevent the gratuitous introduction of irrelevant propositions in an argument or proof. In order to achieve this, the relevant logicians, e.g. Anderson and Belnap (1975), impose certain restrictions. These, together with some techniques to record dependencies due to Martins and Shapiro (1988), will be incorporated in DRML. Briefly, they are:

1. *Actual Use of a Hypothesis and Origin Set*: Within an argument or derivation, each wff is associated with a set containing references to all the hypotheses that were *actually* used in its derivation. This may be called *origin set* (OS). To record the fact that wff *A* is dependent on an origin set α, i.e. OS_{α}, we may write A, α. In the Anderson-Belnap system, the introduction of any new hypothesis is associated with a singleton OS whose element must never appear anywhere before in the derivation.

   A Conditional proof with the implication introduction rule (i.e. $\rightarrow$ I) and discharge rule may illustrate this: if *A* is a hypothesis with OS $\{k\}$, *A* is genuinely used in the derivation of wff *B* (i.e. B, $\alpha \cup \{k\}$ as OS), and both *A* and *B* are in the same subproof, then $A \rightarrow B$ can be deduced. This can be represented schematically thus:

$$\begin{array}{ll} m & A, \{k\} \text{ Hyp.} \\ & \ldots \\ & \ldots \\ n & B, \alpha \cup \{k\} \\ \hline & A \rightarrow B, \alpha \rightarrow I(m,n), \text{OS } \alpha - \{k\} \end{array}$$

2. *Use of Inference Rules and OS*: The use of inference rules must take the OSs into account in order to prevent the introduction of irrelevant propositions. This can be achieved by imposing the following restrictions on the use of inference rule. Whenever a rule of inference is used, the resulting wff must be associated with an OS. The OS could be:

   i) the OS of the parent wff(s), or
   ii) the union of OSs of the parent wffs, or
   iii) the set difference of the OSs of the parent wffs.

   This can be illustrated by the rule of and-introduction (i.e. Λ I): if two wffs *A* and *B* belong to the same OS, then one can deduce *A* Λ *B* and associate it with the same OS. Schematically:

$$\begin{array}{ll} \hline & \ldots \\ & \ldots \\ m & A, \alpha \\ & \ldots \\ & \ldots \\ n & B, \alpha \\ & A \wedge B, \alpha \wedge I(m,n) \\ \hline \end{array}$$

   Similar restrictions are to be imposed on or-introduction (i.e. V I). The violation of these restrictions could result in some of the so-called paradoxes of implication.
3. *Recording Contradictions and Restriction Set*: Although a given defeater *R*, and the *prima facie* reason *P*, are consistent, they might together lead to a belief which is contradictory to the original, intended target belief.

In DRML these contradictions are recorded by associating each wff (and its corresponding origin set) with a set, called the *restriction set* which contains information about which set when unioned with the wff's origin set produces an inconsistent set.

4. *Complexity Problem and Relevance Logic*: A moot point remains. It has already been argued that the complexity of epistemic logic proposed by Hintikka prevents it from being a theory of rationality. Does the complexity of relevant logic fare any better?

   Urquhart (1984) has shown that the decision problem for the main relevant propositional logics are in general undecidable. However, there are important decidable subsystems. In his 1990 paper, Urquhart shows that the decision problem for $R \rightarrow$, the implication fragment of R is exponential space hard. 'This means any Turing machine which solves this decision problem must use an exponential amount of space (relative to the input size) on infinitely many inputs.' Nevertheless, he reports that the first-degree fragment common to E, R, and most other well-known relevant logics has a simple decision procedure, as 'finite matrices are used extensively to prune invalid nodes from the search tree.'

5. *Relevance Logic and a Combination of Forward and Backward System:* In a previous section we considered the ability to take the goal into account in a reasoning process to be an important characteristic of rational behaviour. Most logics deal with forward reasoning (e.g. from the axioms/premises to the theorems/conclusions, which can be viewed as the goal state). However, capturing backward reasoning, i.e. regression from the goal, could be an important feature of a logic as the branching factor is significantly greater going forward from the axioms to the theorems than it is going backward from theorems to axioms. Many mathematicians (Polya, 1957) and the designers of theorem-proving programs (Newell *et al.* 1963) realised that it is a much better strategy to reason backward while trying to prove theorems. It would thus be interesting to see whether this feature of the regression from the goal can be captured in DRML.

   One of the constituents of the DRML, viz. relevant logic, has the notions of intensional conjunction or fusion and intensional disjunction or fission. From these notions one can form *fact expressions* consisting of conjunctions of literals and *goal expressions* consisting of disjunctions of literals using the well-known technique of resolution (Rich 1988; Nilsson 1980). Nilsson (1980, pp. 253–57) develops a production system which combines the forward and backward systems with modified F-rules and B-rules. This could be incorporated in DRML to capture backward reasoning or regression from goals.

One such possible system would proceed not by making successive steps in a single direction but by reducing differences between the current and the goal states and, in effect, sometimes reason backward and sometimes reason in the forward direction.

A Theory of Justification of Belief

A proper account of knowledge will include an account of justification and warrant. The theory of justification of belief must investigate into the processes that in fact occur when an epistemic agent is involved in knowledge acquistion. It would attempt to answer the questions like: what kind of transitions between mental states make for justified belief? What does justification consist in?

A Naturalised Theory of Justification

Kornblith (1980) has criticised the two standard apsychological accounts of justification, viz. foundationalism and the coherence theory, and tried to reconcile the insights of both into a single unified theory of justification. This may be called the naturalized theory of justification. Our account of justified belief endorses this theory.

An account of justification takes a naturalised turn when it is seen to be intimately tied to the question about the sorts of processes responsible for the presence of a true belief. An account of this must include the following:

(*i*) The causal process or mechanism for the formation of belief must be reliable. A belief is justified when it is caused by a reliable process: the reliable account of justification (Goldman 1976, 1979).

(*ii*) Although there could be beliefs such that an account of the process responsible for their presence does not involve other beliefs, an account of what makes the process adequate for justification involves reference to other *relevant* background reasoning process and beliefs. A complete account of justified belief will make reference to those reasoning processes and beliefs.

The Notion of Justification Based on DRML

Our notion of 'initial argument' accounts for (*i*) above, where the presence of a belief is the result of a causal process or mechanism which is reliable. Goldman has already offered a satisfactory account of this reliability.

Our notion of 'interactive argument' suggests how a belief is related to other beliefs. The survival of a belief through a process of defeat and reinstatement precisely describes the nature of reasoning involved in assessing the relevant contending beliefs in counterfactual situations.

Let us illustrate this with a somewhat modified example from Kornblith (1980). The presence of a bowl of fruit cause Joe and Moe, who are looking at it, to believe that there is an apple in front of them. This is perfectly reasonable belief-forming process in the normal circumstances. However, it is not adequate for the justification of the belief unless it leads to the production of true belief in both actual and *relevant* counterfactual situations. Let us suppose that unlike Moe, Joe has an additional belief that he is myopic. Hence, from a distance he would not be able to tell the difference between a real or an artificial apple. This defeater makes the counterfactual situation relevant because the normal situation which makes Joe's belief justified would make it unjustified in the counterfactual situation. Since Moe does not have this belief, the counterfactual situation would not be relevant. At this point, let us further assume that Joe, who was unaware of or momentarily forgot, that he had his glasses on throughout, suddenly becomes aware of it. With this additional relevant belief, his earlier belief in the presence of the apple would be reinstated and justified.

In classical epistemology the notion of justification and the associated logic of justification takes a form which is not compatible with the available facts about human cognition. In DRML, we try to develop a theory of justification which takes these facts into account.

Notes

We are indebted to Daniel Kayser and Daniel Andler for insightful comments on an earlier draft of this paper.

1. Complexity theory, based on the works of Richard M. Karp is concerned with the limit of computation and algorithm. It attempts to analyse and evaluate the size and complexity of a problem or a cognitive task in terms of the steps to be undergone, the time taken and the amount of effort expended.

   It is customary in complexity theory to categorise a cognitive task or problem into solvable and open; then the solvable problems are categorised into decidable and undecidable depending on the availability of algorithm; and, finally, the decidable problems into P-type (polynomial time complexity) and NP type (exponential time complexity).

   Suppose in a given problem n represents the size of the input. The time complexity in terms of the steps to be undergone could be either n(exp 2) [which represents polynomial time complexity] or 2(exp n) [which represents exponential time complexity]. It will be clear from the table given below that when n > 5 the exponential function increases very rapidly:

n	n^2[P]	2^n[NP]
2	4	4
3	9	8
4	16	16
5	25	32
6	36	64
7	49	128
-	-	-
-	-	-
10	100	1024
-	-	-
-	-	-

2. The Minimal Inference Condition states:

   'If A has a particular belief-desire set, A would make some, but not necessarily all, of the sound inferences from the belief set that are apparently appropriate' (Cherniak 1986, p. 10).

3. The Minimal Consistency Condition states:

'If A has a particular belief-desire set, then if any inconsistencies arose in the belief set, A would sometimes eliminate some of them' (Cherniak 1986, p. 16).

4. *P* is a *prima facie* reason for *S* to believe *Q* if *P* is a reason for *S* to believe *Q* and there is an *R* such that *R* is logically consistent with *P* but $(P \wedge R)$ is not a reason for *S* to believe *Q*.
5. *R* is a defeater for *P* as a *prima facie* reason for *Q* if *P* is a reason for *S* to believe *Q* and *R* is logically consistent with *P* but $(P \wedge R)$ is not a reason for *S* to believe *Q*.

References

ALMEDER, ROBERT. 1990. On Naturalizing Epistemology. American Philosophical Quarterly, 27: 4, pp. 263–79.

ANDERSON, A. and N. BELNAP. 1975. *Entailment: The Logic of Relevance and Necessity*. Princeton, N.J.: Princeton University Press.

BARKER, JOHN A. 1989. Relevance Logic and Inferential Knowledge in Jean Norman and Richard Sylvan (eds.), *Direction in Relevant Logic*. Dorderecht: Kluwer Academic Publishers.

CARNAP, R. 1950. *Logical Foundations of Probability*. Chicago: University of Chicago Press.

CHERNIAK, C. 1981a. Minimal Rationality. *Mind*, XC, pp. 161–83.

———. 1981b. Feasible Inferences. *Philosophy of Science*, 48, pp. 248–68.

———. 1984. Computational Complexity and the Universal Acceptance of Logic. *The Journal of Philosophy*, LXXXI: 12, pp. 739–58.

———. 1986. *Minimal Rationality*. Cambridge, MA: M.I.T. Press.

CHISHOLM, R.M. 1957. *Perceiving*. Ithaca, N.Y.: Cornell University Press.

DAVIDSON, DONALD. 1963. Actions, Reasons and Causes, *Journal of Philosophy*, 60, pp. 685–700.

———. 1974. Mental Events. In *Essays on Actions and Events*. Oxford: Clarendon. pp. 207–27.

———. 1976. Psychology as Philosophy. In *Essays on Actions and Events*. Oxford: Clarendon. pp. 229–44.

DENNETT, DANIEL C. 1971. Intentional Systems. *Journal of Philosophy*, 68, pp. 87–106.

DRETSKE, FRED. 1969. *Seeing and Knowing*. Chicago: University of Chicago Press.

FREGE, G. 1983. *The Basic Laws of Arithmetic*, Vol. 1, p. XIV. Translated by M. Furth. Berkeley: University of California Press.

GARFIELD, JAY I. 1990. THE DOG: Relevance and Rationality. In J.M. Dunn and A. Gupta (eds.), *Truth or Consequences*. Dorderecht: Kluwer Academic Publishers. pp. 97–109.

GOLDMAN, A.I. 1976. Discrimination and Perceptual Knowledge. *Journal of Philosophy*, 73, pp. 771–91.

———. 1978. Epistemic: The Regulative Theory of Cognition. In H. Kornblith (ed.), *Naturalized Epistemology*. Cambridge, M.A.: M.I.T. Press, 1985.

GOLDMAN, A.I. 1979. Varieties of Cognitive Appraisal. *Nous*, 13, pp. 23–38.

———. 1985. The Relation between Epistemology and Psychology. In David J. Cole (ed.), *Philosophy, Mind and Cognition*, Dorderecht: Kluwer Academic Publishers, 1990.

———. 1986. *Epistemology and Cognition*. Cambridge: Harvard University Press.

HALPERN, J.Y. 1986. Reasoning about Knowledge: An Overview. In J. Halpern (ed.), *Reasoning about Knowledge*. Palo Alto, CA: Morgan Kaufman, pp. 1–17.

HEMPEL, C.G. 1965. *Aspects of Scientific Explanation*. New York: Free Press.

HINTIKKA, J. 1962. *Knowledge and Belief*. Ithaca: Cornell University Press.

JOHNSON-LAIRD, P.N. 1980. Mental Models in Cognitive Science. *Cognitive Science*, 49, pp. 24–42.

———. 1983. *Mental Models*. Cambridge: Cambridge University Press.

———. 1990. Human Thinking and Mental Models. In K.A. Mohjeldin Said *et al.* (eds.), *Modelling the Mind*. Oxford: Clarendon.

KAHNEMAN, D., P. SLOVIC, and A. TVERSKY. 1982. *Judgement under Uncertainty: Heuristics and Biases*. Cambridge: Cambridge University Press.

KARP, R.M. 1989. 'Reducibility among Combinatorial Problems' in *Complexity of Computer Combutations*. New York: Plenum.

KORNBLITH, HILARY. 1980. Beyond Foundationalism and the Coherence Theory. *The Journal of Philosophy*, LXXII; reprinted in Naturalizing Epistemology, H. Kornblith (ed.), *Naturalizing Epistemology*. Cambridge, M.A.: M.I.T. Press, 1987, pp. 115–28.

KRIPKE, S. 1982. *Wittgenstein on Rules and Private Language*. New York: Basil Blackwell.

LEWIS, D. 1969. Lucas Against Mechanism, *Philosophy*, 44, pp. 231–33.

LUCAS, J.R. 1961. Minds, Machines and Godel. *Philosophy*, 36, pp. 112–27.

MARTINS, J.P. and S.C. SHAPIRO, 1988. A Model for Belief Revision. *Artificial Intelligence*, 35, pp. 25–79.

MEYER, R.K. and M.J. DUNN. 1969. E, R, and Y, *The Journal of Symbolic Logic*, 34, pp. 460–74.

NEWELL, A., J.C. SHAW, and H.A. SIMON. 1963. Empirical Explorations with the Logic Theory Machine: A Case Study in Heuristics in R.A. Feigenbaum and J. Feldman (eds.), *Computers and Thought*. New York: McGraw-Hill.

NILSSON, NILS J. 1980. *Principles of Artificial Intelligence*, Palo Alto, CA: Morgan Kaufmann.

NISBETT, RICHARD and LEE ROSS. 1980. *Human Interface: Strategies and Shortcomings of Social Judgement*. Englewood Cliffs, N.J.: Prentice Hall.

———. 1985. Judgemental Heuristics and Knowledge Structures in H. Kornblith (ed.), *Naturalizing Epistemology*. Cambridge M.A.: M.I.T. Press.

PARIKH, ROHIT. 1987. Knowledge and the Problem of Logical Omniscience, *ISMIS–87*, Amsterdam: North-Holland, pp. 432–39.

POLLOCK, J.I. 1987. Defeasible Reasoning, *Cognitive Science*, II.

POLLOCK, JOHN. 1967. 'Criteria and Our Knowledge of the External World.' *Philosophical Review*, 76, pp. 28–62.

POLYA, G. 1957. *How to Solve*. Princeton, N.J.: Princeton University Press.

PYLYSHYH, Z.W. (ed.) 1987. The Robert's Dilemma: The Frame Problem. In *Artificial Intelligence*, Norwood, New Jersey: Ablex Publishing Corporation.

QUINE, W.V.O. 1969. Epistemology Naturalized. In H. Kornblith (ed.), *Naturalizing Epistemology*. Cambridge M.A.: M.I.T. Press, 1985.

READ, STEPHEN. 1988. *Relevant Logic: A Philosophical Examination of Inference*. Oxford: Basil Blackwell.

RICH, E. 1988. *Artificial Intelligence*. New York: McGraw Hill.

RUSSELL, B. 1971. On Propositions. In R. Marsh (ed.), *Logic and Knowledge*. New York: Putnam.

SIMON, H. 1981. *The Sciences of the Artificial*. Cambridge, MA: M.I.T. Press.

STICH, STEPHEN and RICHARD NISBETT. 1980. Justificational and the Psychology of Human Reasoning. *Philosophy of Science*, 47, pp. 188–202.

STICH, S. 1984. Could Man be an Irrational Animal? Some Notes on the Epistemology of Rationality. In H. Kornblith, *Naturalizing Epistemology*. Cambridge M.A.: M.I.T. Press.

THAGARD, PAUL. 1982. From the Descriptive to the Normative in Psychology and Logic. *Philosophy of Science*, 49, pp. 24–42.

TVERSKY, A. and D. KAHNEMAN. 1974. Judgement under Uncertainty: Heuristics and Biases, *Science*, 185, pp. 1124–34.

URQUHART, ALASDIAR. 1984. The Undecidability of Entailment and Relevant Implication, *Journal of Symbolic Logic*, 49, pp. 1059–73.

——. 1990. The Complexity of Decision Procedures in Relevance Logic. In J.M. Dunn and A. Gupta (eds.), *Truth or Consequence*. Dorderecht: Kluwer Academic Publishers, p. 61.

VON WRIGHT, G.H. 1951. *An Essay in Model Logic*. Amsterdam: North-Holland.

WEBB, J.C. 1980. *Mechanism, Mentalism and Metamathematics*. Dorderecht: Reidel.

WITTGENSTEIN, L. 1958. *The Blue Book*. New York: Harper and Row.

Logic and Rationality in an Artificial Intelligence Perspective

DANIEL KAYSER

Key Issues

In ordinary speech, the adjectives 'rational' and 'logical' denote more or less the same kind of quality, and in several circumstances, they are interchangeable. As it is the vocation of science to develop abstract notions from intuitions in the hope of obtaining general results, it is a genuine scientific task to try and give an account of the underlying notion, i.e. that quality which is referred to by the words 'rational' and 'logical'.

The process of abstraction is often also an idealisation, and there is nothing wrong with this so long as the initial intuition is not lost. However, the abstract notions that have emerged from the intuition of what 'logical' turns out to denote, namely logic as a discipline, provides a satisfactory account of only part of the *logical* behaviours, and seems to have almost completely lost contact with the intuition behind the word *rational*. My position is roughly summarised in the following diagram.

Although we will not develop this issue here, it should be pointed out that the rationality of a behaviour (in its intuitive sense) tends to be evaluated at least relatively to a given *goal*, in a given *culture*, in a given *situation*, while the usual models of rationality (in its formal sense) fail to take these parameters into account. Moreover, in spite of the efforts of mathematicians of this century, logic does not correspond to a single entity ('standard' logic has been challenged by intuitionism, by various modal or many-valued logics, by linear logic, or even by more 'exotic' logics developed for some purpose or another); the boundary of what

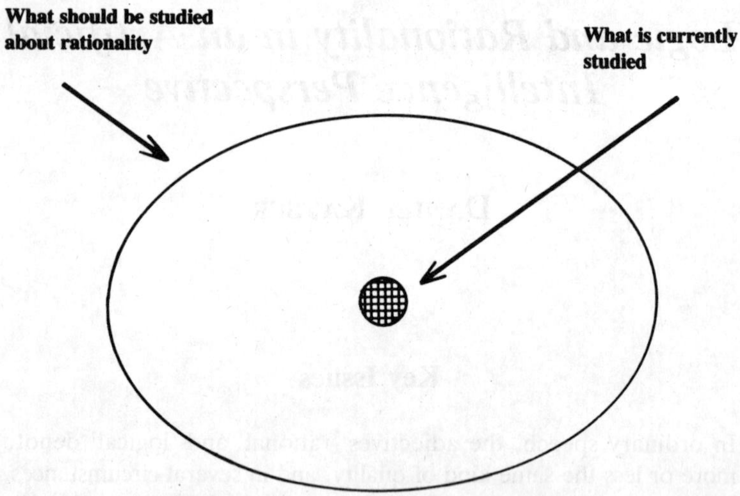

can legitimately be called a logic is still a matter of debate (see Engel 1989).

The bulk of the efforts in logic concerns deductive reasoning, while rational behaviour requires an intensive use of inductive and abductive inferences. The next section will briefly survey some of the tentative answers which Artificial Intelligence (henceforth AI) has brought forth. In the meanwhile, it might be worthwhile mentioning some as yet unsolved issues. Is it rational, when performing *inductive* reasoning, i.e. when building a theory from experimental data, to expect that a unique theory will account for all the data? Alternatively, is it rational to have several, possibly inconsistent theories, to explain the data? When is it rational to accept a probabilistic rather than a deterministic theory, the latter one possibly augmented by some allowances? The data are expressed in a language. Is it rational to have a unique language to describe all the facts? To have a unique formula of the language to express a given observation? Is there a rational policy to select, if any, the *constants*, the *predicates*, the *functions* of the language? To acquire the facts in a principled way?

From the very beginning of AI, the literature contains countless arguments for and against logic as a language for *knowledge representation*. In the latter case, the target is sometimes a specific kind of logic, sometimes any kind of logic (see e.g. Minsky 1975; McDermott 1987). The most fundamental criticisms concern either *expressivity* or *tractability*.

While the issue of *expressivity* has continuously stimulated the invention of new formalisms, the problems remain as to whether this inflation has a limit, or whether we shall always remain under the threat of discovering a kind of knowledge requiring the addition of a new feature to the language; and whether putting together all the fancy embellishments suggested by this or that aspect of knowledge yields a consistent framework. In this article I shall ignore these problems, at least as they are formulated here, since they correspond to rather long term concerns, while more immediate problems need to be solved.

A much more crucial issue is that of *tractability*: borrowing the word from Rabin (1974), we may even say that it is the most serious *impediment* to the project of giving a scientific account of *logical* behaviours in terms of a logic. Along the same lines, Cherniak (1986) provides, in philosophical terms, a very good survey of the reasons why the classical models fail to give an adequate account of *rationality*, and he centres his argumentation primarily on the problem of tractability. But here too, although to a lesser extent, ingenuity helps in designing tractable subclasses of intractable—or even indecidable—problems.

Some AI Answers

Logical and/or *rational* behaviours are, most of the time, analysed according to their representational and inferential aspects. As already mentioned, many other facets of these behaviours should be given an account of. We listed above some problems concerning the construction of theories from experimental data. In AI, Feldman (1972), and more recently Osherson *et al.*, (1985) have tried to give criteria for this kind of enterprise; Gold (1967) and several other authors have shown that under these criteria, the class of learnable theories was rather narrow, questioning the adequacy of the criteria with regard to the intuitive idea of what it is to build rationally a theory 'explaining' some given data.

Abductive reasoning is obviously not a valid form of reasoning; it is therefore overlooked in many models where logic plays the prominent role. Nobody however would deny that rational behaviour must include the ability to suggest possible actions when troubles are detected, and such troubleshooting behaviour is typically abductive (e.g. knowing that *run-down battery* implies *head-lights failure* and observing *head-lights failure* is a rational basis to check whether the battery is all right). Probabilistic accounts of rationality are often invoked in these circumstances (see e.g. probabilistic logics Paass 1988) but ever since the early studies on probability, the deep nature of this notion has been questioned (for recent AI, see e.g. arguments in favour of the epistemic point of view (Cheeseman 1985), and for 'statistical knowledge' (Kyburg 1991). But several psychological studies (Kahnemann *et al.* 1982) show that humans fail to follow probabilistic schemes, while having nonetheless rather good success in several uncertain environments. Other—generally computationally easier—schemes than probabilities have been proposed to cope with uncertainties in the recent AI literature; among the most elaborated are the Dempster-Shafer rules for combining evidence, the 'MYCIN' technique which has been extensively used in expert systems, and the so-called 'possibility theory' (Dubois & Prade 1985). If we acknowledge that tractability must be taken into account in any model of rationality, it is by no means irrational to reject the very sophisticated apparatus of probabilities for a simpler technique, as long as the latter proves its efficiency in real world settings. True enough, ad hoc rules are likely to give very poor results outside of the scope where they have been tested, while probabilities are universal. But human behaviour witnesses that at least one non-probabilistic scheme yields reasonable results in a variety of circumstances. Cognitive modelling may help to implement this behaviour in machines, but other satisfactory solutions are likely to exist, which AI must look for on its own.

Another problem where rationality is obviously involved, and where logic is by itself of no help, is theory *revision*. Two rather different cases are generally considered: the first one consists in updating the 'belief state of a rational agent' after he/she has observed an evolution of the world. The second concerns that updating after an (epistemic) evolution of the agent but with no actual change in the world. In both cases, various logical devices have been designed, but they sometimes conflict with each other, and there seems to be no *rational* basis for preferring one solution over the other. Let us take an example of the second problem. A useful device to extend an epistemic state in the presence of a partial description of the world is provided by Reiter's (1980) so-called 'default rules', i.e. special-purpose inference rules having the general form

$$\frac{A{:}M\ B_i, \ldots, B_n}{C}$$

where A, the B_i, and C are closed formulae of a first-order language.Such a rule is intended to convey a *normative* inferential behaviour and means that whenever an agent believes A (the *prerequisite*) and has no ground to believe that any of the B_i (the *justifications*) is false, then he/she should believe also C (the *consequent*). The technical definition is somehow more involved, in order to avoid oscillations in the belief set when 'pathological' default rules are present. A simple example of 'pathological' default is when the consequent contradicts the justification, e.g.

$$\frac{A{:}\ M\ B}{\neg B}$$

(read: if the agent believes A and does not believe $\neg$B, then he/sheshould precisely believe $\neg$B). In the presence of such 'default rules', various propositions are still possible in order to define what a proper 'belief set' should be: either any one of the fixpoints[1] of a given function (Reiter's [1980] original choice, later known as a 'credulous' point of view) or the intersection of those fixpoints (proposed in a different logical framework by McDermott [1982], later known as the 'skeptical' solution). Moreover, what counts as a fixpoint has also been the subject of various definitions: each applicable default should be applied; or a consistency condition between the 'justification' of the defaults used should be checked (Lukaszewicz 1985); or, according to Poole (1988), only maximal consistent sets obeying certain constraints should be considered, and so on.

The point is that each of these authors—and there are plenty of others, which space limitations prevents from quoting—provides one or a very small amount of (sometimes rather contrived) examples, shows that some previous system yields on these examples a solution which he/she considers as counterintuitive, and goes on to his own solution which provides the desired result for the given examples. Of course, with such a methodology, there is little progress in arriving at a deep understanding of what a *rational* extension of beliefs should look like. Some interesting convergences have been uncovered; for example, starting from an intuition explicitly asserted as different, Moore's (1985) 'auto-epistemic' logic has been later proven, under some conditions, to be equivalent to Reiter's (Konolige 1988). But negative results—proofs that the belief set obtained by one system cannot be obtained by another one—are also widespread in the literature. A potentially more interesting step in recent literature (Kraus *et al.* 1990; Lehmann & Magidor 1992) examines the properties implied by the choice of any inferential scheme obeying simple axioms. But here again, several axioms can be proposed, yielding several norms of rationality; what is noteworthy is the fact that each axiom, considered separately, seems intuitively satisfactory; only after examining the result of its interaction with other equally seemingly rational axioms can it be decided whether it will be maintained or withdrawn from the whole axiomatic system. This shows, in my opinion, that no self-explanatory system has yet been devised which uncontroversially reflects the intuition of what should be done in presence of new information about a stable state of the world. The state-of-the-art of what we have called the first problem, viz. the update in case of a changing world, shows about the same if not higher degree of confusion (see e.g. Somb 1992)!

All the problems discussed so far seem to arise in a context where the description of the world is not put into question. Within an accepted descriptive framework, despite of an impressive body of formal work, we have enumerated some unsolved difficulties in the definition of a *rational* policy for abductive inference or for the update of a belief set. We now turn to what seems to us to be a still more vital challenge for rationality, viz. the choice of a language to describe the world. Of course, philosophy has devoted much effort on conceptual organisation; more recently, psychology has brought some new ideas concerning categorisation, e.g. through Rosch's work on typicality. But very little attention has been paid to an aspect that seems of major importance in rationality: the ability to modify, during the inference itself, the ontological choices.

Whether it makes sense or not to consider a categorisation of objects independently of any consciousness or teleology, is a metaphysical issue which need not be of concern here. But the very concrete choice of a given categorisation, and hence of a descriptive language, for reasoning on some real-world situation is a major prerogative of a rational agent. As a matter of fact, even if we postulate the existence of a pre-existing, immanent as it were, classification of the (concrete and abstract) entities that may intervene in a reasoning, this classification would be of no help in any practical setting. The main reason of this practical impossibility is the combinatorial explosion (another name for intractability) intrinsically caused by the manipulation of a large amount of facts and rules. Hence a principle of ontological parsimony is an absolute necessity, somewhat reminiscent of Occam's razor: introduce in the reasoning only those entities which are relevant to it. This principle entails the choice of a level of granularity (Hobbs 1985).

I shall illustrate this idea with the following example. In the process of planning a trip, the choice of going by plane, train or car depends on the distance, on the fare, on the urgency, which in turn depends on several factors (the distance by car depends on the itinerary, but the choice of the itinerary interacts, because of toll roads, with the expenditure; the fare usually depends on the date and hour; to comply with urgency, an estimate of the duration of flights, ground transportation to the airport, probability of strikes etc. should be available). This list is virtually endless. However, it would sound very irrational to spend more time in evaluating the expected duration of the travel than in travelling, especially if urgency is an important factor in the decision. This shows that reasoning on a coarse—hence approximate—representation is often more rational than using an exact one. But the basis on which to select the 'correct' level of granularity remains unclear.

Moreover—and this point seems to have been overlooked—the selection is part of the reasoning itself. Hence it may happen—and obviously happens frequently—that, as a consequence of the reasoning, a new choice should be considered. As a continuation of the travel example, it may have been decided that a 1 for 1,000,000 scale map was enough to decide a road itinerary, but a close inspection of that map can incite to make the final choice after having had a look to a 1 for 100,000 map. Changing the scale also changes the ontology: the reasoning now takes into account towns, mountains, rivers, which did not even exist when the reasoning was based on the first map. The decision to change the ontology is perceived as belonging to a different level than the plain development of a reasoning.

There is a strong analogy here with the way an operating system works in a computer: either an external event interrupts the normal execution, or control is given at regular intervals—or every time some given configuration appears—to a 'supervisor'; this latter module may decide either to resume the execution or to suspend it and give the control to another task. Several levels of priority are commonly defined, and this allows for a hierarchy of possible causes of interrupts. This solution is the only conceivable one, if the computer is to work in 'real-time', i.e. to react properly to external events within the very short lapses demanded e.g. by process control applications.

In other words, what this analogy suggests amounts to the following. If we imagine that a series of deductive steps in some logic is an adequate model of (abstract) rationality, we are presently adding *into* the model the necessity to interrupt these steps (periodically or whenever some situation dictates) and to perform another series of steps, possibly in a different logic, and definitely on different entities. No single logic, to our knowledge, can incorporate within its proof theory anything like this bi-level reasoning.[2] Therefore, a first interesting achievement of AI is that it can interweave 'object' and 'meta' reasoning, just because it is implemented on computer systems and thus has access to all the tools at the disposal of the users (an early implementation of this approach. was O.G. Selfridge's *Pandemonium*; the modern *Object-Oriented Programming* generalises the idea). Now, as this blending seems a prerequisite to rational behaviour, we believe that AI has a clear advantage over logic as a model of rationality.

But this advantage will soon be lost if no principled way is found to use the tool. Our analogy between rational behaviours and operating systems leads us no further. The parameters of a computer system, e.g. the interval between two interventions of the 'meta' level, the list of the events triggering an intervention, etc. can be determined from a careful study of the computer environment: the characteristics of the programs executed, the 'real-time' constraints, and so on. But since a key requirement for rational behaviour is to react appropriately to unexpected situations, even a good knowledge of the tasks already solved would not suffice to determine the parameters of a good sharing of the resources between 'object' and 'meta' levels.

Even if it is supposed that this sharing has been correctly determined (and perhaps the optimum is not very critical here), what to do at the 'meta' level remains to be decided. Various techniques have been proposed in order to improve the interplay of several 'knowledge sources'; one of the currently most popular ones is the so-called *blackboard architecture*. Another interesting feature, which probably has important philosophical counterparts, concerns *reflexivity*, i.e. the ability for the inferential system to use a representation of itself, such that a modification in the representation corresponds to an actual modification of the behaviour of the system.

Pitrat (1990) forcefully argues in favour of 'spying' on the object level, and to apply strategic rules that operate on the result provided by the 'spies'. The strategic rules decide either the resumption of the interrupted process, or more or less drastic changes. In the case where a change is decided, advantage must be taken of those results already obtained which the change under consideration does not undermine. This problem has strong similarities with the revision issues mentioned earlier, and also with the well-known 'frame problem'.

The current trend in AI is thus to model rationality by (standard or non-standard) logical tools at the object level, by non-standard logics at the meta level, and to forsake the idea of a logical account of the switch between the levels.

A Proposal for One of the Issues

I have tried to develop another paradigm—variable-depth reasoning—to provide a more principled account of *one* of the difficulties encountered, namely the variation of ontology during the reasoning (see e.g. Kayser 1988) and will try to give a logical account of this possibility. The key ideas are that:

- *contradictions* are the triggers to an ontological change
- *existentials* provide a lever to switch the reasoning to another level
- *non-monotonic logics* provide a convenient framework for keeping as many results as can consistently be done when proceeding from level to level.

More precisely, I shall consider a (partial) description of the world in terms of closed literals in a first-order language (the *facts*), which will give us both strict rules (i.e. material implications) and default rules (in Reiter's sense). Rules have as their premises a universally quantified conjunction of literals: in *flat* rules, the consequence is also a conjunction of literals involving the same variables; in *existential* rules, existentially quantified new variables appear in the literals of the conclusion. The predicates appearing in the justifications of the defaults are special predicates, partially ordered, and no 'greater' predicate can conclude to the negation of a 'lesser' predicate, which ensures the existence of extensions (Etherington 1987).

The variable-depth reasoner operates as follows (Grandemange 1992): every *flat* rule whose premises match the facts creates facts of the same 'depth'; every *flat* default rule having its prerequisite matching the facts, and none of its justification deniable from the facts, creates its consequence as facts of the same 'depth'. Several sets of facts (extensions) may be derived at this stage; each fact is therefore labelled by the set of defaults which allow its derivation, by an ATMS-like technique (DeKleer 1986; Levy 1989). *Existential* rules, the premises of which match the facts, generate *shells*, i.e. formulae with a free variable. When every fact deducible in some extension at the current depth has been created, the control is passed to a supervisor which may either decide to stop at that depth, or to pursue at a deeper level. In the latter case, the free variables in the shells are non-deterministically filled with constants of the language, looking for possible contradictions. Every non-contradictory instantiation has the same status as a default, i.e. the facts derived assuming it are labelled by it; as several possible instantiations may exist, we may get several extensions. Shells which do not have non-contradictory instantiation on the existing constants are called *mandatory*. At the end of the process, the supervisor once more gets control. If it decides to go further, a new step in depth is performed, and *new constants are created* in the language in order to complete the mandatory shells.

It may thus be seen that, as a result of the reasoning itself, the ontology of the universe on which it operates is expanded. The technique described ensures an incremental minimality of the universe (but not an overall minimality); minimality of the set of objects considered is consonant with a principle of economy which seems to be basic in all rational achievements.

Example: Suppose that we have the following fact:

P(a)

and the rules:

R1	$(\forall x, y)(P(x) \wedge R(x,y) \supset \neg Q(x))$
R2	$(\forall x)(P(x) \supset (\exists y)(Q(y)))$
R3	$(\forall x)(Q(x) \supset (\exists y)(R(x,y)))$

All the rules are strict. R1 is flat; R2 and R3 are existential. The fact does not match the premise of R1, so no new fact is generated. It matches the premise of R2, so the *shell* Q(y) is created. Nothing more can be done at this depth (let us call it depth 0).

If the supervisor decides to go deeper, at the next level (call it depth 1), the shell is instantiated on the only constant of the language, and fact Q(a) is added with a label L1 explaining that it depends on the instantiation of shell Q(y) on (a). Rule R3 becomes applicable and creates a new shell R(a,y) with the same label L1.

If the reasoning goes deeper (depth level 2), this new shell is instantiated with the same constant, yielding the fact R(a,a); this fact bears labels L1 and L2 where L2 describes the instantiation of shell R(a,y) on (a). Rule R1 then fires, and concludes ¬Q(a); it is then noticed that the conjunction of L1 and L2 is inconsistent. We cancel the most recent instantiation; as(a) was the only solution, the shell R(a,y) is now considered as mandatory.

Therefore, at depth level 3, a new object, say(b), is created, and the shell is transformed into the fact R(a,b), labelled by L1 and L3 where L3 expresses that shell R(a,y) is instantiated on(b). But R1 fires again, concludes ¬Q(a), and the conclusion L1 and L3 is seen to be inconsistent. Whenever the instantiation of a mandatory shell creates an inconsistency, an older instantiation must be changed. We therefore reconsider the instantiation of shell Q(y), and create the fact Q(b) with label L4. Rule R2 builds a new shell R(b,y).

At depth level 4, we have two solutions: instantiate the shell on (a) (label L5) or instantiate it on(b) (label L6). This generates two extensions, one with fact R(b,a) labelled by L4 and L5, the other with facts R(b,b) labelled by L4 and L6. Nothing more is deducible. This entire process is summed up in the following table:

Depth	Extension 1	Extension 2 (if applicable)
0	P(a); shell Q(y)	
1	P(a); Q(a) {L1}; shell R(a,y)	
2	P(a); Q(a) {L1}; mandatory shell R(a,y)	
3	P(a); Q(b) {L4}; shell R(b,y)	
4	P(a); Q(b) {L4}; R(b,a) {L4,L5}	P(a),Q(b) {L4}; R(b,b) {L4,L6}

Notes: L1: instantiate shell Q(y) with y =a
L4: instantiate shell Q(y) with y = b
L5: instantiate shell R(b,y) with y = a
L6: instantiate shell R(b,y) with y = b
for the sake of completeness: L2: instantiate shell R(a,y) with y = a
L3: instantiate shell R(a,y) with y = b

What appears clearly from this above table is that object (*a*) has property Q at levels 1 and 2 and loses this property at deeper levels, and that object (*b*) does not exist until depth level 3.

This very artificial and contrived example shows how, inside a logical framework, the universe can be modified during the reasoning. More realistic examples—including the use of default rules—can be found in Grandemange 1992. Although Default Logic is undecidable (Reiter 1980), the fragment we use—and the procedural way which we use in it—makes it decidable, but the number of inference steps and the size of the set of labels grow exponentially with the size of the problem.

Conclusion

Further work will show whether this kind of approach is helpful for the difficult and important problem of making the universe of discourse evolve during the reasoning. Whatever it may be, other large areas of rational behaviour are still poorly dealt with in AI Take for example the problem of the 'meta-level' decision: on which basis should it be decided that there is no hope to solve a task in a given framework, and which action if any, should then be taken among the following:

- change the description of the situation (not only of the universe, but maybe also the predicates, the modalities present in the description language, etc.);

- learn more about the world (by observing? by experimenting? which parameters to be considered as relevant? which solution space?);
- relax some constraints (e.g. accept default conclusions, analogies . . .).

Partial and empirical solutions have been adopted for some of these issues, but whether a unifying principle can justify a well-founded answer to these problems is still an open question.

Similarly, we have seen that several developments in computer science have provided interesting answers to issues concerning rationality which had been overlooked in theoretical studies: the notions of supervisor and interrupts introduced in operating systems; the idea, developed in early AI works, of programming with 'daemons' watching for given configurations to become active; more recently techniques such as non-monotonicity, meta-knowledge, blackboard, reflexivity, reason-maintenance seem able to meet some long-lasting demands emanating from models of rationality. But, despite the sophistication of these answers, it is still not clear on which principles of rationality they are grounded: is there an underlying logic lurking behind these techniques, or is it a necessity that any model of rationality contains an intrinsically a-logical part?

Notes

Professor Amitabha Gupta, from the Indian Institute of Technology at Bombay, made invaluable comments on an earlier version of this paper, on which Frank Döring, from LIPN and CREA, provided interesting remarks. Both of them are gratefully thanked.

1. A fixpoint of a function f(x) is a solution of the equation x = f(x). It may happen—and it is the case here—that no general method is known in order to compute the solution of the equation. However, in several practically useful particular cases, efficient methods can be found.
2. Although it makes perfect sense indeed to use two different logics, one for each level. But then articulating together the two levels, which is crucial for the study of rationality, would remain out of the scope of any logic.

References

CHEESEMAN, PETER. 1985. In defence of probability Proc. 9th Int. Joint Conf. on Artificial Intelligence, pp. 1002–9, Los Angeles, Aug.

CHERNIAK, CHRISTOPHER. 1986. *Minimal Rationality*. Cambridge M.A.: M.I.T. Press.

DEKLEER, JOHAN. 1986. *Problem Solving with the ATMS*, A.I. Journal, Vol. 28, pp. 197–224.

DUBOIS, DIDIER and PRADE, HENRI (with HENRI FARRENY, ROGER MARTIN CLOUAIRE and CLAUDETTE TESTEMALE). 1985. *Théorie des possibilitiés. Applications à la représentation des connaissances en Informatique*. Paris: Masson.

ENGEL, PASCAL. 1989. *La Norme du Vrai: philosophie de la logique*. Paris: Gallimard.

ETHERINGTON, DAVID W. 1987. *Formalizing Nonmonotonic Reasoning Systems*, A.I. Journal 31(1): 41–85, January.

FELDMAN, JEROME. 1972. *Some Decidability Results on Grammatical Inference and Complexity*. Information & Control, 20(3): 244–62, April.

GOLD, E.M. 1967. *Language Identification in the Limit*. Information and Control, 10: 447–74.

GRANDEMANGE, PHILIPPE. 1992. *Représentation des connaissances et profondeur variable: Une implantation*. Thèse d'Informatique, Université Paris–Nord, 12 February.

HOBBS, JERRY R. 1985. Granularity. Proceedings of the 9th International Joint Conference on Artificial Intelligence. Los Angeles: pp. 432–35, August.

KAHNEMANN, DANIEL, SLOVIC PAUL and TVERSKY AMOS (eds.). 1982. *Judgement Under Uncertainty: Heuristics and Biases*. Cambridge: Cambridge University Press.

KAYSER, DANIEL. 1988. *Le raisonnement à profondeur variable* Proc. 2èmes Journées Nationales du PRC-GRECO Intelligence Artificielle, Editions Teknea: Toulouse pp. 109–36.

KONOLIGE, KURT. 1988. *On the Relationship between Default Logic and Autoepistemic Logic*. A.I. Journal 35(3): 343–82. July.

KRAUS, SARIT, LEHMANN, DANIEL and MAGIDOR MENAHEM. 1990. *Non-monotonic Reasoning, Preferential Models, and Cumulative Logics*. A.I. Journal 44 (1–2): 167–207, July.

KYBURG, HENRY E. 1991. Contributions to the section *Quantified Uncertainty*, special issue on 'Uncertainty, Conditionals, and Non-Monotonicity', Journal of Applied Non-Classical Logics 1(2): 105–97.

LEHMANN, DANIEL, and MAGIDOR, MENACHEM. 1992. *What does a conditional knowledge base entail?* A.I. Journal, 55(1): 1–60, May.

LÉVY, FRANÇOIS. 1989. *Contribution à la réalisation d'un raisonneur à profondeur variable: le système de maintenance* Thèse d'Informatique, Université Paris–Nord, 8 February.

LUKASZEWICZ, WITOLD. 1985. *Two results on Default Logic*. Proc. 9th Int. Joint Conf. on Artificial Intelligence, Los Angeles: pp. 459–61, August.

McDermott, Drew V. 1982. *Non-Monotonic Logic II: Non-Monotonic Modal Theories*. Journal of the A.C.M. 29(1): 33–57, January.

———. 1987. *A Critique of Pure Reason*. Computational Intelligence, 3: 151–160.

Minsky, Marvin. 1975. *A Framework for Representing Knowledge* in P.H. Winston (ed), *The Psychology of Computer Vision*, pp. 211–77, McGraw Hill.

Moore, Robert C. 1985. *Semantical Considerations on Non-Monotonic Logic*. A.I. Journal 25(1): 75–94, January.

Osherson, Daniel N. Stob Michael, and Weinstein Scott. 1985. *Systems that Learn*. M.I.T. Press.

Paass, Gerhard. *Probabilistic Logic* in Smets, Ph. *et al*. (eds.) *Non-standard Logics for Automated Reasoning*. Academic Press: pp. 213–51.

Pitrat, Jacques. 1990. *Métaconnaissance, futur de l'Intelligence Artificielle*, Paris: Hermès.

Poole, David. 1988. *A Logical Framework for Default Reasoning*. A.I. Journal 36: 27–47.

Rabin, Michael O. 1974. *Theoretical Impediments to Artificial Intelligence*. Proceedings of the IFIP Congress preprints: pp. 615–19, Stockholm, August.

Reiter, Raymond. 1980. *A Logic for Default Reasoning*. A.I. Journal 13(1–2): 81–132, April.

Sombé Léa, (pseudonym for Besnard Philippe, Cholvy Laurence, Cordier Marie-Odile, Dubois Didier, Farinas Luis del Cerro, Froidevaux Christine, Lévy François, Moinard Yves, Prade Henri, Schwind Camilla, and Siegel Pierre). 1992. *Révision de bases de connaissances*. Proc. 4èmes Journées Nationales du PRC-GDR Intelligence Artificielle, pp. 207–38. Toulouse: Éditions Teknea.

Characterising Coherence in Multimodal Discourse

S. BANDYOPADHYAY

Introduction

A text cannot be characterised in terms of mere collection of sentences. A text has an organisational framework and the sentences in a text bind themselves together to constitute a 'unit' that itself has a meaningful interpretation. Text linguistics, therefore, argue that texts constitute an object to be modelled differently from sentences in isolation (Van Dijk 1985). This is due to the occurrence of a linguistic phenomenon, generally termed as *text coherence* (Hobbs 1983; Reichman 1978) which establishes textuality above the sentence level.

The same analysis is true for a sequence of pictures, for example in silent movies or cartoons where the pictures bind themselves together in order that a definite meaningful interpretation of the whole is possible. A sequence of pictures in silent movie is, therefore, different from a collection of pictures in that the former are 'coherently' sequenced.

When we talk about multimodal presentation of information (in terms of a combination of text and pictures or audio-visuals in general), discourse segments are no longer structurally uniform. Each discourse segment in this mode is either a text segment (or audio segment) or a picture segment (visual segment) or a combination of both. Again, the notion of 'coherence' is the binding force which links the multimodal discourse segments together to generate an overall meaningful presentation i.e. a presentation which is *valid* (adjacent discourse segments are properly connected at the surface level), *structured* (have a well-formed thematic organisation) and *effective* (capable of meeting the intended goal of the discourse from the perspective of the receptor of the discourse).

In this study, a proper characterisation of coherence is attempted in a multimodal framework which can ensure validity, thematic well-formedness and effectiveness of a multimodal discourse.

Characterising Coherence

What is Coherence? Some General Observations

According to Rescher (1973), the coherence of a propositional set is understood as requiring not simply the obvious minimum of consistency but also the feature of being connected in some special ways. He argues that the *special* connection required links consistency pragmatically with our affective life. Samet (1984) has also depicted coherence as connectedness of various sorts that plays a part in building the receptors' mental representation.

Margolis (1984) defines coherence in terms of a model of the rationality of human thought and action in order to satisfy the condition for the special connection as proposed by Rescher (1973). According to him, coherence is a function of the rationality of human thought and action with self-consistency as its necessary and minimal condition. Incoherence is a relative discrepancy of some sort in the speech, thought, behaviour or work of human beings in terms of their products or acts or deeds. But at the same time, incoherence cannot be a mere formal property of sentences or statements or analogous elements detached from the human activity by which they are generated; it must be *linked* to the intention and purposive life of the human being. The idea of coherence is therefore context dependent: what is coherent in one context may appear incoherent in some other context (Margolis 1984).

Based on all this, we can say that:

- A minimum requirement for coherence is self-consistency.
- Coherence is connectedness (or, togetherness) which reflects the rationality of an agent who tends to maintain coherence to fulfil his rational purpose. There are two implications of

this. First, as coherence can be seen as connectedness, it is more often a matter of degree than a clearcut presence or absence. Second, as connectedness in this context is a consequence of rationality, the connections are often to be explained in terms of intention and purposive life of human beings.

- Coherence is context-sensitive. For example, the following three sentences are coherent in the context of the question 'Tell me what thoughts are crossing your mind'; 'It is snowing in Tibet. The moon is round. Cats eat rats.' (Ziff 1984)

Coherence in Discourse: A Three-Level Description

Coherence in Discourse: Related Research

Some significant work has been devoted to the study of coherence in texts—dialog or monolog (Enkvist 1978, Grimes 1975, Hobbs 1979, Hobbs 1983, Reichman 1978 and others). But unfortunately, no work of any significance has been done in the area of coherence in picture sequence or in multimodal discourse composed of text and pictures (or audio-visuals in general).

Hobbs (1983) describes three kinds of coherence: the *global coherence* of an utterance is its relation to the speaker's overall plan (Litman [1986] also adopts the same basis for analysis); the *local coherence* of an utterance or larger segment of discourse is its relation to adjacent segments in the discourse in terms of satisfying certain textual goals; and *thematic coherence* is a coherence pattern observable, for example, in different writings of the same author. Hoobs defined a set of coherence relations and stated that a text will strike one as coherent to a degree that varies inversely with the degree of 'difficulty' in inferring certain coherence relations.

Lindeberg's *Cohesion, Coherence Pattern and EFL Essay Evaluation* (1985), is an exploratory study designed to find and describe coherence patterns which distinguish good expository essays from poor ones. He emphasises degree of coherence rather than coherence/incoherence in its absolute sense and shows that it is possible to arrive at coherence patterns, based on sequences of rhetorical functions, that are frequent in good essays and less frequent in poor essays. He segments each text into functional units and identifies the functional role of each unit (which expresses the rhetorical purpose of the unit in the discourse, e.g., assert, specify, contrast, result, cause and so on). Some combinations of functional roles generate patterns which are less coherent than other combinations. For example, the combination 'assert-assert-assert' (i.e. three consecutive functional units each making a major assertion by introducing a new topic or an aspect of a topic not covered before) will appear to be less coherent than the combination 'assert-specify-cause' or 'assert-cause-specify'.

Samet and Schank (1984) analyse coherence in terms of connectivity of various sorts that plays a part in the hearer's evolving a mental representation of the discourse. They identified two types of connectivities: *internal*, which is connectivity within the discourse among discourse segments; and *external*, which concerns the integratability of this unified representation into the hearer's pre-existing conceptual framework. A text is coherent for a reader if and only if the reader has constructed an adequate conceptualisation for it which involves building a conceptualisation (internal connectivity) and integrating it with one's own world-view (external connectivity). They defined a set of hierarchical connections similar to coherent relations as defined by Hobbs (1983).

Most of these works concentrate mainly on coherence detection in texts in terms of identifying certain coherence relations or coherence patterns in the process of understanding a text. Although the notion of global coherence (Hobbs 1983) and external connectivity (Samet 1984) are introduced to capture the role of readers' interpretive ability in detecting coherence, they received less attention in the overall discussion.

Efforts have also been made to generate coherent text based on rhetorical relations (Moore 1988, McKeown 1985). McKeown did not try to characterise coherence; rather, coherence is a consequence of her methodology in a restricted sense. While it would be interesting to explore the possibility of using rhetorical relations as described in RST (Mann 1987), in characterising coherence in multimodal discourse, the present study is restricted to characterising coherence which is primarily motivated by Hobbs (1983).

The next section is an attempt to characterise coherence in a more general framework of multimodal presentation of information involving text and pictures (or audio-visuals in general), where a discourse segment is either a text (or audio) segment, or a picture (or visual segment) or a combination of both. Coherence is a function both of features in the discourse and of the addressees' interpretive ability which depends on various factors like knowledge of the subject, inferencing ability, purpose and motivation (Grice 1975). We therefore attempt to characterise coherence (*a*) *at the discourse level* which concerns the validity and thematic well-formedness of the discourse, and (*b*) *at the user level* which concerns the effectiveness of the discourse in the specific context.

A Three-Level Description of Coherence in Discourse

Coherence can be characterised at three levels: at the syntactic level (or *syntactic coherence*), at the semantic level (or *semantic coherence*) and coherence at the pragmatic level (or *pragmatic coherence*). The first two levels together constitute the *discourse level* and the third is the *user level* characterisation of coherence.

Syntactic coherence in a discourse is a surface-level phenomenon that deals with the immediate connectivity among adjacent segments using some rules or conventions of connectivity. Maintaining syntactic coherence ensures the validity of the presentation at the syntactic level and at the same time helps the addressee capture the coherence at the semantic level. Syntactic coherence is the necessary and minimal condition for maintaining coherence but it alone does not ensure the coherence of the discourse. Its role in a discourse is similar to the role of grammaticality in a sentence in the sense that grammatical correctness is necessary for easy understanding of the sentence but it alone does not ensure the meaningfulness of the sentence as such.

Semantic coherence concerns the content and global structuring of a discourse. It ensures a well-formed thematic organisation of a discourse so that the discourse can be conceived as a unified whole. The discourse segments are coupled together through some semantic ties (coherence relation [Hobbs 1983] or conceptual connectivity [Samet 1984] which in turn generate a thematic structure of the discourse. Semantic coherence presupposes syntactic coherence; but in certain cases, even if the syntactic clues from a semantically coherent discourse are removed, one can still make sense out of it.

While coherence at the semantic level concerns what to communicate and how to organise it, it does not address the issue of the extent of communication required to suit the addressees' interpretive ability. In other words, semantic coherence deals with the relevance of a discourse segment with respect to the entire discourse but it does not deal with the adequacy of presentation. As a result, a semantically coherent discourse may appear incoherent to an addressee whose interpretive ability does not match with that assumed in the discourse. Instead of calling it incoherent, we will call it *pragmatically* incoherent with respect to that addressee.

Pragmatic coherence, therefore, concerns the *effectiveness* of a discourse. A discourse is pragmatically coherent to an addressee or a group of addressees if it is compatible with their interpretive ability. It presupposes semantic coherence.

The purpose of discourse is communication which can be understood. The generator of the discourse generates some messages and the receptor applies an active inference process to understand the messages and to link what is new in them to what he already knows. It is thus part of the generator's job to provide the necessary linkage and to try to manipulate the receptor's inference process to lead him to make a proper interpretation (Hobbs 1979). Pragmatic coherence thus deals with the manipulation of the receptor's inference process. A discourse which is pragmatically incoherent to a receptor can be sometimes made pragmatically coherent with the addition of more information in terms of clarification, elaboration, examples and so on which would make the presentation adequate for that receptor. But these additions must be relevant to the discourse in order to preserve its semantic coherence.

Another aspect of pragmatic coherence is the discourse typology. A discourse should be coherent with its purpose and situation in order to make the communication effective. An effective love letter should look different from an effective business letter. A discourse should therefore, be context dependent and be capable of *linking* itself to the relevant purpose and situation for effective communication.

To summarise, syntactic coherence concerns the linking mechanism of the discourse segments at the surface level in order to make the presentation valid; semantic coherence concerns the linking of discourse segments through some semantic ties in order to generate a well-formed thematic organisation of the entire presentation; and pragmatic coherence concerns effective presentation through linking the discourse with the addressees' pre-existing conceptual framework by making it compatible with their interpretive ability, and linking the discourse with the purpose and situation by selecting a proper discourse typology.

Coherence in Text

Syntactic Coherence

Syntactic coherence in text concerns the immediate connectivity among adjacent text segments through some surface markers, cohesive devices or overt linguistic mechanisms such as anaphora, cataphora, ellipsis and conjunctions which serve to link text segments to each other. This surface-level, language dependent linking mechanism is often termed *text cohesion* (Halliday 1976). Hobbs (1979) pointed out that cohesion does not ensure text coherence. Lindeberg (1985) also showed that a text may have a fair amount of cohesive tightness but may not be coherent. Still, if a text is to be well-formed, it must have semantic coherence as well as sufficient signals of surface cohesion to enable the reader to capture the coherence. McKeown (1985) indicated that certain choices at the surface level are critical in order to produce a coherent text. She pointed out that the generator must be able to make reasoned decisions about when to use pronominal reference and about the syntactic construction that should be used. The following is an example of lexical choice (bought vs. sold) where (*b*) is a text sequence *incoherent* and the *syntactic level* (ibid):

(*a*) Jane was in a hurry to finish her shopping. It was a chore she particularly despised. First, Jane *bought* $3.00 worth of bobby socks from Michael.

(*b*) Jane was in a hurry to finish her shopping. It was a chore she particularly despised. First, Michael *sold* $3.00 worth of bobby socks to Jane.

Semantic Coherence

Text coherence at the semantic level concerns the global structuring of the text in terms of a well-formed thematic organisation. Instances of global text structuring through text coherence can be illustrated by regular patterns of thematic progression in a text (Danes 1974) or by various additional functional coherence relations such as elaborations, parallels, contrasts, occasion, compatibility and the like (Hobbs 1979, 1983; Reichman 1978).

Danes (1974) describes text coherence patterns which follow either the principle of a constant theme, or linear thematization of themes, or derived themes. According to Werlich (1976) coherence is created whenever the generator uses some linguistic unit as a text base unit with a theme and then expands this in linear progression in conventionally ordered and completed sequences of linguistic units. For instance, Hahn (1986) attempts to create a text graph using this notion of text coherence pattern in a expository text.

Coherence relations describe a set of semantic ties which in turn defines the coherent structure of a text. Although Samet (1984) has shown that these coherence relations alone cannot ensure the semantic coherence of a text (the notion of causal connectivity plays an important role which is beyond the coherence relation in a narrative text), coherence relations are an important first step towards describing semantic ties among text segments. Tucker (1986) also uses coherence relations in analysing coherence in expository texts.

Pragmatic Coherence

The notion of pragmatic coherence in a text takes care of the readers' interpretive ability and adequacy of presentation in order to make the text effective. In other words, a text would be pragmatically coherent if it is compatible with the readers' interpretive ability and is adequate to satisfy the reader.

As an example, suppose that a computer salesman describes a product of his company in the following way: 'The PC 2000 is equipped with a 8086 cpu as opposed to the 8085 of the previous model. The standard amount of dynamic RAM is 256 Kbytes. One of the two RS–232C ports also serves as a higher speed RS–422 port . . .' This description is coherent and effective to a computer engineer who has a sufficient background knowledge of computer hardware. But the same description will appear incoherent to a person outside the computer engineering profession. A more effective description to him would be: 'The personal computer PC 2000 is more powerful than its previous model. It has a standard amount of memory and an appropriate mechanism to link it with another computer for communication at a higher speed . . . .' Again, this description is not effective enough to satisfy a computer engineer because it is *inadequate*.

Another aspect of pragmatic coherence concerns the pragmatic appropriateness of a text (Hovy 1987) in terms of its suitability for the hearer, the conversational setting and their interpersonal goal.

Selection of a proper text typology or stylistics and the use of a proper pragmatic particle (Enkvist 1985) can make a communication far more effective. On the other hand, a improper use of these can render the communication inappropriate and incoherent at the pragmatic level.

Coherence in Picture-Sequence

Syntactic Coherence

The pictures in a picture-sequence are not autonomous; their task is to contribute to the flow of information transmitted by the picture-sequence, to link up with what went before and with what comes after. Each picture represents a set of objects or event and contains a single focus of attention. Syntactic coherence in a picture-sequence concerns the immediate cor·ectivity of adjacent pictures. The conventions of this connectivity at the surface level are termed as *continuity*. Three type of continuities are identified:

(*i*) Continuity in perspective
(*ii*) Continuity of action
(*iii*) Continuity in focus of attention

To illustrate this, we will take examples from the instruction manual for an expresso coffee machine (adopted from Philips Espresso Machine HD 5649).

Continuity in perspective: Examples of continuity in perspectives are spatial continuity, continuity in viewpoint, continuity in colour, etc.

Spatial continuity is said to be maintained when there is no *sudden* change in the relative position or relative size of the objects (or parts of objects) between two consecutive pictures in a picture-sequence having *same focus of attention*. For example, the pictures A and B in Figure 1 are coherently illustrating the process of filling water in the water-container of a coffee machine. A *spatial discontinuity* is observed if one moves from picture B to picture C in the sequence, where the position of the hole is shifted from left (in B) to right (in C). A spatial discontinuity due to a sudden change in size of the switch is also observable from B to C.

Continuity in *viewpoint* is said to be maintained when there is no change in viewpoint between two consecutive pictures in a picture-sequence having *same focus of attention*. For example, a continuity in viewpoint is observed in Figure 1. If the same machine is viewed from a different angle as compared to that in A in the sequence, a discontinuity in viewpoint is observable.

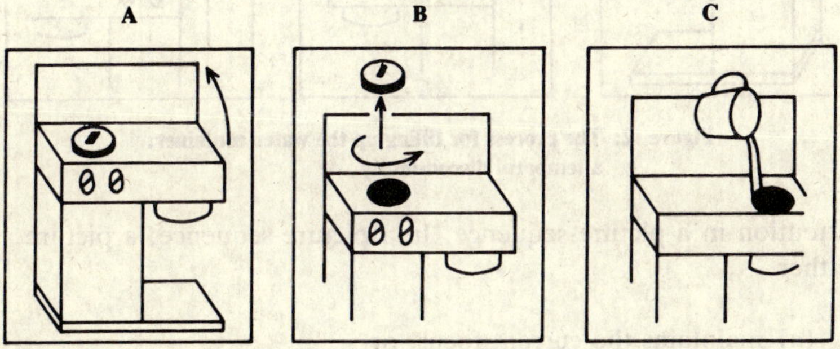

Figure 1: Filling up the water container of a coffee machine

Continuity in *colour* is said to be maintained when there is no *sudden* change in the relative colours of the objects (or parts of objects) between two consecutive pictures in a picture-sequence having the *same focus of attention*.

Continuity of action, also called *temporal* continuity, is said to be maintained when temporally sequenced pictures in a picture-sequence are continuous in the time domain. In other words, when a picture-sequence depicts a certain action or sequence of actions, a picture in the sequence needs to maintain continuity of action with respect to the picture preceding it in the sequence, unless there is a total shift in the focus of attention. For example, take Figure 2, where pictures A, B and C illustrate the process of filling up the water-container. A temporal *discontinuity* is observed, since the cover is shown in its position in C even after its removal in B.

Continuity in focus of attention: Sidner (1979) showed how the immediate focus can shift or be maintained in a text. The same constraints can be used to understand the continuity in focus of

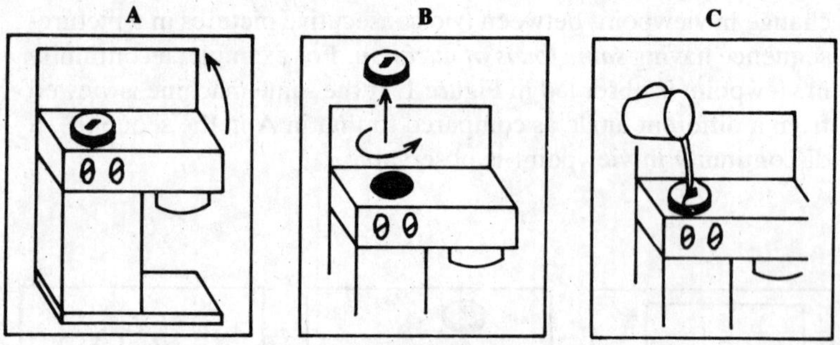

Figure 2: The process for filling up the water container: a temporal discontinuity

attention in a picture-sequence. In a picture-sequence, a picture either

(*a*) maintains the current focus, or
(*b*) shifts focus to an item just introduced, or
(*c*) returns to a previous focus, or
(*d*) focuses on an item *implicity related* to the current focus.

(The implicit relation mentioned in (*d*) is essentially a semantic relation which will be illustrated in the context of semantic coherence.)

Semantic Coherence

The discourse structure of a picture-sequence can be described by defining certain coherence relations. These coherence relations, derived from Hobbs (1979, 1983), are the semantic ties that can exist between two pictures in a sequence, not necessarily adjacent. These relations are:

(*a*) elaboration
(*b*) occasion (cause or enablement)
(*c*) expansion
  (*c.1*) specification
  (*c.2*) generalisation
  (*c.3*) parallel

(*c.4*) contrast

(*d*) temporal relation

(*d.1*) parallel

(*d.2*) sequential

a) Elaboration: A picture p1 is an elaboration of a picture p0 in a sequence if p1 expresses the *same* thought as in p0 from a different perspective. p1 has either the same focus of attention as p0, or has a focus of attention directly related to p0. Figure 3 is an example of this where picture B elaborates the process of opening the cover as depicted in picture A.

b) Occasion (Cause or Enablement): p1 is said to be occasioned by p0 when the event depicted in p0 produces a state that is consumed (Balzer 1977) by the event depicted in p1. When the first event causes the second event to occur, they are causally related; when the first event sets up the situation for the second, p1 is said to be enabled by p0. For example, in Figure 4, changing the position of the switch causes the steam to come out of the nozzle (i.e. A causes B) and this enables one to hold the coffee cup at the exhaust (i.e. C is enabled by B).

c) Expansion. *Specification*: When p1 specifies something illustrated in p0 in more detail, it is said to be a specification relation. The use of close-up or zoom is a method for this in a picture-sequence. For example, if p0 shows an object and p1 zooms in on a part of the same object to show it in greater detail, they are said to be related through a specification relation.

Generalisation: When p0 indicates some entity or a set of entities and p1 indicates the class to which they belong, p1 is a generalisation of p0. It is sometimes the reverse of a specification relation.

Parallel: A parallel relation involves moving from a predication about some set of entities illustrated in p0 to the *same* predication about a similar set of entities illustrated in p1. For example, if p0 shows a machine consisting of three parts and p1, p2 and p3 show one of those parts, each then p1, p2 and p3 are said to be related parallelly. At the same time, p1, p2 and p3 are each related to p0 through the specification relation.

Contrast: p0 and p1 are said to be contrastive if the depiction in p1 is either dissimilar or opposite from that depicted in p0.

d) Temporal relation. *Parallel:* p0 and p1 are said to be temporally parallel when the event depicted in p0 and that depicted in p1

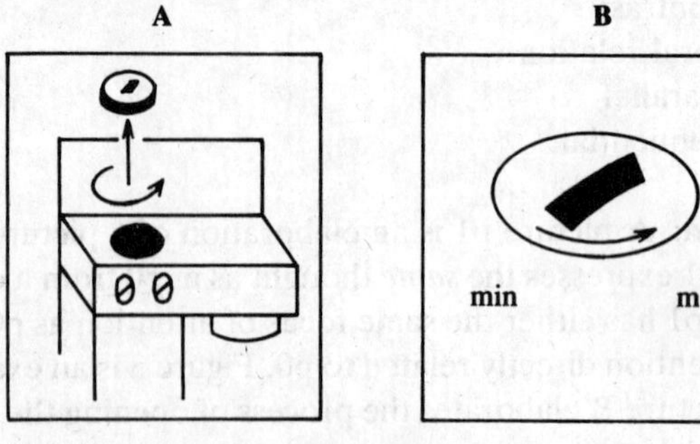

Figure 3: Process of removing the cover in a coffee machine

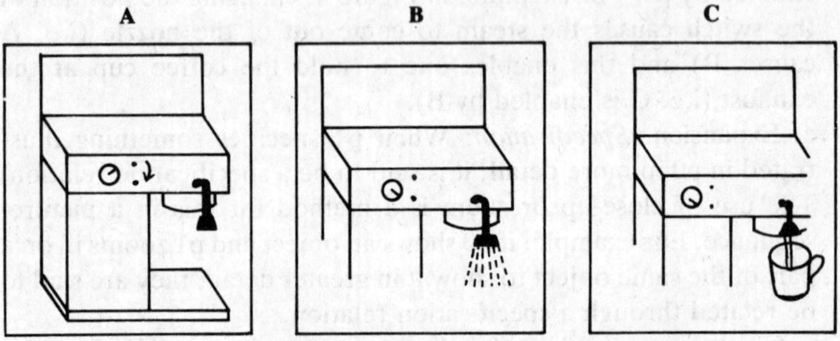

Figure 4: Preparing espresso coffee

occurs simultaneously. Two simultaneous events are shown in a picture-sequence through a mechanism known as inter-cut where two simultaneous events are broken into sub-events and presented in an alternate sequence of pictures.

Sequential: p0 and p1 are said to be temporally sequential when they depict a sequence of actions of an action-sequence. Figure 5 presents a sequence which is incoherent due to temporal non-sequentiality.

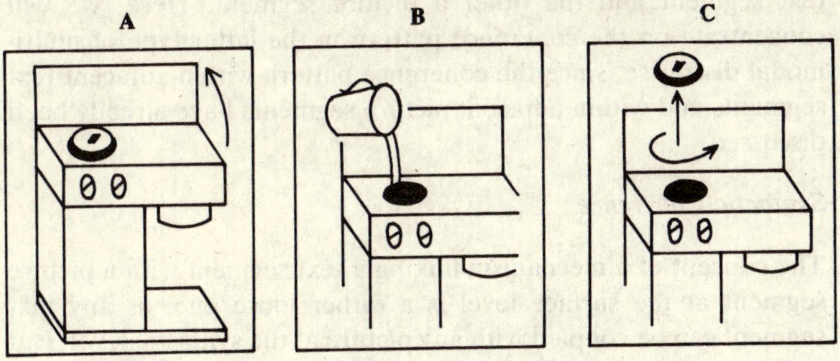

Figure 5: Filling up the water container: an incoherent sequence

Pragmatic Coherence

A message can be conveyed through a coherent sequence of pictures in a multiple number of ways. But all of them may not be pragmatically coherent with respect to the intended addressees. The number of pictures and the content of each picture in a picture-sequence are the determining factors in generating a pragmatically coherent picture-sequence.

Moreover, a picture can be taken from different angles (e.g. high or low), from different viewpoints (e.g. front, rear and side view) and from different distances (e.g. close-up, medium or long distance). The effectiveness of a picture in a picture-sequence often depends on the proper selection of the angle, viewpoint and distance.

The purpose and situation also determine the style of presentation of a picture-sequence. For example, building up a fast, suspenseful sequence in a movie is different from building up a slow, romantic sequence. The first one usually comprises a larger number of shots, each of a short duration; the second one usually uses fewer shots, each of a longer duration.

Coherence in Multimodal Discourse

In multimodal discourse, two adjacent discourse segments can both be text segments or picture segments, or one of them can be a text segment and the other a picture segment. Here, we will concentrate on the coherence pattern in the latter type of multimodal discourse, since the coherence pattern within adjacent text segments and within adjacent picture segments have already been discussed.

Syntactic Coherence

The concept of a mechanism linking a text segment with a picture segment at the surface level is a rather loose one, as any text segment can be coupled with any picture at the syntactic level. But one important consideration in the presentation of a multimodal document is the positioning of picture with respect to the text segment referring to that picture. If the picture is too far away from the relevant text segment or comes after some other pictures, it will lead to a surface-level incoherence.

The notion of syntactic coherence is more important in a movie where this linking mechanism between the audio segment and the corresponding visual segment is termed as *synchronisation*. Synchronisation of voice with picture (with lip and movement) is one way of maintaining syntactic coherence in a movie.

Semantic Coherence

A set of coherence relations similar to that described in section dealing with coherence in a picture-sequence, with an additional relation termed as *equivalence* can be described to illustrate the semantic tie between a text segment and a picture segment or vice versa.

Equivalence: A picture and a text segment can never be truly equivalent in the formal sense of the term. However, a picture is said to be equivalent with a text segment if they convey a conceptually identical thought from the same perspective with reference to a common focus of attention. The simplest example is the relation between a picture and its caption (usually).

Figure 6 gives an example of the coherence relations between a picture and text segments.

Pragmatic Coherence

Pragmatic coherence in multimodal discourse concerns the factors illustrated in pragmatic coherence in texts and pragmatic coherence

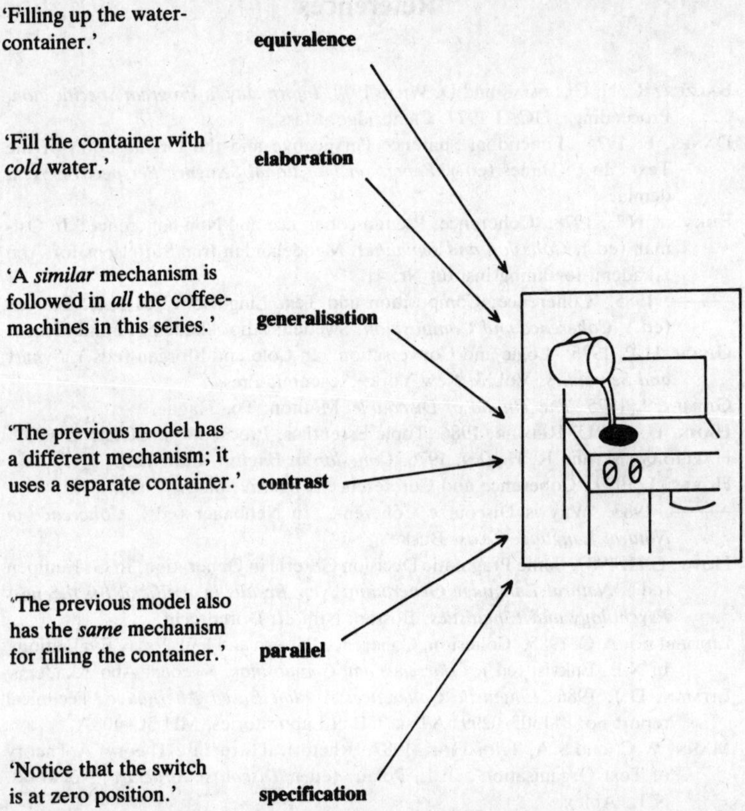

Figure 6: Illustrating coherence relations between text and picture segment

in picture-sequences. In addition, it concerns the *selection of modality* of presentation, that is, the decisions such as when to use a picture and when to use text and to what extent. At the same time, the selection of the most suitable graphic presentation for a given object among several possibilities needs to be done in a specific context to make the presentation more effective.

References

BALZER, R., N. GOLDMAN and D. WILE. 1977. *Informality in Program Specification.* Proceedings, IJCAI 1977. Cambridge, Mass.

DANES, F. 1974. 'Functional Sentence Perspective and the Organisation of the Text.' In F. Danes (ed.), *Papers on Functional Sentence Perspective.* Academia.

ENKVIST, N.E. 1978. 'Coherence, Pseudo-coherence and Non-coherence.' In Östman (ed.), *Cohesion and semantics*. Meddelanden fran Stiftelsens för Abo Akademi forskningsinstitut Nr. 41.

———. 1985. 'Coherence, Composition and Text Linguistics.' In N.E. Enkvist (ed.), *Coherence and Composition*. Sweden: Abo Akademi Foundation.

GRICE, H.P. 1975. 'Logic and Conversation.' In Cole and Morgan (eds.), *Syntax and Semantics*, Vol. 3. New York: Academic Press.

GRIMES, J. 1975. *The Thread of Discourse.* Mouton: The Hague.

HAHN, U. and U. REIMER. 1986. Topic Essentials. Proceedings. COLING 1986.

HALLIDAY, M. and R. HASAN. 1976. *Cohesion in English*. Longman.

HOBBS, J. 1979. 'Coherence and Coreference' *Cognitive Science* 3(1).

———. 1983. 'Why is Discourse Coherent?' In Neubauer (ed.), *Coherence in Natural Language Texts*. Buske.

HOVY, E.H. 1987. Some Pragmatic Decision Criteria in Generation. In G. Kempen (ed.), *Natural Language Generation: New Results in Artificial Intelligence, Psychology and Linguistics*. Boston: Nijhoff: Doredrecht.

LINDEBERG, A.C. 1985. 'Cohesion, Coherence Pattern and EFL Essay Evaluation.' In N.E. Enkvist (ed.), *Coherence and Composition*. Sweden: Abo Akademi.

LITMAN, D.J. 1986. *Linguistic Coherence: A Plan Based Alternative.* Technical report no. 311305–0299. AT & T Bell Laboratories, MH 3C–408A.

MANN, W.C. and S.A. THOMPSON. 1987. 'Rhetorical Structure Theory: A Theory of Text Organisation'. In L. Polany (ed.), *Discourse Structure*. Norwood, N.J.: Ablex.

MARGOLIS J. 1984. 'The Locus of Coherence.' *Liguistics & Philosophy*, 7(1).

MCKEOWN, K.R. 1985. 'Discourse Strategies for Generating Natural Language Text.' *Artificial Intelligence*, 27.

MOORE, J.D. and C.L. PARIS , 1988. *Constructing Coherent Text Using Rhetorical Relations*. Proceedings, Tenth Annual Conference of the Cognitive Science Society.

REICHMAN, R. 1978. 'Conversational Coherency.' *Cognitive Science* 2(4).

RESCHER, N. 1973. *The Coherence Theory of Truth*. Oxford: Clarendon.

SAMET, J. and R. SCHANK. 1984. 'Coherence and Connectivity'. *Linguistics & Philosophy*, 7(1).

SIDNER, C.L. 1979. *Towards a Computational Theory of Definite Anaphora Comprehension in English Discourse*. Ph.d. Dissertation. Cambridge. Mass.: MIT.

TUCKER, A.B., S. NIRENBURG and V. RASKIN. 1986. Discourse and Cohesion in Expository Text. Proceedings. COLING 1986.

VAN DIJK, T.A. (ed.). 1985. *Handbook of Discourse Analysis*. U.K: Academic Press.

WERLICH, E. 1976. *A Text Grammar of English*. Heidelberg: Quelle & Meyer.

ZIFF, P. 1984. 'Coherence'. *Linguistics & Philosophy*, 7(1).

Notes on Contributors

Daniel Andler is Professor of Philosophy at the Université Paris X-Nanterre and heads a research group in cognitive science at CREA, Ecole Polytechnique and CNRS, Paris. He holds doctorates in mathematics from Berkeley and Paris. Professor Andler has papers on model theory, and on philosophy and foundations of cognitive science to his name. He has edited several books and journal issues on cognitive sciences.

S Bandyopadhyay teaches at the Department of Computer Science and Engineering, Indian Institute of Technology, Bombay. His main area of interest is natural language processing, discourse and representations.

Mohua Banerjee is a Senior Research Fellow at the Department of Pure Mathematics, University of Calcutta. Her main areas of interest are category theory and mathematical logic.

Parthasarathi Banerjee is attached to the National Institute of Science, Technology and Development Studies, New Delhi. His main areas of interest include theories of politics, economics and cognition.

Kalyan S Basu is at the Department of Computer Science and Engineering, Indian Institute of Technology, Kanpur. He is deeply interested in algebra-logic-computer science, philosophy of cognition and language, philosophy of social sciences, and direct-reference theory.

S M Bhave teaches at Nowrosjee Wadia College, Pune. His main areas of interest are philosophy of mathematics, and philosophy of science and Nyaya. Dr. Bhave was a post-doctoral fellow of the ICPR between 1985 and 1988.

Laurent Bochereau studied at the Ecole Polytechnique, Paris, and has a doctorate in computer science and neural networks.

Danièle Bourcier is presently Researcher at the National Centre for Scientific Research, Paris, and also heads a research team on computer science, law and linguistics. She works mainly on computational aspects of law.

Paul Bourgine teaches computer science and cognitive science in Paris. He is the Head of the AL & Artificial Intelligence Laboratory of CEMAG-REF.

M K Chakraborty teaches mathematics at the Department of Pure Mathematics, University of Calcutta. Interested in mathematical logic, fuzzy set theory and related areas, he is the founder secretary of the Indian Society for Fuzzy Mathematics and Information Processing (IS-FUMIP).

Mahasweta Chaudhury teaches philosophy at Calcutta University. Her main areas of interest include philosophy of sociology, epistemology, and philosophy of knowledge.

Jean-Claude Gardin is Director of Research at the National Centre for Scientific Research, Paris, and teaches at the School of Higher Studies in Social Sciences, Paris. His main area of interest is the analysis of the intellectual processes in archaeology and other human sciences.

Allan Gibbard received his doctorate from Harvard University, and is currently Professor of philosophy at the University of Michigan, Ann Arbor. His main area of interest is ethics.

Olivier Guillaume is at present Director of the French Institute, St. Petersburg, Russia. Prior to this, he was Director, Centre de Sciences Humaines, Embassy of France in India. Dr. Guillaume holds a Ph.D. in classical archaeology.

Amitabha Gupta is Professor at the Department of Humanities and Social Sciences, Indian Institute of Technology, Bombay. He is well-versed in problems of logic, theory of knowledge and cognitive science, as also European and Indian traditions of philosophy.

Pierre Jacob holds a doctorate in philosophy from Harvard University, and is presently Researcher at the National Centre for Scientific Research, Paris. He is also a member of the Centre for Research in Applied Epistemology, Ecole Polytechnique, Paris. He has published extensively in the philosophy of mind and the philosophy of science.

Satish K Jain is Professor of Economics at the Centre for Economic Studies and Planning, Jawaharlal Nehru University, New Delhi. Social choice theory and welfare economics are his areas of specialisation.

V N Jha is Professor of Sanskrit and Director of the Centre of Advanced Study in Sanskrit, Pune University. A scholar as well-versed in Western philosophy, he is an exponent of the Nyaya tradition and has a number of books to his credit.

Daniel Kayser is Professor of Computer Science at the University of Paris-Nord, and Director of a research team in artifical intelligence at the Institute Galilee for Computer Science. He is primarily interested in artifical intelligence, logic, language and reasoning.

Serge-Christophe Kolm is at the Institute for Advanced Studies in Paris. His main area of specialisation is foundations of economics.

A Raghuramaraju teaches philosophy at the University of Hyderabad. His research interests include social and political philosophy and contemporary Indian philosophy.

François Recanati is Director of Research at the National Centre for Scientific Research, Paris. His areas of interest are philosophy of language, pragmatics and philosophy of mind.

T K Sarkar is Professor at the Department of Philosophy, Jadavpur University. A prolific writer, he is an expert on both Western and Indian logic, philosophy and theories of knowledge.

Wagish Shukla teaches mathematics at the Indian Institute of Technology, Delhi. His wideranging interests include category theory, number theory, Indian philosophy and Persian poetry.

Nilesh Sutaria obtained his B.Tech. degree in Computer Science and Engineering from the Indian Institute of Technology, Bombay in 1993. He is currently pursuing graduate studies in the University of Connecticut, USA.

John Vattanky, Professor of Indian Philosophy and Sanskrit at Jnena-Deepa Vidyapeeth, Pune, is a Nyaya scholar who was educated at Oxford and Vienna. Among his major publications are *Gangesa's Philosophy of God* and *Nyaya Philosophy of Language.*

